Graduate Texts in Mathematics **165**

Springer
New York
Berlin
Heidelberg
Barcelona
Budapest
Hong Kong
London
Milan
Paris
Santa Clara
Singapore
Tokyo

Graduate Texts in Mathematics

1 TAKEUTI/ZARING. Introduction to Axiomatic Set Theory. 2nd ed.
2 OXTOBY. Measure and Category. 2nd ed.
3 SCHAEFER. Topological Vector Spaces.
4 HILTON/STAMMBACH. A Course in Homological Algebra.
5 MAC LANE. Categories for the Working Mathematician.
6 HUGHES/PIPER. Projective Planes.
7 SERRE. A Course in Arithmetic.
8 TAKEUTI/ZARING. Axiomatic Set Theory.
9 HUMPHREYS. Introduction to Lie Algebras and Representation Theory.
10 COHEN. A Course in Simple Homotopy Theory.
11 CONWAY. Functions of One Complex Variable I. 2nd ed.
12 BEALS. Advanced Mathematical Analysis.
13 ANDERSON/FULLER. Rings and Categories of Modules. 2nd ed.
14 GOLUBITSKY/GUILLEMIN. Stable Mappings and Their Singularities.
15 BERBERIAN. Lectures in Functional Analysis and Operator Theory.
16 WINTER. The Structure of Fields.
17 ROSENBLATT. Random Processes. 2nd ed.
18 HALMOS. Measure Theory.
19 HALMOS. A Hilbert Space Problem Book. 2nd ed.
20 HUSEMOLLER. Fibre Bundles. 3rd ed.
21 HUMPHREYS. Linear Algebraic Groups.
22 BARNES/MACK. An Algebraic Introduction to Mathematical Logic.
23 GREUB. Linear Algebra. 4th ed.
24 HOLMES. Geometric Functional Analysis and Its Applications.
25 HEWITT/STROMBERG. Real and Abstract Analysis.
26 MANES. Algebraic Theories.
27 KELLEY. General Topology.
28 ZARISKI/SAMUEL. Commutative Algebra. Vol.I.
29 ZARISKI/SAMUEL. Commutative Algebra. Vol.II.
30 JACOBSON. Lectures in Abstract Algebra I. Basic Concepts.
31 JACOBSON. Lectures in Abstract Algebra II. Linear Algebra.
32 JACOBSON. Lectures in Abstract Algebra III. Theory of Fields and Galois Theory.
33 HIRSCH. Differential Topology.
34 SPITZER. Principles of Random Walk. 2nd ed.
35 WERMER. Banach Algebras and Several Complex Variables. 2nd ed.
36 KELLEY/NAMIOKA et al. Linear Topological Spaces.
37 MONK. Mathematical Logic.
38 GRAUERT/FRITZSCHE. Several Complex Variables.
39 ARVESON. An Invitation to C^*-Algebras.
40 KEMENY/SNELL/KNAPP. Denumerable Markov Chains. 2nd ed.
41 APOSTOL. Modular Functions and Dirichlet Series in Number Theory. 2nd ed.
42 SERRE. Linear Representations of Finite Groups.
43 GILLMAN/JERISON. Rings of Continuous Functions.
44 KENDIG. Elementary Algebraic Geometry.
45 LOÈVE. Probability Theory I. 4th ed.
46 LOÈVE. Probability Theory II. 4th ed.
47 MOISE. Geometric Topology in Dimensions 2 and 3.
48 SACHS/WU. General Relativity for Mathematicians.
49 GRUENBERG/WEIR. Linear Geometry. 2nd ed.
50 EDWARDS. Fermat's Last Theorem.
51 KLINGENBERG. A Course in Differential Geometry.
52 HARTSHORNE. Algebraic Geometry.
53 MANIN. A Course in Mathematical Logic.
54 GRAVER/WATKINS. Combinatorics with Emphasis on the Theory of Graphs.
55 BROWN/PEARCY. Introduction to Operator Theory I: Elements of Functional Analysis.
56 MASSEY. Algebraic Topology: An Introduction.
57 CROWELL/FOX. Introduction to Knot Theory.
58 KOBLITZ. p-adic Numbers, p-adic Analysis, and Zeta-Functions. 2nd ed.
59 LANG. Cyclotomic Fields.
60 ARNOLD. Mathematical Methods in Classical Mechanics. 2nd ed.

continued after index

Melvyn B. Nathanson

Additive Number Theory

Inverse Problems and the Geometry of Sumsets

Melvyn B. Nathanson
Department of Mathematics
Lehman College of the
City University of New York
250 Bedford Park Boulevard West
Bronx, NY 10468-1589 USA

Mathematics Subject Classifications (1991): 11-01, 11P99

Library of Congress Cataloging-in-Publication Data
Nathanson, Melvyn B. (Melvyn Bernard), 1944–
Additive number theory:inverse problems and the geometry of sumsets/Melvyn B. Nathanson.
p. cm. – (Graduate texts in mathematics;165)
Includes bibliographical references and index.
ISBN 0-387-94655-1 (hardcover:alk. paper)
1. Number theory. I. Title. II. Series.
QA241.N3468 1996
512′.73 – dc20 96-12929

Printed on acid-free paper.

Production managed by Hal Henglein; manufacturing supervised by Joe Quatela.
Camera-ready copy prepared from the author's LaTeX files.
Printed and bound by R.R. Donnelley & Sons, Harrisonburg, VA.
Printed in the United States of America.

9 8 7 6 5 4 3 2 1

ISBN 0-387-94655-1 Springer-Verlag New York Berlin Heidelberg SPIN 10490825

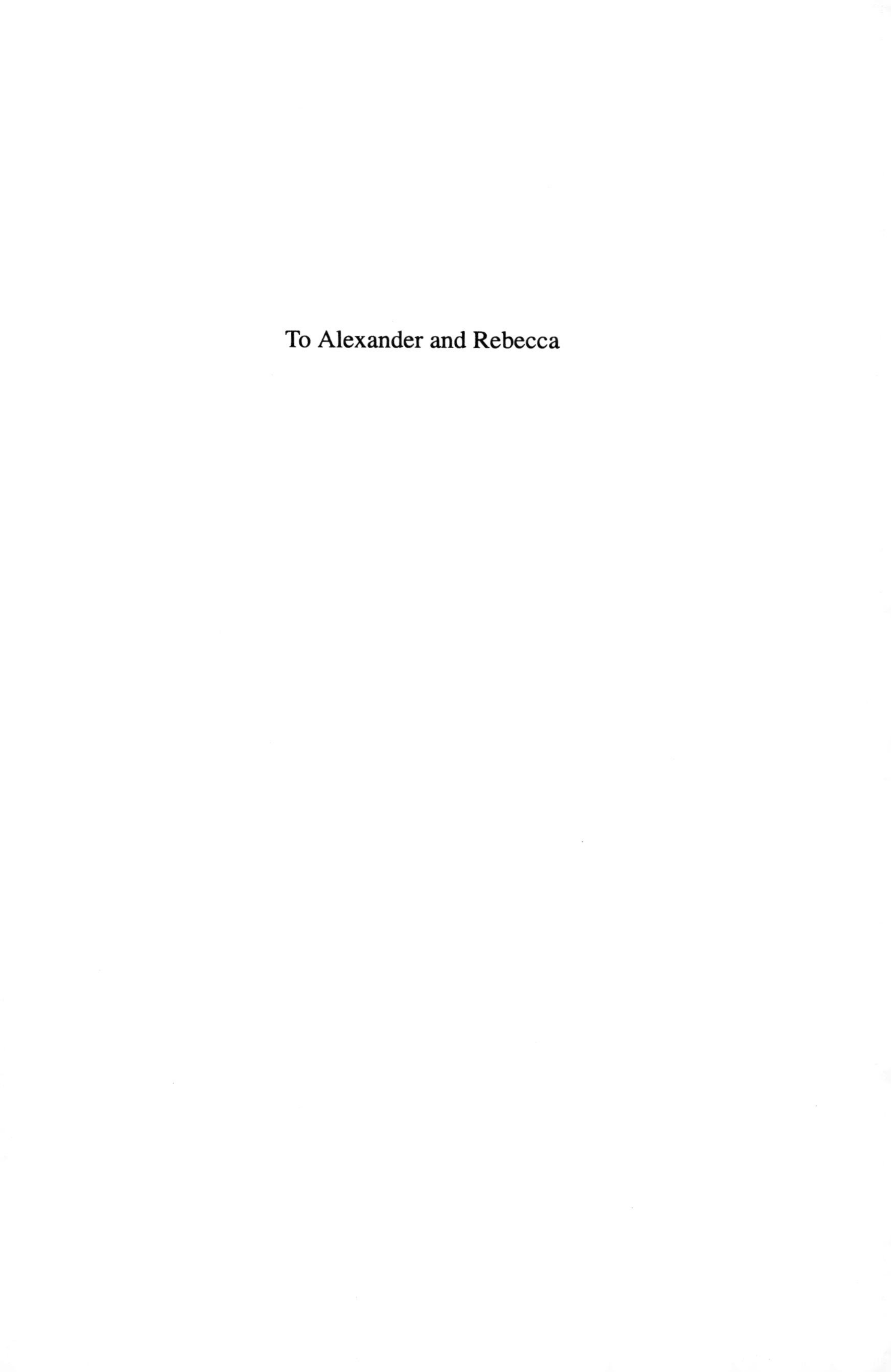
To Alexander and Rebecca

Preface

> Il est vrai que M. Fourier avait l'opinion que le but principal des mathématiques était l'uilité publique et l'explication des phénomènes naturels; mais un philosophe comme lui aurait dû savoir que le but unique de la science, c'est l'honneur de l'esprit humain, et que sous ce titre, une question de nombres vaut autant qu'une question du système du monde.[1]
>
> C. G. J. Jacobi [71, vol. I, p. 454]

The classical problems in additive number theory are *direct problems*, in which we start with a set A of integers and proceed to describe the h-fold sumset hA, that is, the set of all sums of h elements of A. In an *inverse problem*, we begin with the sumset hA and try to deduce information about the underlying set A. In the last few years, there has been remarkable progress in the study of inverse problems for finite sets in additive number theory. There are important inverse theorems due to Freiman, Kneser, Plünnecke, Vosper, and others. In particular, Ruzsa recently discovered a new method to prove a generalization of Freiman's theorem. One goal of this book is to present Ruzsa's beautiful proof.

The prerequisites for this book are undergraduate courses in elementary number theory, algebra, and analysis. Beyond this, the volume is self-contained. I include

[1]It is true that Fourier believed that the principal goal of mathematics was the public welfare and the understanding of nature, but as a philosopher he should have understood that the only goal of science is the honor of the human spirit, and, in this regard, a problem in number theory is as important as a problem in physics.

complete proofs of results from exterior algebra, combinatorics, graph theory, and the geometry of numbers that are used in the proofs of the Erdős–Heilbronn conjecture, Plünnecke's inequality, and Freiman's theorem. Indeed, a second goal of the book is to introduce different methods that have been used to obtain results in this field.

This is the second of several books on additive number theory. It is independent of the related volume *Additive Number Theory: The Classical Bases* [96], which is a study of the direct problems that are historically at the center of this subject. I had originally planned to write one short and comprehensive book on additive problems, but the project has become a long and complex enterprise. I am grateful to my publisher, Springer-Verlag, for its interest in and understanding of this work.

I wish to thank Antal Balog, Gregory Freiman, Yahya Ould Hamidoune, Vsevolod F. Lev, Öystein Rödseth, Imre Z. Ruzsa, and Endre Szemerédi, who provided me with preprints of their papers on additive number theory and made helpful comments on preliminary versions of this book. I also benefited greatly from a conference on Freiman's work that was organized by Jean-Marc Deshouillers at CIRM Marseille in June, 1993, and from a workshop on combinatorial number theory that was held at the Center for Discrete Mathematics and Theoretical Computer Science (DIMACS) of Rutgers University in February, 1996. Much of this book was written while I was on leave at the School of Mathematics of The Institute for Advanced Study, and at DIMACS. I am especially grateful to Henryk Iwaniec and the late Daniel Gorenstein for making it possible for me to work at Rutgers.

I have taught additive number theory at Southern Illinois University at Carbondale, Rutgers University—New Brunswick, and the Graduate Center of the City University of New York. I am grateful to the students and colleagues who participated in my graduate courses and seminars.

This work was supported in part by grants from the PSC-CUNY Research Award Program and the National Security Agency Mathematical Sciences Program.

I would very much like to receive comments or corrections from readers of this book. My e-mail addresses are nathansn@alpha.lehman.cuny.edu and nathanson@worldnet.att.net. A list of errata will be available on my homepage at http://www.lehman.cuny.edu or http://math.lehman.cuny.edu/nathanson.

Melvyn B. Nathanson
Maplewood, New Jersey
June 18, 1996

Contents

Notation

$\mathbf{N}$	The positive integers $1, 2, 3, \ldots$
$\mathbf{N}_0$	The nonnegative integers $0, 1, 2, \ldots$
$\mathbf{Z}$	The integers $0, \pm 1, \pm 2, \ldots$
$\mathbf{R}$	The real numbers
$\mathbf{R}^n$	n-dimensional Euclidean space
$\mathbf{Z}^n$	The integer lattice points in $\mathbf{R}^n$
$\mathbf{C}$	The complex numbers
$\|z\|$	The absolute value of the complex number z
$\Re z$	The real part of the complex number z
$\Im z$	The imaginary part of the complex number z
$[x]$	The integer part of the real number x
$\{x\}$	The fractional part of the real number x
$\\|x\\|$	The distance from the real number x to the nearest integer, that is, $\\|x\\| = \min(\{x\}, 1 - \{x\})$.
$(a_1, a_2, \ldots, a_k)$	The greatest common divisor of the integers $a_1, a_2, \ldots, a_k$
$[a_1, a_2, \ldots, a_k]$	The least common multiple of the integers $a_1, a_2, \ldots, a_k$
$[a, b]$	The interval of integers n such that $a \leq n \leq b$ (Context will always make clear whether $[a, b]$ denotes an interval of integers or the least common multiple of two integers.)
$Q(q_0; q_1, \ldots, q_n; l_1, \ldots, l_n)$	An n-dimensional arithmetic progression of integers
$G(V, E)$	A graph G with vertex set V and edge set E
$\|X\|$	The cardinality of the set X
hA	The h-fold sumset, consisting of all sums of h elements of A
$h^\wedge A$	The set of all sums of h distinct elements of A.
$A - B$	The difference set, consisting of all elements $a - b$ with $a \in A$ and $b \in B$
$hA - kA$	The difference set formed from the sumsets hA and kA
$\lambda * A$	The set of all elements of the form λa with $a \in A$

$f \ll g$	$\lvert f(x)\rvert \le c\lvert g(x)\rvert$ for some absolute constant c and all x in the domain of f
$f \ll_{a,b,\ldots} g$	$\lvert f(x)\rvert \le c\lvert g(x)\rvert$ for some constant c that depends on $a, b, \ldots$ and for all x in the domain of f

1

Simple inverse theorems

1.1 Direct and inverse problems

Additive number theory is the study of sums of sets of integers. Let $h \geq 2$, and let $A_1, A_2, \ldots, A_h$ be sets of integers. The *sumset*

$$A_1 + A_2 + \cdots + A_h$$

is the set of all integers of the form $a_1 + a_2 + \cdots + a_h$, where $a_i \in A_i$ for $i = 1, 2, \ldots, h$. If A is a set of integers and $A_i = A$ for $i = 1, 2, \ldots, h$, then we denote the sumset $A_1 + A_2 + \cdots + A_h$ by hA. Thus, the h-fold sumset hA is the set of all sums of h elements of A, with repetitions allowed.

Sumsets can also be defined in any abelian group and, indeed, in any set in which there is a binary operation. For example, we shall consider sumsets in the group $\mathbf{Z}/m\mathbf{Z}$ of congruence classes modulo m, and in the group $\mathbf{Z}^n$ of integer lattice points in $\mathbf{R}^n$.

A *direct problem* in additive number theory is a problem in which we try to determine the structure and properties of the h-fold sumset hA when the set A is known. An example of a direct theorem, indeed, the archetypical theorem in additive number theory, is Lagrange's theorem that every nonnegative integer can be written as the sum of four squares. Thus, if A is the set of all nonnegative squares, then the sumset $4A$ is the set of all nonnegative integers.

There is a simple and beautiful solution of the direct problem of describing the structure of the h-fold sumset hA for any finite set A of integers and for all sufficiently large h. We require the following notation.

Let A and B be sets of integers. Let $|A|$ denote the cardinality of A. We define the *difference set*

$$A - B = \{a - b : a \in A \text{ and } b \in B\}.$$

For any integers c and q, we define the sets

$$c + A = \{c\} + A,$$

$$c - A = \{c\} - A,$$

and

$$q * A = \{qa \mid a \in A\}.$$

Then $q * (A + B) = q * A + q * B$.

Denote by $(a_1, \ldots, a_k)$ the greatest common divisor of the integers $a_1, \ldots, a_k$. If $A = \{a_0, a_1, \ldots, a_{k-1}\}$ is a finite set of integers such that $a_0 < a_1 < \cdots < a_{k-1}$, we define

$$d(A) = (a_1 - a_0, a_2 - a_0, \ldots, a_{k-1} - a_0).$$

Let $a_i' = (a_i - a_0)/d(A)$ for $i = 0, 1, \ldots, k-1$, and let

$$A^{(N)} = \{a_0', a_1', \ldots, a_{k-1}'\}.$$

Clearly,

$$0 = a_0' < a_1' < \cdots < a_{k-1}',$$

$$d(A^{(N)}) = (a_1', \ldots, a_{k-1}') = 1,$$

$$A = a_0 + d * A^{(N)},$$

and

$$hA = \{ha_0\} + d(A) * hA^{(N)}.$$

It follows that

$$|hA| = |hA^{(N)}|. \tag{1.1}$$

The set $A^{(N)}$ is called the *normal form* of the set A.

Let $[a, b]$ denote the *interval of integers* n such that $a \leq n \leq b$.

For example, if $A = \{8, 29, 71, 92\}$ and $h = 2$, then $d(A) = 21$, $A^{(N)} = \{0, 1, 3, 4\}$, $2A^{(N)} = [0, 8]$, and $2A = \{16 + 21n : n \in [0, 8]\}$.

Lemma 1.1 *Let $k \geq 2$ and let $a_1, \ldots a_{k-1}$ be positive integers such that*

$$(a_1, \ldots, a_{k-1}) = 1.$$

If

$$(a_{k-1} - 1)\sum_{i=1}^{k-2} a_i \leq n \leq ha_{k-1} - (k-2)(a_{k-1} - 1)a_{k-1},$$

then there exist nonnegative integers $u_1, \ldots, u_{k-1}$ such that

$$n = u_1 a_1 + \cdots + u_{k-1} a_{k-1}$$

and

$$u_1 + \cdots + u_{k-1} \leq h.$$

Proof. Since $(a_1, \ldots, a_{k-1}) = 1$, there exist integers $x_1, \ldots, x_{k-1}$ such that

$$n = x_1 a_1 + \cdots + x_{k-1} a_{k-1}.$$

For $i = 1, \ldots, k-2$, let u_i be the least nonnegative residue of x_i modulo a_{k-1}. Then

$$\begin{aligned} n &\equiv x_1 a_1 + \cdots + x_{k-2} a_{k-2} \pmod{a_{k-1}} \\ &\equiv u_1 a_1 + \cdots + u_{k-2} a_{k-2} \pmod{a_{k-1}}, \end{aligned}$$

and so there exists an integer u_{k-1} such that

$$n = u_1 a_1 + \cdots + u_{k-2} a_{k-2} + u_{k-1} a_{k-1}.$$

Since $0 \leq u_i \leq a_{k-1} - 1$ for $i = 1, \ldots, k-2$, it follows that

$$u_{k-1} a_{k-1} = n - (u_1 a_1 + \cdots + u_{k-2} a_{k-2}) \geq n - (a_{k-1} - 1) \sum_{i=1}^{k-2} a_i \geq 0,$$

and so $u_{k-1} \geq 0$. Similarly,

$$u_{k-1} a_{k-1} \leq n \leq h a_{k-1} - (k-2)(a_{k-1} - 1) a_{k-1}$$

and

$$u_{k-1} \leq h - (k-2)(a_{k-1} - 1).$$

It follows that

$$u_1 + \cdots + u_{k-2} + u_{k-1} \leq (k-2)(a_{k-1} - 1) + u_{k-1} \leq h.$$

This completes the proof.

By (1.1), the structure of the sumset hA is completely determined by the structure of the sumset $hA^{(N)}$, and so it suffices to consider only finite sets in normal form.

Theorem 1.1 (Nathanson) *Let $k \geq 2$ and let $A = \{a_0, a_1, \ldots, a_{k-1}\}$ be a finite set of integers such that*

$$0 = a_0 < a_1 < \cdots < a_{k-1}$$

and

$$(a_1, \ldots, a_{k-1}) = 1.$$

Then there exist integers c and d and sets $C \subseteq [0, c-2]$ and $D \subseteq [0, d-2]$ such that

$$hA = C \cup \left[c, ha_{k-1} - d\right] \cup (ha_{k-1} - D) \tag{1.2}$$

for all $h \geq \max(1, (k-2)(a_{k-1} - 1) a_{k-1})$.

Proof. If $k = 2$, then $a_1 = 1$, $A = \{0, 1\}$, $hA = [0, h]$, and the theorem holds with $c = d = 0$ for all $h \geq 1$.

Let $k \geq 3$. Then $a_{k-1} \geq 2$. We define

$$h_0 = (k-2)(a_{k-1} - 1)a_{k-1}. \tag{1.3}$$

Then

$$h_0 \geq (a_{k-1} - 1)\left(1 + \sum_{i=1}^{k-2} a_i\right) \tag{1.4}$$

and

$$\begin{aligned} h_0 a_{k-1} &\geq 2h_0 \\ &\geq (k-2)(a_{k-1} - 1)a_{k-1} + a_{k-1} - 1 \\ &\quad +(a_{k-1} - 1)\sum_{i=1}^{k-2} a_i. \end{aligned} \tag{1.5}$$

The theorem is proved by induction on $h \geq h_0$. Choose integers c and d such that $[c, h_0 a_{k-1} - d]$ is the largest interval of integers satisfying

$$\begin{aligned} \left[(a_{k-1} - 1)\sum_{i=1}^{k-2} a_i, h_0 a_{k-1} - (k-2)(a_{k-1} - 1)a_{k-1}\right] &\subseteq [c, h_0 a_{k-1} - d] \\ &\subseteq h_0 A. \end{aligned}$$

Lemma 1.1 implies that this maximal interval exists. It follows that $c - 1 \notin h_0 A$ and $h_0 a_{k-1} - (d - 1) \notin h_0 A$. Moreover,

$$c \leq (a_{k-1} - 1)\sum_{i=1}^{k-2} a_i < h_0 \leq h \tag{1.6}$$

and

$$d \leq (k-2)(a_{k-1} - 1)a_{k-1}. \tag{1.7}$$

It follows that

$$\begin{aligned} c + d &\leq (a_{k-1} - 1)\sum_{i=1}^{k-2} a_i + (k-2)(a_{k-1} - 1)a_{k-1} \\ &\leq h_0 a_{k-1} - a_{k-1} + 1, \end{aligned}$$

and so

$$[c, c + a_{k-1} - 1] \subseteq [c, h_0 a_{k-1} - d]. \tag{1.8}$$

Let C and D be the finite sets of integers defined by

$$C = h_0 A \cap [0, c-2]$$

and

$$h_0 a_{k-1} - D = h_0 A \cap [h_0 a_{k-1} - (d-2), h_0 a_{k-1}].$$

Then $D \subseteq [0, d-2]$ and

$$h_0 A = C \cup [c, h_0 a_{k-1} - d] \cup (h_0 a_{k-1} - D).$$

Thus, (1.2) holds for h_0.

Suppose that (1.2) is true for some $h \geq h_0$. Let

$$\begin{aligned} B &= C \cup [c, (h+1)a_{k-1} - d] \cup ((h+1)a_{k-1} - D) \\ &= C \cup [c, c + a_{k-1} - 1] \cup [c + a_{k-1}, (h+1)a_{k-1} - d] \\ &\quad \cup ((h+1)a_{k-1} - D). \end{aligned}$$

The second equality follows from (1.8).

Since $0 \in A$, we have $hA \subseteq (h+1)A$ and so

$$\begin{aligned} C \cup [c, c + a_{k-1} - 1] &\subseteq C \cup [c, h_0 a_{k-1} - d] \\ &\subseteq hA \\ &\subseteq (h+1)A. \end{aligned}$$

Since $a_{k-1} \in A$, it follows that $a_{k-1} + hA \subseteq (h+1)A$ and

$$\begin{aligned} [c + a_{k-1}, (h+1)a_{k-1} - d] &\subseteq a_{k-1} + [c, ha_{k-1} - d] \\ &\subseteq a_{k-1} + hA \\ &\subseteq (h+1)A. \end{aligned}$$

Similarly,

$$\begin{aligned} (h+1)a_{k-1} - D &= a_{k-1} + (ha_{k-1} - D) \\ &\subseteq (h+1)A. \end{aligned}$$

Therefore, $B \subseteq (h+1)A$.

Let $b \in (h+1)A$. If $b < c$, then (1.6) implies that b cannot be the sum of $h+1$ nonzero elements of A, so $b \in hA$, hence $b \in C \subseteq B$. If $c \leq b \leq c + a_{k-1} - 1$, then $b \in [c, c + a_{k-1} - 1] \subseteq B$.

Suppose that $b \in (h+1)A$ and $b \geq c + a_{k-1}$. If $b - a_{k-1} \notin hA$, then b is the sum of $h+1$ elements of A that are all strictly less than a_{k-1}, and so

$$b \leq (h+1)(a_{k-1} - 1). \tag{1.9}$$

Since $[c, ha_{k-1} - d] \subseteq hA$, the conditions $b - a_{k-1} \geq c$ and $b - a_{k-1} \notin hA$ imply that

$$b - a_{k-1} > ha_{k-1} - d \geq ha_{k-1} - (k-2)(a_{k-1} - 1)a_{k-1}. \tag{1.10}$$

Combining inequalities (1.9) and (1.10), we obtain

$$h + 1 < (k-2)(a_{k-1} - 1)a_{k-1} = h_0 \leq h$$

which is absurd. Therefore, $b - a_{k-1} \in hA$. By the induction hypothesis, either

$$b \in a_{k-1} + [c, ha_{k-1} - d] = [c + a_{k-1}, (h+1)a_{k-1} - d] \subseteq B$$

or

$$b \in a_{k-1} + (ha_{k-1} - D) = ((h+1)a_{k-1} - D) \subseteq B.$$

Therefore, $(h+1)A \subseteq B$. This completes the proof.

An *inverse problem* in additive number theory is a problem in which we attempt to deduce properties of the set A from properties of the sumset hA. For example, if A is a finite set of integers and if the cardinality of the h-fold sumset hA is small, what can we conclude about the structure of the set A?

The following result is the simplest inverse theorem in additive number theory.

Theorem 1.2 *If A is a set of k integers, then $|2A| \geq 2k - 1$. If A is a set of k integers and if $|2A| = 2k - 1$, then A is an arithmetic progression.*

Proof. Let

$$A = \{a_0, a_1, a_2, \ldots, a_{k-1}\},$$

where

$$a_0 < a_1 < a_2 < \cdots < a_{k-1}.$$

Then the sumset $2A$ contains the k integers $2a_i$ for $i = 0, 1, \ldots, k-1$, and the $k-1$ integers $a_{i-1} + a_i$ for $i = 1, \ldots, k-1$. Since

$$2a_{i-1} < a_{i-1} + a_i < 2a_i$$

for $i = 1, \ldots, k-1$, it follows that $|2A| \geq 2k - 1$.

If $|2A| = 2k - 1$, then every element of $2A$ is of the form $2a_i$ or $a_{i-1} + a_i$. Since

$$a_{i-1} + a_i < a_{i-1} + a_{i+1} < a_i + a_{i+1}$$

and

$$a_{i-1} + a_i < 2a_i < a_i + a_{i+1}$$

for $i = 1, \ldots, k-2$, it follows that

$$2a_i = a_{i-1} + a_{i+1}$$

or, equivalently,

$$a_i - a_{i-1} = a_{i+1} - a_i$$

for $i = 1, \ldots, k-2$. Therefore, A is an arithmetic progression. This completes the proof.

The most important inverse theorems in this book are due to Freiman, Kneser, Plünnecke, and Vosper. Vosper's theorem is an inverse to the Cauchy–Davenport theorem (Theorem 2.2), which states that if A and B are nonempty sets of congruence classes modulo p, then

$$|A + B| \geq \min(p, |A| + |B| - 1).$$

The Cauchy–Davenport theorem is a direct theorem in additive number theory. Vosper's inverse theorem (Theorem 2.7) describes the structure of the "critical pairs" A and B such that $A + B \neq \mathbf{Z}/p\mathbf{Z}$ and $|A + B| = |A| + |B| - 1$. In particular, if $|A| \neq 1$ and $|2A| < 2|A| < p$, then A is an arithmetic progression in the group $\mathbf{Z}/p\mathbf{Z}$.

A *finite n-dimensional arithmetic progression* is a set of the form

$$\{q_0 + x_1 q_1 + \cdots + x_n q_n : 0 \leq x_i < l_i \text{ for } i = 1, \ldots, n\}.$$

In 1964, Freiman [53] discovered a deep and beautiful fact (Theorem 8.10) about the structure of finite sets of integers with small sumsets. Let $c \geq 2$. If A is a finite set of integers such that $|A| = k$ and

$$|2A| \leq ck,$$

then A is a subset of an n-dimensional arithmetic progression Q, where $|Q| \leq c'k$ and n and c' are constants that depend only on c. Ruzsa has extended this result to sumsets of the form $A + B$, where A and B are finite subsets of any torsion-free abelian group.

Nothing is known, however, about the structure of the finite set A if, for example, $|A| = k$ and

$$|2A| \leq k^{1+\delta}$$

for some $\delta > 0$, or even if

$$|2A| \leq ck \log k.$$

Nor is anything known about the structure of A if, for some $h \geq 3$,

$$|hA| \leq ck^{h-1}$$

or even

$$|hA| \leq ck^2.$$

These are important unsolved inverse problems.

1.2 Finite arithmetic progressions

Let k and q be positive integers, and let a_0 be an integer. An *arithmetic progression* of length k with difference q and initial term a_0 is a set of the form

$$\{a_0, a_0 + q, a + 2q, \ldots, a_0 + (k-1)q\} = a_0 + q * [0, k-1].$$

Let $A_1, \ldots, A_h$ be nonempty, finite sets of integers, and let $|A_i| = k_i$ for $i = 1, \ldots, k$. We shall prove that

$$|A_1 + \cdots + A_h| \geq |A_1| + \cdots + |A_h| - (h-1)$$

and that this lower bound is attained if the sets A_i are arithmetic progressions with the same common difference. This is a direct theorem. The corresponding inverse theorem, Theorem 1.5, states that the lower bound is attained only for arithmetic progressions with the same common difference. The proof is easy, but much of the rest of this book will be devoted to proving an inverse theorem of Freiman (Theorem 8.10) that generalizes this result.

The following are simple lower and upper bounds for the cardinality of sums of finite sets of integers.

Theorem 1.3 *Let $h \geq 2$. Let A be a finite set of integers with $|A| = k$. Then*

$$hk - (h-1) \leq |hA| \leq \binom{k+h-1}{h} = \frac{k^h}{h!} + O(k^{h-1}).$$

Proof. Let $A = \{a_0, a_1, \ldots, a_{k-1}\}$, where $a_0 < a_1 < \cdots < a_{k-1}$. Then

$$hA \supseteq \{ha_0\} \cup \bigcup_{j=1}^{k-1} \{(h-i)a_{j-1} + ia_j : i \in [1,h]\}.$$

Since

$$ha_{j-1} < (h-1)a_{j-1} + a_j < \cdots < a_{j-1} + (h-1)a_j < ha_j,$$

for $j = 1, \ldots, k-1$, it follows that

$$|hA| \geq 1 + (k-1)h = hk - (h-1).$$

This gives the lower bound.

The upper bound follows from the combinatorial fact (Exercise 5) that the number of expressions of the form $a_{i_1} + \cdots + a_{i_h}$ with $a_{i_j} \in A$ for $j = 1, 2, \ldots, h$ and $0 \leq i_1 \leq \cdots \leq i_h \leq k-1$ is exactly

$$\binom{k+h-1}{h} = \frac{k(k+1)(k+2)\cdots(k+h-1)}{h!}.$$

Theorem 1.4 *Let $h \geq 2$, and let $A_1, A_2, \ldots, A_h$ be finite sets of integers. Then*

$$|A_1| + \cdots + |A_h| - (h-1) \leq |A_1 + \cdots + A_h| \leq |A_1| \cdots |A_h|.$$

Proof. We shall prove the lower bound by induction on h. Let $h = 2$, and let $A_1 = \{a_0, a_1, \ldots, a_{k-1}\}$ and $A_2 = \{b_0, b_1, \ldots, b_{l-1}\}$, where $a_0 < a_1 < \cdots < a_{k-1}$ and $b_0 < b_1 < \cdots < b_{l-1}$. Suppose that $|A_1| = k \leq l = |A_2|$. The sumset $A_1 + A_2$ contains the distinct elements

$$\begin{aligned} a_0 + b_0 &< a_0 + b_1 < a_1 + b_1 < a_1 + b_2 < \cdots \\ &< a_i + b_i < a_i + b_{i+1} < a_{i+1} + b_{i+1} < \cdots \\ &< a_{k-1} + b_{k-1} < a_{k-1} + b_k < \cdots < a_{k-1} + b_{l-1}, \end{aligned}$$

and so

$$|A_1 + A_2| \geq (2k - 1) + (l - k) = |A_1| + |A_2| - 1.$$

Let $h \geq 3$, and suppose that the lower bound holds for the sum of any $h - 1$ finite sets of integers. Then

$$\begin{aligned}
|A_1 + \cdots + A_{h-1} + A_h| &= |(A_1 + \cdots + A_{h-1}) + A_h| \\
&\geq |A_1 + \cdots + A_{h-1}| + |A_h| - 1 \\
&\geq |A_1| + \cdots + |A_{h-1}| - (h - 2) + |A_h| - 1 \\
&= |A_1| + \cdots + |A_{h-1}| + |A_h| - (h - 1).
\end{aligned}$$

The upper bound is a consequence of the fact that the number of expressions of the form $a_1 + \cdots + a_h$ with $a_i \in A_i$ for $i = 1, 2, \ldots, h$ is exactly $|A_1| \cdots |A_h|$. This completes the proof.

Lemma 1.2 *Let A and B be finite sets of integers with $|A| = |B| = k$. If $|A + B| = |A| + |B| - 1$, then A and B are arithmetic progressions with the same common difference.*

Proof. Let $A = \{a_0, a_1, \ldots, a_{k-1}\}$ and $B = \{b_0, b_1, \ldots, b_{k-1}\}$, where $a_0 < a_1 < \cdots < a_{k-1}$ and $b_0 < b_1 < \cdots < b_{k-1}$. The sumset $A + B$ contains the following strictly increasing sequence of $2k - 1$ integers:

$$\begin{aligned}
a_0 + b_0 &< a_0 + b_1 < a_1 + b_1 < a_1 + b_2 \cdots \\
&< a_{i-1} + b_i < a_i + b_i < a_i + b_{i+1} < a_{i+1} + b_{i+1} \\
&< \cdots < a_{k-1} + b_{k-1}.
\end{aligned}$$

Since $|A + B| = 2k - 1$, it follows that this sequence of integers includes all the integers in $A + B$. Since

$$a_{i-1} + b_i < a_i + b_i < a_i + b_{i+1}$$

and

$$a_{i-1} + b_i < a_{i-1} + b_{i+1} < a_i + b_{i+1},$$

it follows that

$$a_{i-1} + b_{i+1} = a_i + b_i$$

or, equivalently,

$$a_i - a_{i-1} = b_{i+1} - b_i \tag{1.11}$$

for $i = 1, \ldots, k - 2$. Similarly, the inequalities

$$a_{i-1} + b_{i-1} < a_{i-1} + b_i < a_i + b_i$$

and

$$a_{i-1} + b_{i-1} < a_i + b_{i-1} < a_i + b_i$$

imply that

$$a_{i-1} + b_i = a_i + b_{i-1}$$

or, equivalently,

$$a_i - a_{i-1} = b_i - b_{i-1} \tag{1.12}$$

for $i = 1, \ldots, k-1$. Equations (1.11) and (1.12) imply that the positive integer $q = a_1 - a_0$ satisfies

$$a_i - a_{i-1} = b_i - b_{i-1} = q$$

for $i = 0, \ldots, k-2$. This completes the proof.

Lemma 1.3 *Let A and B be finite sets of integers with $|A| = k \geq 2$ and $|B| = \ell \geq 2$. If $|A+B| = k+\ell-1$, then A and B are arithmetic progressions with the same common difference.*

Proof. Let $A = \{a_0, a_1, \ldots, a_{k-1}\}$ and $B = \{b_0, b_1, \ldots, b_{\ell-1}\}$, where $a_0 < \cdots < a_{k-1}$ and $b_0 < \cdots < b_{\ell-1}$. Suppose that $k \leq \ell$, and let $0 \leq t \leq \ell - k$. Let $B = B_0^{(t)} \cup B_1^{(t)} \cup B_2^{(t)}$, where

$$\begin{aligned}
B_0^{(t)} &= \{b_0, b_1, \ldots, b_{t-1}\}, \\
B_1^{(t)} &= \{b_t, b_{t+1}, \ldots, b_{t+k-1}\}, \\
B_2^{(t)} &= \{b_{t+k}, b_{t+k+1}, \ldots, b_{\ell-1}\}.
\end{aligned}$$

Then

$$A + B \supseteq \left(a_0 + B_0^{(t)}\right) \cup \left(A + B_1^{(t)}\right) \cup \left(a_{k-1} + B_2^{(t)}\right). \tag{1.13}$$

The three sumsets on the right side of (1.13) are pairwise disjoint, since

$$\begin{aligned}
a_0 + B_0^{(t)} &\subseteq [a_0 + b_0, a_0 + b_{t-1}], \\
A + B_1^{(t)} &\subseteq [a_0 + b_t, a_{k-1} + b_{t+k-1}], \\
a_{k-1} + B_2^{(t)} &\subseteq [a_{k-1} + b_{t+k}, a_{k-1} + b_{\ell-1}].
\end{aligned}$$

Moreover,

$$\begin{aligned}
|a_0 + B_0^{(t)}| &= t, \\
|A + B_1^{(t)}| &\geq |A| + |B_1^{(t)}| - 1 = 2k - 1, \\
|a_{k-1} + B_2^{(t)}| &= \ell - t - k.
\end{aligned}$$

It follows that

$$\begin{aligned}
k + \ell - 1 &= |A + B| \\
&\geq |a_0 + B_0^{(t)}| + |A + B_1^{(t)}| + |a_{k-1} + B_2^{(t)}| \\
&\geq t + (2k - 1) + (\ell - t - k) \\
&= k + \ell - 1,
\end{aligned}$$

and so

$$|A + B_1^{(t)}| = 2k - 1$$

for $t = 0, 1, \ldots, \ell - k$. By Lemma 1.2, there exists a positive integer q such that the set A is an arithmetic progression with difference q and the sets $B_1^{(t)}$ are arithmetic progressions with difference q for $t = 0, 1, \ldots, \ell - k$. Therefore, B is an arithmetic progression with difference q.

Theorem 1.5 *Let $h \geq 2$, and let $A_1, A_2, \ldots, A_h$ be h nonempty finite sets of integers. Then*

$$|A_1 + \cdots + A_h| = |A_1| + \cdots + |A_h| - (h - 1) \tag{1.14}$$

if and only if the sets $A_1, \ldots, A_h$ are arithmetic progressions with the same common difference.

Proof. Let $|A_i| = k_i$ for $i = 1, \ldots, h$. We can assume without loss of generality that $k_i \geq 2$ for all i. If $A_i = [0, k_i - 1]$ for $i = 1, \ldots, h$, then

$$A_1 + \cdots A_h = [0, k_1 + \cdots + k_h - h]$$

and so

$$|A_1 + \cdots + A_h| = k_1 + \cdots + k_h - h + 1 = |A_1| + \cdots + |A_h| - (h - 1).$$

Let the sets A_i be arithmetic progressions with the same common difference q. Then there exist integers $a_{0,i}$ such that

$$A_i = a_{0,i} + q * [0, k_i - 1].$$

It follows that

$$A_1 + \cdots A_h = (a_{0,1} + \cdots + a_{0,h}) + q * [0, k_1 + \cdots + k_h - h].$$

This implies (1.14).

We shall prove that equation (1.14) implies that the sets A_i are arithmetic progressions with the same common difference. The proof is by induction on h. The case $h = 2$ is Lemma 1.3 with $A = A_1$ and $B = A_2$. Now let $h \geq 3$, and assume that the theorem holds in the case of $h - 1$ sets. For $j = 1, \ldots, h$, consider the sumset

$$B_j = A_1 + \cdots + A_{j-1} + A_{j+1} + \cdots + A_h.$$

Then

$$|B_j| \geq \sum_{\substack{i=1 \\ i \neq j}}^{h} |A_i| - (h - 2).$$

Since

$$\begin{aligned}\sum_{i=1}^{h}|A_i| - (h-1) &= |A_j + B_j| \\ &\geq |A_j| + |B_j| - 1 \\ &\geq \sum_{i=1}^{h}|A_i| - (h-1),\end{aligned}$$

it follows that

$$|B_j| = \sum_{\substack{i=1 \\ i\neq j}}^{h} |A_i| - (h-2),$$

and so the $h-1$ sets $A_1, \ldots, A_{j-1}, A_{j+1}, \ldots, A_h$ are arithmetic progressions with the same common difference for $j =, 1, \ldots, h$. This completes the proof.

Theorem 1.6 *Let $h \geq 2$. Let A be a finite set of integers with $|A| = k$. Then $|hA| = hk - (h-1)$ if and only if A is a k-term arithmetic progression.*

Proof. This follows from Theorem 1.5 by letting $A_i = A$ for $i = 1, \ldots, h$.

Theorem 1.7 *Let A be a finite set of integers with $|A| = k$. Then $|hA| = hk-(h-1)$ if and only if $A^{(N)} = [0, k-1]$.*

Proof. This follows from Theorem 1.6 and the fact that the set A is an arithmetic progression if and only if its normal form $A^{(N)}$ is an interval of integers.

Let $o(h)$ denote an arithmetic function such that $\lim_{h\to\infty} o(h) = 0$.

Theorem 1.8 *Let A be a finite set of integers with $|A| = k$. If*

$$|hA| = hk - (h-1) + o(h)$$

for infinitely many h, then A is a k-term arithmetic progression.

Proof. Let $A^{(N)}$ be the normal form of A. Then $A^{(N)} = \{a'_0, a'_1, \ldots, a'_{k-1}\}$, where

$$0 = a'_0 < a'_1 < \cdots < a'_{k-1}$$

and

$$(a'_1, \ldots, a'_{k-1}) = 1.$$

Also,

$$k - 1 \leq a'_{k-1}.$$

Theorem 1.1 implies that

$$|hA^{(N)}| = ha'_{k-1} + 1 - r$$

for some integer $r = r(A) \geq 0$ and all $h \geq h_0$. If

$$ha'_{k-1} + 1 - r = |hA^{(N)}| = |hA| = hk - (h-1) + o(h)$$

for infinitely many h, then

$$a'_{k-1} = k - 1 + \frac{r + o(h)}{h} = k - 1,$$

and so $A^{(N)} = [0, k - 1]$.

Theorem 1.6 is a simple example of an inverse theorem in additive number theory: If A is a finite set of integers and if, for some $h \geq 2$, the cardinality of the sumset hA is as small as possible, then A must be an arithmetic progression. By Theorem 1.7, the normal form $A^{(N)}$ must be an interval, and so A is the affine image of an interval of consecutive integers. The set $A^{(N)}$ can also be described as the set of lattice points contained inside some convex subset of the real line. This geometrical point of view is important in connection with Freiman's inverse theorem.

An important and general inverse problem in additive number theory can be stated as follows: Let A be a finite set of integers. Suppose that the sumset hA is "small." What does this imply about the arithmetic or geometric structure of A? In Section 1.5, we shall prove that if $|A| = k$ and $|2A| \leq 3k - 4$, then A is a "large" subset of a "small" arithmetic progression.

1.3 An inverse problem for distinct summands

Let $A = \{a_0, a_1, \ldots, a_{k-1}\}$ be a nonempty finite set of integers, where

$$|A| = k$$

and

$$a_0 < a_1 < \cdots < a_{k-1}.$$

For $h \geq 1$, let $h^\wedge A$ denote the set of all sums of h distinct elements of A. If $h > k$, then $h^\wedge A = \emptyset$. We define $0^\wedge A = \{0\}$.

The direct problem for $h^\wedge A$ is to find a lower bound for $|h^\wedge A|$. The inverse problem for $h^\wedge A$ is determine the structure of the finite sets A of integers for which $|h^\wedge A|$ is minimal. In this section, we solve these two problems.

Let A' be a subset of A. We define the *subset sum*

$$s(A') = \sum_{a \in A'} a.$$

In particular, $s(\emptyset) = 0$. Then

$$h^\wedge A = \{s(A') : A' \subseteq A, |A'| = h\}$$

and so $0^\wedge A = \{0\}$. If $A' \subseteq A$ and $|A'| = h$, then $|A \setminus A'| = k - h$ and

$$s(A') + s(A \setminus A') = s(A).$$

This identity establishes a natural bijection

$$\Phi : h^\wedge A \to (k-h)^\wedge A$$

defined by

$$\Phi(s(A')) = s(A) - s(A').$$

It follows that

$$|h^\wedge A| = |(k-h)^\wedge A| \tag{1.15}$$

for $h = 0, 1, \ldots, k$.

Theorem 1.9 *Let A be a set of k integers, and let $1 \leq h \leq k$. Then*

$$|h^\wedge A| \geq hk - h^2 + 1 = h(k-h) + 1. \tag{1.16}$$

This lower bound is best possible.

Proof. Let $A = \{a_0, a_1, \ldots, a_{k-1}\}$ be a finite set of integers, where

$$a_0 < a_1 < \cdots < a_{k-1}.$$

For $i = 0, 1, \ldots, k-h-1$ and $j = 0, 1, \ldots, h$, we define

$$s_{i,j} = \sum_{\substack{l=0 \\ l \neq h-j}}^{h} a_{i+l}. \tag{1.17}$$

Let

$$s_{k-h,0} = \sum_{l=0}^{h-1} a_{k-h+l}. \tag{1.18}$$

Each of these numbers is a sum of h distinct elements of A, and so $s_{i,j} \in h^\wedge A$ for all i and j. Moreover, for $i = 0, 1, \ldots, k-h-1$, we have

$$s_{i,h} = \sum_{l=1}^{h} a_{i+l} = \sum_{l=0}^{h-1} a_{i+1+l} = s_{i+1,0}.$$

For $j = 0, 1, \ldots, h-1$, we have

$$s_{i,j+1} - s_{i,j} = a_{i+h-j} - a_{i+h-j-1} > 0.$$

It follows that

$$s_{i,0} < s_{i,1} < s_{i,2} < \cdots < s_{i,h-1} < s_{i,h} = s_{i+1,0}$$

and so

$$|h^\wedge A| \geq h(k-h) + 1 = hk - h^2 + 1.$$

This proves (1.16). Let $A = [0, k-1]$. Since

$$h^\wedge A = \left[\binom{h}{2}, hk - \binom{h+1}{2}\right] = \binom{h}{2} + [0, hk - h^2],$$

it follows that the lower bound in Theorem 1.9 is best possible.

An inverse problem for $h^\wedge A$ is to determine the extremal sets A such that

$$|h^\wedge A| = hk - h^2 + 1 = h(k-h) + 1. \tag{1.19}$$

If a_0 and q are integers and $q \neq 0$, then

$$h^\wedge(a_0 + q * A) = ha_0 + q * h^\wedge A,$$

and so

$$|h^\wedge(a_0 + q * A)| = |h^\wedge A|. \tag{1.20}$$

This means that the function $|h^\wedge A|$ is an *affine invariant* of the set A. Since every interval A of length k satisfies condition (1.19), it follows from (1.20) that every k-term arithmetic progression also satisfies (1.19). Let $|A| = k$ and $h \in [0, k]$. The symmetry (1.15) implies that if A is an arithmetic progression whenever $|h^\wedge A|$ satisfies (1.19), then A is also an arithmetic progression whenever $|(k-h)^\wedge A|$ satisfies (1.19).

Not all extremal sets are arithmetic progressions. Here are some examples:

(i) Let A be any set of k integers. If $h = 0$ or $h = k$, then $h(k-h) + 1 = 1$ and

$$|0^\wedge A| = |h^\wedge A| = 1.$$

(ii) Let A be any set of k integers. If $h = 1$ or $h = k-1$, then $h(k-h) + 1 = k$ and

$$|1^\wedge A| = |(h-1)^\wedge A| = k.$$

(iii) If $h = 2$ and $k = 4$, then $h(k-h) + 1 = 5$. Let

$$A = \{a_0, a_1, a_2, a_3\}$$

be a set of integers such that $a_0 < a_1 < a_2 < a_3$. Then

$$2^\wedge A = \{a_0 + a_1, a_0 + a_2, a_0 + a_3, a_1 + a_2, a_1 + a_3, a_2 + a_3\}$$

and so $|2^\wedge A| = 5$ or 6. Since

$$a_0 + a_1 < a_0 + a_2 < a_0 + a_3 < a_1 + a_3 < a_2 + a_3$$

and

$$a_0 + a_2 < a_1 + a_2 < a_1 + a_3,$$

it follows that $|2^\wedge A| = 5$ if and only if

$$a_1 - a_0 = a_3 - a_2.$$

Thus, $\{a_0, a_1, a_2, a_2 + a_1 - a_0\}$ is an extremal set for all $a_0 < a_1 < a_2$.

We shall prove that these three examples are the only examples of extremal sets that are not arithmetic progessions.

Theorem 1.10 *Let $k \geq 5$, and let $2 \leq h \leq k-2$. If A is a set of k integers such that*

$$|h^\wedge A| = hk - h^2 + 1,$$

then A is an arithmetic progression.

Proof. Let $A = \{a_0, a_1, \ldots, a_{k-1}\}$, where

$$a_0 < a_1 \cdots < a_{k-1}.$$

It follows from the proof of Theorem 1.9 that the set $h^\wedge A$ consists precisely of the numbers $s_{i,j}$ defined in (1.17) and (1.18). Let $i = 0, 1, \ldots, k-h-2$ and $j = 2, 3, \ldots, h$. Then

$$s_{i,j} = \left(\sum_{\substack{l=0 \\ l \neq h-j}}^{h-1} a_{i+l} \right) + a_{i+h}$$

and

$$s_{i,1} < s_{i,2} < s_{i,3} < \cdots < s_{i,h} = s_{i+1,0} < s_{i+1,1}.$$

Consider the integers

$$u_{i,j} = \left(\sum_{\substack{l=0 \\ l \neq h+1-j}}^{h-1} a_{i+l} \right) + a_{i+h+1} \in S_h(A).$$

Since

$$s_{i,1} < u_{i,2} < u_{i,3} < \cdots < u_{i,h} < s_{i+1,1},$$

it follows that

$$s_{i,j} = u_{i,j}$$

and so

$$a_{i+h-j+1} + a_{i+h} = a_{i+h-j} + a_{i+h+1}.$$

Thus, for each $i = 0, 1, \ldots, k-h-2$, we have

$$a_{i+h-j+1} - a_{i+h-j} = a_{i+h+1} - a_{i+h} \tag{1.21}$$

for $j = 2, 3, \ldots, h$ or, equivalently,

$$\begin{aligned} a_{i+1} - a_i &= a_{i+2} - a_{i+1} = \cdots = a_{i+h-2} - a_{i+h-3} \\ &= a_{i+h-1} - a_{i+h-2} = a_{i+h+1} - a_{i+h}. \end{aligned}$$

We must show that

$$a_{i+h} - a_{i+h-1} = a_{i+1} - a_i.$$

Suppose that $3 \leq h \leq k-3$. If $1 \leq i \leq k-h-2$, then

$$\begin{aligned} a_{i+h} - a_{i+h-1} &= a_{i-1+(h+1)} - a_{i-1+h} \\ &= a_{i-1+(h-1)} - a_{i-1+(h-2)} \\ &= a_{i+h-2} - a_{i+h-3} \\ &= a_{i+1} - a_i. \end{aligned}$$

If $i = 0$, then

$$\begin{aligned} a_h - a_{h-1} &= a_{1+(h-1)} - a_{1+(h-2)} \\ &= a_{1+(h-2)} - a_{1+(h-3)} \\ &= a_{h-1} - a_{h-2} \\ &= a_1 - a_0. \end{aligned}$$

Therefore,

$$a_{i+1} - a_i = a_1 - a_0$$

for all $i = 1, \ldots, k-2$, and so A is an arithmetic progression.

Suppose that $h = 2$ and $|2^\wedge A| = 2k - 3$. It follows from equation (1.21) that

$$a_{i+1} - a_i = a_{i+3} - a_{i+2}$$

for $i = 0, 1, \ldots, k-4$. It suffices to prove that

$$a_1 - a_0 = a_4 - a_3. \tag{1.22}$$

Since $k \geq 5$, the six smallest elements of the set $2^\wedge A$ are

$$a_0 + a_1 < a_0 + a_2 < a_1 + a_2 < a_1 + a_3 < a_2 + a_3 < a_2 + a_4.$$

Since

$$a_0 + a_3 = a_1 + a_2$$

and

$$a_1 + a_4 = a_2 + a_3,$$

it follows that

$$a_1 + a_2 = a_0 + a_3 < a_0 + a_4 < a_1 + a_4 = a_2 + a_3,$$

and so

$$a_0 + a_4 = a_1 + a_3.$$

This proves (1.22).

Finally, if $h = k - 2$, then

$$|2^\wedge A| = |(k-2)^\wedge A| = 2(k-2) + 1 = 2k - 3$$

by (1.15), and so A is again an arithmetic progression. This completes the proof.

It would be interesting to prove a more general inverse theorem for distinct summands: Let A be a set of k integers such that $|h^\wedge A|$ is "small." Is A a "large" subset of some arithmetic progression?

1.4 A special case

Let $k \geq 3$, and let b be a nonnegative integer. Let $a_i = i$ for $i = 0, 1, \ldots, k-2$ and

$$a_{k-1} = k - 1 + b.$$

We shall consider the finite set

$$A = \{a_0, a_1, \ldots, a_{k-2}, a_{k-1}\} = [0, k-2] \cup \{k - 1 + b\}.$$

For fixed $h \geq 2$, we shall examine how the cardinality of the sumset hA increases as the largest element a_{k-1} increases. We shall find that $|hA|$ is a strictly increasing, piecewise-linear function of $b = a_{k-1} - (k-1)$ for $0 \leq b \leq (h-1)(k-2)$ and that $|hA|$ is constant for $b \geq (h-1)(k-2)$. For $h = 2$ and $h = 3$, the graphs of $|hA|$ as a function of b are the following:

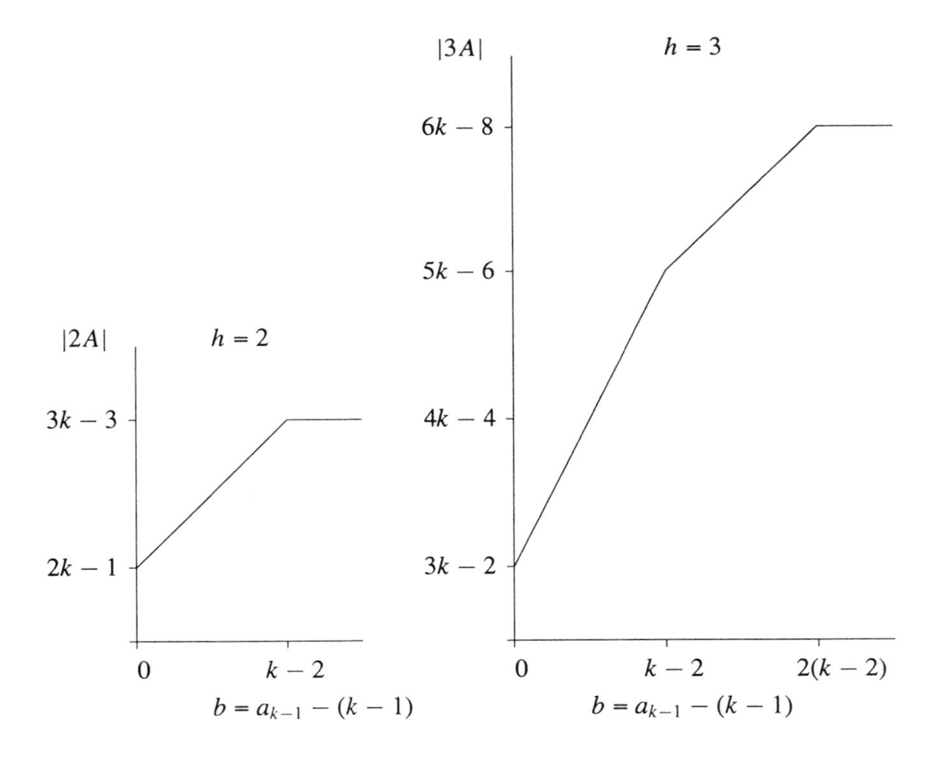

Theorem 1.11 *Let $h \geq 2$ and $k \geq 3$. For $b \geq 0$, let*

$$A = [0, k-2] \cup \{k - 1 + b\}$$

and

$$b = q(k-2) + r,$$

where $q \geq 0$ *and* $0 \leq r \leq k-3$. *If* $b \leq (h-1)(k-2)$, *then*

$$|hA| = hk - (h-1) + \frac{q(2h-q-1)(k-2)}{2} + (h-q-1)r.$$

If $b \geq (h-1)(k-2)$, *then*

$$|hA| = hk - (h-1) + \frac{h(h-1)(k-2)}{2}.$$

Proof. If $b = 0$, then $q = r = 0$ and $A = [0, k-1]$. It follows that $hA = [0, hk-h]$, and $|hA| = hk - h + 1 = hk - (h-1)$.

Let $b \geq 1$. The sumset hA is a union of $h+1$ not necessarily disjoint intervals:

$$\begin{aligned} hA &= \bigcup_{\ell=0}^{h} ([0, (h-\ell)(k-2)] + \{\ell(k-1+b)\}) \\ &= \bigcup_{\ell=0}^{h} [\ell(k-1+b), \ell(k-1+b) + (h-\ell)(k-2)] \\ &= \bigcup_{\ell=0}^{h} [\ell(k-1+b), h(k-2) + \ell(b+1)] \\ &= \bigcup_{\ell=0}^{h} I_\ell, \end{aligned}$$

where

$$\begin{aligned} I_\ell &= [\ell(k-1+b), h(k-2) + \ell(b+1)] \\ &= [\ell(k-1+b), \ell(k-1+b) + (h-\ell)(k-2)] \end{aligned}$$

and

$$|I_\ell| = (h-\ell)(k-2) + 1.$$

Notice that the intervals I_ℓ "move to the right" in the sense that, as ℓ increases from 0 to h, the sequence of right-hand endpoints of the intervals I_ℓ is strictly increasing, and the sequence of left-hand endpoints of the intervals I_ℓ is strictly increasing.

For $\ell = 1, \ldots, h$, the set $I_{\ell-1} \cup I_\ell$ will be the interval

$$[(\ell-1)(k-1+b), \ell(k-1+b) + (h-\ell)(k-2)]$$

if and only if

$$\ell(k-1+b) \leq (\ell-1)(k-1+b) + (h-\ell+1)(k-2) + 1,$$

which is equivalent to

$$b \leq (h-\ell)(k-2).$$

If $1 \leq b \leq (h-1)(k-2)$, then there exists a unique $t \in [1, h-1]$ such that

$$(h-t-1)(k-2) < b \leq (h-t)(k-2).$$

It follows that $I_{\ell-1} \cup I_\ell$ is an interval for $\ell = 1, \ldots, t$, and so

$$J = \bigcup_{\ell=0}^{t} I_\ell = [0, h(k-2) + t(b+1)].$$

If $t+1 \leq \ell \leq h$, then $h - \ell \leq h - t - 1$ and

$$(h-\ell)(k-2) \leq (h-t-1)(k-2) < b.$$

It follows that the intervals $J, I_{t+1}, I_{t+2}, \ldots, I_h$ are pairwise disjoint. Therefore,

$$\begin{aligned}
|hA| &= \left| \bigcup_{\ell=0}^{h} I_\ell \right| \\
&= |J| + \bigcup_{\ell=t+1}^{h} |I_\ell| \\
&= h(k-2) + t(b+1) + 1 + \sum_{\ell=t+1}^{h} ((h-\ell)(k-2)+1) \\
&= h(k-2) + t(b+1) + (h-t+1) + (k-2)\sum_{\ell=0}^{h-t-1} \ell \\
&= h(k-2) + tb + h + 1 + \frac{(h-t)(h-t-1)(k-2)}{2} \\
&= hk - (h-1) + tb + \frac{(h-t)(h-t-1)(k-2)}{2}.
\end{aligned}$$

If $r = 0$, then $b = q(k-2)$, so $q = h - t$ and

$$\begin{aligned}
|hA| &= hk - (h-1) + (h-q)q(k-2) + \frac{q(q-1)(k-2)}{2} \\
&= hk - (h-1) + \frac{q(2h-q-1)(k-2)}{2}.
\end{aligned}$$

If $r \geq 1$, then

$$q(k-2) < b = q(k-2) + r < (q+1)(k-2),$$

so $q = h - t - 1$ and

$$\begin{aligned}
|hA| &= hk - (h-1) + (h-q-1)(q(k-2)+r) + \frac{q(q+1)(k-2)}{2} \\
&= hk - (h-1) + (h-q-1)r + \frac{q(2h-q-1)(k-2)}{2}.
\end{aligned}$$

If $b > (h-1)(k-2)$, then

$$(h-\ell)(k-2) \leq (h-1)(k-2) < b$$

for $\ell = 1, 2, \ldots, h$, and so the intervals $I_0, I_1, \ldots, I_h$ are pairwise disjoint. Therefore,

$$\begin{aligned} |hA| &= \left| \bigcup_{\ell=0}^{h} I_\ell \right| = \bigcup_{\ell=0}^{h} |I_\ell| \\ &= \sum_{\ell=0}^{h} ((h-\ell)(k-2)+1) \\ &= \frac{h(h+1)(k-2)}{2} + h + 1 \\ &= hk - (h-1) + \frac{h(h-1)(k-2)}{2}. \end{aligned}$$

This completes the proof.

Theorem 1.12 *Let $k \geq 3$, and let*

$$A = [0, k-2] \cup \{k-1+b\}.$$

If $0 \leq b \leq k-3$, then

$$|2A| = 2k - 1 + b \leq 3k - 4.$$

If $b \geq k-2$, then

$$|2A| = 3k - 3.$$

1.5 Small sumsets: The case $|2A| \leq 3k - 4$

We proved that the set A must be an arithmetic progression if A is a finite set of integers whose twofold sumset is as small as possible. That is, if $|A| = k$ and $|2A| = 2k - 1$, then $A^N = [0, k-1]$. In this section, we shall show that if $|A| = k$ and $|2A| \leq 3k - 4$, then A is a subset of a short arithmetic progression. More precisely, we shall prove that if $|A| = k$ and $|2A| = 2k - 1 + b \leq 3k - 4$, then $A^{(N)} \subseteq [0, k-1+b]$. Theorem 1.12 shows that these lower bounds are best possible.

Theorem 1.13 *Let $k \geq 3$. Let $A = \{a_0, a_1, \ldots, a_{k-1}\}$ be a set of integers such that*

$$0 = a_0 < a_1 < \cdots < a_{k-1} \leq 2k - 3.$$

Let $a_{k-1} = k - 1 + r$, where $r \in [0, k-2]$. Then

$$|2A| \geq 2k - 1 + r = k + a_{k-1}.$$

Proof. Consider the set

$$S = A \cup (a_{k-1} + A) \subseteq 2A \subseteq [0, 2a_{k-1}].$$

Then S consists of the $2k-1$ integers

$$\begin{aligned} 0 &< a_1 < a_2 < \cdots < a_{k-2} < a_{k-1} \\ &< a_{k-1} + a_1 < a_{k-1} + a_2 < \cdots < 2a_{k-1}. \end{aligned}$$

Let $W = [1, a_{k-1}] \setminus A$. Then

$$|W| = a_{k-1} - (k-1) = r.$$

For $w \in W$, let

$$S(w) = \{w, a_{k-1} + w\} \subseteq [1, 2a_{k-1}].$$

The $r+1$ sets S and $\{S(w)\}_{w \in W}$ are pairwise disjoint, and

$$[0, 2a_{k-1}] = S \cup \bigcup_{w \in W} S(w).$$

Therefore,

$$2A = S \cup \bigcup_{w \in W} (S(w) \cap 2A).$$

It suffices to prove that

$$|S(w) \cap 2A| \geq 1$$

for all $w \in W$.

For each $w \in W = [1, a_{k-1}] \setminus A$, there exists a unique $t \in [1, k-1]$ such that $a_{t-1} < w < a_t$. Define the sets I, Y, and Z by

$$\begin{aligned} I &= [w + 1, w + a_{k-1} - 1] \\ Y &= I \cap S = \{a_t, a_{t+1}, \ldots, a_{k-1}, a_{k-1} + a_1, \ldots, a_{k-1} + a_{t-1}\} \\ Z &= \{w + a_{k-1} - a_j \mid j = 1, 2, \ldots, k-2\}. \end{aligned}$$

Then $Y \subseteq I$ and $Z \subseteq I$. Also, $|Y| = k-1$ and $|Z| = k-2$. Since

$$|I| = a_{k-1} - 1 \leq 2k - 4 < 2k - 3 = (k-1) + (k-2) = |Y| + |Z|,$$

it follows that $Y \cap Z \neq \emptyset$. Therefore, there exist $i \in [1, k-1]$ and $j \in [1, k-2]$ such that either

$$a_i = w + a_{k-1} - a_j$$

or

$$a_{k-1} + a_i = w + a_{k-1} - a_j.$$

In the first case, we have

$$w + a_{k-1} = a_i + a_j \in 2A,$$

and in the second case we have

$$w = a_i + a_j \in 2A.$$

In both cases, $|S(w) \cap 2A| \geq 1$. This completes the proof.

Theorem 1.14 *Let $k \geq 3$. Let $A = \{a_0, a_1, \ldots, a_{k-1}\}$ be a finite set of integers in normal form, that is,*

$$0 = a_0 < a_1 < \cdots < a_{k-1}$$

and

$$d(A) = (a_1, \ldots, a_{k-1}) = 1.$$

If $a_{k-1} \geq 2k - 3$, then

$$|2A| \geq 3k - 3.$$

Proof. If $a_{k-1} = 2k - 3$, then $|2A| \geq 3k - 3$ by Theorem 1.13. Therefore, we can assume that

$$a_{k-1} \geq 2k - 2.$$

The proof will be by induction on $k = |A|$.

Let $k = 3$. Then $A = \{0, a_1, a_2\}$, and $a_2 \geq 4$. We must show that $|2A| \geq 6$. Since

$$2A = \{0, a_1, a_2, 2a_1, a_1 + a_2, 2a_2\}$$

and

$$0 < a_1 < a_2 < a_1 + a_2 < 2a_2,$$

it follows that $|2A| = 5$ or 6. Since

$$a_1 < 2a_1 < a_1 + a_2,$$

we see that $|2A| = 5$ if and only if $a_2 = 2a_1$, which implies that

$$1 = (a_1, a_2) = (a_1, 2a_1) = a_1$$

and so $a_2 = 2a_1 = 2$. This is impossible since $a_2 \geq 4$. Therefore, $|2A| = 6$.

Let $k \geq 4$, and assume that the theorem holds for sets of cardinality $k - 1$. Let

$$A' = A \setminus \{a_{k-1}\} = \{0, a_1, \ldots, a_{k-2}\}$$

and

$$d' = d(A') = (a_1, a_2, \ldots, a_{k-2}).$$

If $d' = d(A') \geq 2$, then d' divides all elements of the sumset $2A'$. Since $d(A) = 1$, it follows that $(a_{k-1}, d') = 1$ and so $(a_{k-1} + a_i, d') = 1$ for $i = 0, 1, \ldots, k - 2$. Therefore, $2A' \cap (A' + a_{k-1}) = \emptyset$. Also,

$$2a_{k-1} > \max\left(\max(2A'), \max(A' + a_{k-1})\right).$$

By Theorem 1.3,

$$|2A'| \geq 2(k - 1) - 1 = 2k - 3.$$

Since

$$2A' \cup (A' + a_{k-1}) \cup \{2a_{k-1}\} \subseteq 2A,$$

it follows that

$$|2A| \geq |2A'| + |A' + a_{k-1}| + 1 \geq (2k-3) + (k-1) + 1 = 3k - 3.$$

Therefore, we can assume that $d(A') = 1$, and so A' is in normal form. There are several cases to consider.

Case 1. Suppose that $a_i < 2i$ for all $i = 1, 2, \ldots, k-2$. Then $0 < a_1 < 2$ and so $a_1 = 1$. Let

$$C = [0, 2k-4] \setminus A'.$$

Then

$$|C| = (2k-3) - (k-1) = k-2.$$

If $c \in C$, then $c > a_1 = 1$ and there exists a unique $t \in [1, k-2]$ such that

$$a_t < c < a_{t+1}.$$

Consider the sets

$$D_1 = \{a_i : i \in [1, t]\}$$

and

$$D_2 = \{c - a_j : j \in [1, t]\}.$$

Then $|D_1| = |D_2| = t$ and

$$D_1 \cup D_2 \subseteq [1, c-1].$$

If $t < k-2$, then

$$c < a_{t+1} < 2(t+1) = 2t+2$$

and so $c \leq 2t$ or, equivalently, $c - 1 < 2t$. If $t = k-2$, then $c \leq 2k-4$ implies that

$$c - 1 \leq 2k - 5 < 2k - 4 = 2t.$$

In both cases, it follows that

$$D_1 \cap D_2 \neq \emptyset,$$

and so there exist $i, j \in [1, t] \subseteq [1, k-2]$ such that

$$a_i = c - a_j.$$

Therefore, $c = a_i + a_j \in 2A'$ and

$$A' \cup C = [0, 2k-4] \subseteq 2A' \subseteq 2A.$$

Since

$$2k - 4 < 2k - 2 \leq a_{k-1} \leq a_{k-1} + a_i$$

for $i = 0, 1, \ldots, k-1$, it follows that

$$A' \cup C \cup (a_{k-1} + A) = [0, 2k-4] \cup (a_{k-1} + A) \subseteq 2A$$

and

$$|2A| \geq (2k - 3) + k = 3k - 3.$$

Case 2. Suppose that $a_{k-2} < 2(k-2)$, but $a_{i-1} \geq 2(i-1)$ for some $i \in [2, k-2]$. Choose $s \in [2, k-2]$ such that $a_j < 2j$ for $j = s, s+1, \ldots, k-2$, and $a_{s-1} \geq 2(s-1)$. Then

$$2s - 2 \leq a_{s-1} < a_s < 2s,$$

and so $a_s = 2s - 1$ and $a_{s-1} = 2s - 2$. Define the sets A_1 and A_2 by

$$A_1 = \{a_0, a_1, \ldots, a_{s-1}, a_s\}$$

and

$$A_2 = \{a_{s-1}, a_s, a_{s+1}, \ldots, a_{k-2}, a_{k-1}\}.$$

Since $a_s - a_{s-1} = 1$, it follows that $d(A_1) = d(A_2) = 1$. Let

$$k_1 = |A_1| = s + 1.$$

Then

$$3 \leq k_1 \leq k - 1$$

and

$$a_{k_1-1} = a_s = 2s - 1 = 2k_1 - 3 = (k_1 - 1) + (k_1 - 2).$$

It follows from Theorem 1.13 that

$$|2A_1| \geq (2k_1 - 1) + (k_1 - 2) = 3k_1 - 3 = 3s.$$

Define the set A_2^* by

$$A_2^* = A_2 - \{a_{s-1}\} = \{0, 1, a_{s+1} - a_{s-1}, \ldots, a_{k-1} - a_{s-1}\}.$$

This set is also in normal form. Let

$$k_2 = |A_2^*| = |A_2| = k - s + 1.$$

Then

$$3 \leq k_2 \leq k - 1.$$

The largest element of A_2^* is $a_{k-1} - a_{s-1}$. Since $a_{s-1} = 2s - 1$, we have the inequality

$$a_{k-1} - a_{s-1} \geq (2k - 2) - (2s - 2) = 2(k - s) = 2k_2 - 2.$$

It follows from the induction hypothesis that

$$|2A_2| = |2A_2^*| \geq 3k_2 - 3 = 3k - 3s.$$

Since

$$2A_1 \cup 2A_2 \subseteq 2A$$

and

$$2A_1 \cap 2A_2 = \{2a_{s-1}, a_{s-1} + a_s, 2a_s\}$$

(see Exercise 7), it follows that

$$|2A| \geq |2A_1| + |2A_2| - 3 \geq 3s + (3k - 3s) - 3 = 3k - 3.$$

This proves that the theorem holds if $a_{k-2} < 2k - 4$.

Case 3. We can now assume that

$$a_{k-2} \geq 2k - 4 = 2(k - 2).$$

It follows from the induction hypothesis that $|2A'| \geq 3(k-1) - 3 = 3k - 6$. To complete the proof the theorem, it suffices to show that $|2A \setminus 2A'| \geq 3$. The two largest elements of $2A'$ are $a_{k-2} + a_{k-3}$ and $2a_{k-2}$. Since

$$\{a_{k-1} + a_{k-3}, a_{k-1} + a_{k-2}, 2a_{k-1}\} \subseteq 2A,$$

it follows that $|2A \setminus 2A'| \geq 3$ unless $a_{k-1} + a_{k-3} = 2a_{k-2}$. Therefore, we can assume that the numbers a_{k-3}, a_{k-2}, and a_{k-1} are in arithmetic progression.

If $a_{k-1} - a_{k-2} = m \geq 2$, then

$$a_{k-1} \equiv a_{k-i} \pmod{m}$$

for $i = 1, 2, 3$. Suppose that

$$a_{k-1} \equiv a_{k-i} \pmod{m}$$

for all $i = 1, 2, \ldots, k$. It follows that $a_{k-1} \equiv a_0 \equiv 0 \pmod{m}$, and so m divides each a_i, which contradicts the condition that $d(A) = 1$. Therefore, there exists an integer t such that $4 \leq t \leq k$ and

$$a_{k-1} \equiv a_{k-i} \pmod{m}$$

for $i = 1, \ldots, t - 1$, but

$$a_{k-1} \not\equiv a_{k-t} \pmod{m}.$$

Moreover, $a_{k-1} + a_{k-t} \in 2A$. If $a_{k-1} + a_t \in 2A'$, then there exist integers r and s such that $1 < r \leq s < t$ and

$$a_{k-r} + a_{k-s} = a_{k-1} + a_{k-t} < a_{k-1} + a_{k-3}.$$

This implies that

$$a_{k-1} - a_{k-t} = (a_{k-1} - a_{k-r}) + (a_{k-1} - a_{k-s}) \equiv 0 \pmod{m},$$

which is false. Therefore,

$$\{a_{k-1} + a_{k-t}, a_{k-1} + a_{k-2}, 2a_{k-1}\} \subseteq 2A \setminus 2A',$$

and $|2A \setminus 2A'| \geq 3$.

It follows that we can assume that

$$a_{k-1} - a_{k-2} = a_{k-2} - a_{k-3} = 1.$$

Consider the set

$$\begin{aligned} A^* &= \{a_{k-1} - a_i : i \in [0, k-1]\} \\ &= \{0, 1, 2, a_{k-1} - a_{k-4}, \ldots, a_{k-1} - a_2, a_{k-1} - a_1, a_{k-1}\}. \end{aligned}$$

Then $2A^* = \{2a_{k-1} - b : b \in 2A\}$, and so $|2A^*| = |2A|$. It follows from the preceding analysis that if $|2A^*| < 3k - 3$, then $a_1 = 1$ and $a_2 = 2$. Therefore, we can assume that $a_0 = 0, a_1 = 1, a_2 = 2, a_{k-3} = a_{k-1} - 2$, and $a_{k-2} = a_{k-1} - 1$. Since $a_{k-1} \geq 2k - 2 = 2(k-1)$, it follows that $a_i \geq 2i$ for $i = k-1, k-2, k-3$. Let ℓ be the least positive integer such that

$$a_\ell \geq 2\ell.$$

Then

$$3 \leq \ell \leq k - 3$$

and

$$a_i < 2i \quad \text{for } i = 1, 2, \ldots, \ell - 1.$$

Define the sets A_1 and A_2 by

$$A_1 = \{a_0, a_1, \ldots, a_{\ell-1}, a_\ell\}$$

and

$$A_2 = \{a_{\ell-1}, a_\ell, \ldots, a_{k-2}, a_{k-1}\}.$$

Then $d(A_1) = 1$. Moreover,

$$4 \leq k_1 = |A_1| = \ell + 1 \leq k - 2$$

and

$$a_{k_1-1} = a_\ell \geq 2\ell = 2k_1 - 2.$$

The induction hypothesis implies that

$$|2A_1| \geq 3k_1 - 3 = 3\ell.$$

Define the set A_2^* by

$$A_2^* = A_2 - \{a_{\ell-1}\} = \{0, a_\ell - a_{\ell-1}, \ldots, a_{k-2} - a_{\ell-1}, a_{k-1} - a_{\ell-1}\}.$$

Then $d(A_2^*) = 1$. Since $\ell \in [3, k-3]$, we have

$$4 \leq k_2 = |A_2| = |A_2^*| = k - \ell + 1 \leq k - 2$$

and

$$a_{k-1} - a_{\ell-1} > (2k - 2) - (2\ell - 2) = 2(k - \ell) = 2k_2 - 2.$$

Again, the induction hypothesis implies that

$$|2A_2| = |2A_2^*| \geq 3k_2 - 3 = 3k - 3\ell.$$

Since

$$2A_1 \cup 2A_2 \subseteq 2A$$

and

$$2A_1 \cap 2A_2 = \{2a_{\ell-1}, a_{\ell-1} + a_\ell, 2a_\ell\}$$

(see Exercise 7), it follows that

$$|2A| \geq |2A_1| + |2A_2| - 3 \geq 3\ell + (3k - 3\ell) - 3 = 3k - 3.$$

This completes the proof of the theorem.

Theorem 1.15 *Let $k \geq 3$. Let $A = \{a_0, a_1, \ldots, a_{k-1}\}$ be a finite set of integers in normal form. Then*

$$|2A| \geq \min(3k - 3, k + a_{k-1}).$$

Proof. If $a_{k-1} \leq 2k-3$, then $|2A| \geq k+a_{k-1}$ by Theorem 1.13. If $a_{k-1} \geq 2k-3$, then $|2A| \geq 3k - 3$ by Theorem 1.14.

Theorem 1.16 (Freiman) *Let A be a set of integers such that $|A| = k \geq 3$. If*

$$|2A| = 2k - 1 + b \leq 3k - 4,$$

then A is a subset of an arithmetic progression of length $k + b \leq 2k - 3$.

Proof. Let $A^{(N)} = \{a_0, a_1, \ldots, a_{k-1}\}$ be the normal form of A. Since

$$|2A^{(N)}| = |2A| \leq 3k - 4,$$

it follows from Theorem 1.14 that $a_{k-1} \leq 2k - 4$. Theorem 1.13 implies that

$$k + a_{k-1} \leq |2A^{(N)}| = 2k - 1 + b,$$

ans so $a_{k-1} \leq k - 1 + b$. It follows that

$$A^{(N)} \subseteq [0, k - 1 + b],$$

and A is a subset of an arithmetic progression with $k + b$ terms.

1.6 Application: The number of sums and products

Let A be a nonempty, finite set of positive integers. Let

$$2A = \{a + a' \mid a, a' \in A\}$$

denote the twofold *sumset* of A, and let

$$A^2 = \{aa' \mid a, a' \in A\}$$

denote the twofold *product set* of A. We let

$$E_2(A) = 2A \cup A^2$$

denote the set of all integers that can be written as the sum or product of two elements of A. If $|A| = k$, then

$$|2A| \leq \binom{k+1}{2}$$

and

$$|A^2| \leq \binom{k+1}{2},$$

and so the number of sums and products of two elements of A is

$$|E_2(A)| \leq k^2 + k.$$

Erdős and Szemerédi [38, 44] made the beautiful conjecture that a finite set of positive integers cannot have simultaneously few sums and few products. More precisely, they conjectured that for every $\varepsilon > 0$ there exists an integer $k_0(\varepsilon)$ such that, if A is a finite set of positive integers and

$$|A| = k \geq k_0(\varepsilon),$$

then

$$|E_2(A)| \gg_\varepsilon k^{2-\varepsilon}.$$

We shall use Theorem 1.16 to prove this conjecture in the special case that the number of sums of two elements of the set A is small in the sense that $|2A| \leq 3k-4$. This is the only case in which the conjecture has been proven.

For any set A of positive integers, let $\rho_A(n)$ denote the number of representations of n in the form $n = aa'$, where $a, a' \in A$, and let $d_A(n)$ denote the number of positive divisors of n that belong to the set A. Clearly, for every integer n,

$$\rho_A(n) \leq d_A(n).$$

If Q is a set of positive integers that contains A, then

$$\rho_A(n) \leq \rho_Q(n)$$

and

$$d_A(n) \leq d_Q(n).$$

Let $d(n)$ denote the usual divisor function, that is, the number of positive divisors of n. We shall use the estimate

$$d(n) \ll_\varepsilon n^{\varepsilon/4}$$

for every $\varepsilon > 0$.

Lemma 1.4 *Let Q be a set of positive integers that is an arithmetic progression of length l. For any $\varepsilon > 0$,*

$$\rho_Q(n) \ll_\varepsilon l^\varepsilon. \tag{1.23}$$

Proof. Let $Q = \{r + xq : x = 0, 1, \ldots, l-1\}$, and let $\varepsilon > 0$. We can assume without loss of generality that $(r, q) = 1$. If the integer n has an essentially unique representation as the product of two elements of Q, then

$$\rho_q(n) \leq 2 \ll l^\varepsilon.$$

By Exercise 18, we have $\{x, y\} = \{u, v\}$ if and only if $x + y = u + v$ and $xy = uv$. If n has at least two essentially distinct representations, then there exist integers $0 \leq x, y, u, v < l$ such that

$$\{x, y\} \neq \{u, v\} \tag{1.24}$$

and

$$(r + xq)(r + yq) = (r + uq)(r + vq).$$

Then

$$(x + y)r + xyq = (u + v)r + uvq, \tag{1.25}$$

and so $x + y = u + v$ if and only if $xy = uv$. It follows from (1.24) that

$$x + y \neq u + v$$

and

$$xy \neq uv.$$

Since $(r, q) = 1$, it follows from (1.25) that

$$x + y \equiv u + v \pmod{q}$$

and

$$xy \equiv uv \pmod{r}.$$

Therefore,

$$q \leq |(x + y) - (u + v)| < 2l$$

and

$$r \leq |xy - uv| < l^2.$$

Consequently,

$$1 \leq r + xq, r + yq < l^2 + 2l^2 = 3l^2$$

and so

$$1 \leq n < 9l^4.$$

It follows that

$$\rho_Q(n) \leq d_Q(n) \leq d(n) \ll_\varepsilon n^{\varepsilon/4} \ll_\varepsilon l^\varepsilon.$$

Theorem 1.17 *Let A be a set of k positive integers such that*

$$|2A| \leq 3k - 4,$$

and let $\varepsilon > 0$. Then

$$|A^2| \gg_\varepsilon k^{2-\varepsilon}.$$

Proof. By Theorem 1.16, the set A is a subset of an arithmetic progression Q of length $l < 2k$. Let $\rho_Q(n)$ denote the number of representations of n in the form $n = qq'$, where $q, q' \in Q$. By Lemma 1.4, we have

$$\rho_A(n) \leq \rho_Q(n) \ll_\varepsilon l^\varepsilon,$$

and so

$$\begin{aligned} k^2 &= \sum_{n \in A^2} \rho_A(n) \\ &\leq \sum_{n \in A^2} \rho_Q(n) \\ &\ll_\varepsilon |A^2| l^\varepsilon \\ &\ll_\varepsilon |A^2| k^\varepsilon. \end{aligned}$$

Therefore,

$$|A^2| \gg_\varepsilon k^{2-\varepsilon}.$$

This completes the proof.

1.7 Application: Sumsets and powers of 2

Let $n \geq 1$, and let B^* be the set of all multiples of 3 contained in the interval $[1, n]$. Then $|B^*| \leq n/3$, and every sum of elements of B^* is divisible by 3. Certainly, no such sum is a power of 2. This set B^* is the extremal case: We shall prove that if B is any subset of $[1, n]$ such that $|B| > n/3$, then some power of 2 can be written as the sum of at most four (not necessarily distinct) elements of B.

Lemma 1.5 *Let $m \geq 1$, and let C be a subset of $[0, m]$ such that*

$$|C| \geq \frac{m}{2} + 1.$$

Then some power of 2 is either an element of C or the sum of two distinct elements of C.

Proof. The proof is by induction on m. It is easy to check that the result is true for $m = 1, 2, 3$, and 4. Let $m > 4$, and assume that the result holds for all positive integers $m' < m$. Choose $s \geq 2$ such that

$$2^s \leq m < 2^{s+1},$$

and let

$$r = m - 2^s \in [0, 2^s - 1].$$

Let

$$C' = C \cap [0, 2^s - r - 1]$$

and

$$C'' = C \cap [2^s - r, 2^s + r].$$

Then C is the disjoint union of C' and C'', and

$$|C| = |C'| + |C''|.$$

Suppose that the lemma is false for the set C. Then $|C| \geq m/2+1$, but no power of 2 either belongs to C or is the sum of two distinct elements of C. It follows that $2^s \notin C''$ and, for each $i = 1, \ldots, r$, the set C'' contains at most one of the two integers $2^s - i, 2^s + i$. Therefore,

$$|C''| \leq r.$$

If $m = 2^{s+1} - 1$, then $r = 2^s - 1$ and $C' \subseteq \{0\}$; thus

$$|C'| \leq 1.$$

It follows that

$$\frac{m}{2} + 1 \leq |C| \leq 1 + r = 2^s = \frac{m+1}{2},$$

which is impossible.

Similarly, if $2^s \leq m < 2^{s+1} - 1$, then $0 \leq r < 2^s - 1$ and $m' = 2^s - r - 1 \geq 1$. Since the set C contains C', it follows that no power of 2 either belongs to C' or is the sum of two distinct elements of C'. By the induction hypothesis, we have

$$|C'| < \frac{m'}{2} + 1 = \frac{2^s - r - 1}{2} + 1,$$

and so

$$\frac{m}{2} + 1 \leq |C| = |C'| + |C''| < \frac{2^s - r - 1}{2} + 1 + r = \frac{m+1}{2},$$

which is also impossible.

Theorem 1.18 *Let $n \geq 1$, and let B be a set of integers contained in the interval $[1, n]$. If $|B| > n/3$, then there is a power of 2 that can be written as the sum of at most four elements of B.*

Proof. Since $B \subseteq [1, n]$, we have $|B| > n/3 \geq \max(B)/3$. Thus, we can assume that $\max(B) = n$. Let d be the greatest common divisor of the elements of B. The number of multiples of d in the interval $[1, n]$ is $[n/d]$, so

$$\frac{n}{3} < |B| \leq \frac{n}{d}.$$

Therefore, $d = 1$ or2.

If $d = 2$, we can consider the set

$$B' = \{b/2 :\in B\} \subseteq \left[1, \frac{n}{2}\right].$$

The greatest common divisor of the elements of B' is 1. The set B' also satisfies the hypotheses of the theorem. If the theorem holds in the case $d = 1$, then there exists $h \leq 4$ and there exist integers $b'_1, \ldots, b'_h \in B'$ such that $b'_1 + \cdots + b'_h = 2^s$. It follows that $2b'_1, \ldots, 2b'_h \in B$ and $2b'_1 + \cdots + 2b'_h = 2^{s+1}$. Therefore, we can assume that $d = 1$.

Let $A = \{0\} \cup B$. Then $d(A) = 1$, $\max(A) = \max(B) = n$, and

$$k = |A| = |B| + 1 > \frac{n}{3} + 1.$$

It follows from Theorem 1.15 that

$$|2A| \geq \min(3k - 3, k + n) \geq n + 1.$$

Since $2A \subseteq [0, 2n]$, we can apply Lemma 1.5 with $C = 2A$ and $m = 2n$. Since $2C = 4A$, it follows that some power of 2 can be written as the sum of at most four elements of A. This completes the proof.

In Exercise 19, we construct examples of finite sets $B \subseteq [1, n]$ such that $|B| > n/3$, but no power of 2 can be written as the sum of three elements of B. This shows that Theorem 1.18 is best possible.

1.8 Notes

The principal result in this chapter is Theorem 1.16, which was proved by Freiman [49, 54, 55]. Freiman [52] has extended this to sumsets of the form $A + B$. Steinig [121] has an expanded version of Freiman's proof. In Chapter 4 I give a different proof, discovered by Lev and Smeliansky [81], that uses a theorem of Kneser on sumsets in abelian groups. Freiman's monograph *Foundations of a Structural Theory of Set Addition* [54] is devoted to Freiman's work on inverse theorems.

Theorem 1.1 is due to Nathanson [91]. For some related results, see Lev [80]. The simple inverse theorem for hA (Theorem 1.5) is probably ancient, but I have not seen it in print. The inverse theorems for the sets $h^\wedge A$ are due to Nathanson [95].

Very little is known about Erdős's conjecture that $|E_2(A)| \gg_\varepsilon k^{2-\varepsilon}$. Erdős and Szemerédi [44] have shown that there exists a real number $\delta > 0$ such that

$$|E_2(A)| \gg k^{1+\delta},$$

and Nathanson [97] proved that

$$|E_2(A)| \geq ck^{32/31},$$

where $c = 0.00028\ldots$. Ford [47] has improved the exponent to $16/15$. The proof of Lemma 1.4 is due to Erdős and Pomerance (personal communication). Using a theorem of Vinogradov and Linnik [125], Nathanson and Tenenbaum [99] strengthened this result: They proved that if Q is an arithmetic progression of length l, then $d_Q(m) \ll l^2(\log l)^3$ for all $m \in Q^2$. This implies that if $|A| = k$ and $|2A| \leq 3k-4$, then $|A^2| \gg k^2/(\log k)^3$.

Erdős and Freud had conjectured that if $A \subseteq [1, n]$ and $|A| > n/3$, then some power of 2 can be written as the sum of distinct elements of A. This was proved by Erdős and Freiman [39] (with an unbounded number of summands), and Nathanson and Sárkőzy [98] (with a bounded number of summands). Theorem 1.18, due to Lev [79], improves results of Nathanson and Sárkőzy [98] and of Freiman [58].

Closely related to the "structural" inverse problems is another class of inverse problems in additive number theory that we can call *recognition problems* or *decomposition problems*. We write $A \sim B$ if A and B are sets of integers that coincide from some point on. If we are given a finite or infinite set B of integers, can we determine whether B is a sumset or even asymptotically a sumset? This means the following: Let $h \geq 2$. Does there exist a set A such that $hA = B$? More generally, do there exist sets $A_1, \ldots, A_h$ such that $|A_i| \geq 2$ for $i = 1, 2, \ldots, h$ and $A_1 + \cdots + A_h = B$? Do there exist sets $A_1, \ldots, A_h$ such that $|A_i| \geq 2$ for $i = 1, 2, \ldots, h$ and $A_1 + \cdots + A_h \sim B$? Ostmann [100] introduced this class of inverse problems.

Sets of integers that decompose into sumsets are rare. Let us associate to each set A of nonnegative integers the real number

$$\sum_{a \in A} 2^{-a-1} \in [0, 1].$$

Wirsing [128] proved that the Lebesgue measure of the set of real numbers that correspond to sets A such that $A \sim B + C$ for some B and C is zero.

An important decomposition problem is the following: Do there exist infinite sets A and B of nonnegative integers such that the sumset $A + B$ and the set $\mathbf{P}$ of odd prime numbers eventually coincide, that is,

$$\mathbf{P} \sim A + B. \tag{1.26}$$

The answer is almost surely no, but there is no proof. Hornfeck [69, 70] proved that (1.26) is impossible if the set A is finite and $|A| \geq 2$.

There are other kinds of recognition problems: Does the set B contain a sumset? Given sets A and B, does there exist a set C such that $B + C \subseteq A$? The twin prime conjecture is a special case of this inverse problem: Let **P** be the set of odd prime numbers. Does there exist an infinite set A such that

$$A + \{0, 2\} \subseteq \mathbf{P}?$$

Practically nothing is known about these questions.

We do not consider "partition problems" in this book. A *partition* of a positive integer N is a representation of N as the sum of an unrestricted number of elements taken from a fixed set of positive integers. A good reference for the classical approach to partitions is Andrews's monograph *The Theory of Partitions* [4]. For interesting examples of inverse theorems for partitions, see the papers of Cassels [15], Erdős [35], Erdős, Gordon, Rubel, and Straus [41], and Folkman [46].

An early version of this chapter appeared in Nathanson [92].

In this book we investigate only h-fold sumsets of finite sets. A subsequent volume (Nathanson [90]) will examine sums of infinite sets of integers in additive number theory. For example, it includes a deep and beautiful inverse theorem of Kneser [76] concerning the asymptotic density of sumsets, and an important recent improvement of this due to Bilu [10].

For a comprehensive treatment of many of the most important results on Waring's problem and the Goldbach conjecture, see Nathanson, *Additive Number Theory: The Classical Bases* [96]. There is no other recent book on additive number theory.

1.9 Exercises

1. Compute the sumset $2A$ for each of the following sets of integers:

   (a) $A = \{0, 1, 3, 4\}$.

   (b) $A = \{0, 1, 3, 7, 15, 31\}$.

   (c) $A = \{0, 1, 4, 9, 16, 25\}$.

   (d) $A = \{3, 5, 7, 11, 13, 17, 19, 23, 29\}$.

   (e) $A = \{2x_1 + 7x_2 \mid 0 \leq x_1 < 4, 0 \leq x_2 < 3\}$.

2. Let $A = \{0, 2, 3, 6\}$. Compute the sumsets hA for all $h \geq 1$.

3. In Theorem 1.1, show that $c = 0$ if and only if $a_1 = 1$, and $d = 0$ if and only if $a_{k-1} - a_{k-2} = 1$.

4. Let $A = \{a_0, a_1, \ldots, a_{k-1}\}$ be a finite set of integers such that

$$0 = a_0 < a_1 < \cdots < a_{k-1}$$

   and

$$(a_1, \ldots, a_{k-1}) = 1.$$

Define the integer h_0 by (1.3). Prove that

$$|hA| - |(h-1)A| = a_{k-1}$$

for all $h > h_0$.

5. Let A be a set of integers with $|A| = k$. Prove that the number of expressions of the form $a_1 + \cdots + a_h$ with $a_i \in A$ for $i = 1, 2, \ldots, h$ and $a_1 \leq \cdots \leq a_h$ is exactly $\binom{k+h-1}{h}$.

6. Let $A = \{a_0, a_1, \ldots, a_{k-1}\}$ be a set of integers such that $a_i > ha_{i-1}$ for $i = 1, \ldots, k-1$. Prove that $|hA| = \binom{k+h-1}{h}$.

7. Let $a_0, a_1, \ldots, a_{k-1}$ be a strictly increasing sequence of integers, and let $1 \leq s \leq k-2$. Let

$$A_1 = \{a_0, a_1, \ldots, a_{s-1}, a_s\}$$

and

$$A_2 = \{a_{s-1}, a_s, \ldots, a_{k-1}\}.$$

Prove that

$$2A_1 \cap 2A_2 = \{2a_{s-1}, a_{s-1} + a_s, 2a_s\}.$$

8. Let $k \geq 3$ and

$$A = [0, k-2] \cup \{k-1+r\}.$$

Show that

$$|2^\wedge A| = 2k - 3 + r$$

for $0 \leq r \leq k-4$, and

$$|2^\wedge A| = 3k - 6$$

for $r \geq k-3$.

9. Let $k \geq 4$ and $A = \{a_0, a_1, \ldots, a_{k-1}\}$, where

$$0 = a_0 < a_1 \cdots < a_{k-1} \leq 2k - 5.$$

Define r by $a_{k-1} = k - 1 + r$. Prove that

$$|2^\wedge A| = 2k - 3 + r.$$

10. Let $k \geq 4$, and let $A = \{a_0, a_1, \ldots, a_{k-1}\}$, where

$$0 = a_0 < a_1 \cdots < a_{k-2} < a_{k-1}.$$

Let $A' = A \setminus \{a_{k-1}\}$. Suppose that $d' = d(A') > 1$ and $(a_{k-1}, d') = 1$. Prove that

$$|2^\wedge A| \geq 3k - 6.$$

11. The *subset sum* of a finite set A' of integers is defined by

$$s(A') = \sum_{a \in A'} a.$$

For any finite set A of positive integers, define

$$S(A) = \{s(A') \mid A' \subseteq A, A' \neq \emptyset\}.$$

Prove that if A is a set of k positive integers, then

$$|S(A)| \geq \binom{k+1}{2}.$$

12. Let A be a set of k positive integers such that

$$|S(A)| = \binom{k+1}{2}.$$

Prove that there exists a positive integer m such that

$$A = m * [1, k] = \{m, 2m, 3m, \ldots, km\}.$$

13. For $k \geq 3$, let $f_k(n)$ denote the number of sets $A \subseteq [0, n-1]$ such that $|A| = k$ and $|2A| < \binom{k+1}{2}$. Prove that

$$\lim_{n \to \infty} \frac{f_k(n)}{\binom{n}{k}} = 0.$$

14. Let θ be a positive real number, and let $f_\theta(n)$ denote the number of sets $A \subseteq [0, n-1]$ such that $|A| = [n^\theta]$ and $|2A| < \binom{[n^\theta]+1}{2}$. Prove that there exists $\theta > 0$ such that

$$\lim_{n \to \infty} \frac{f_\theta(n)}{\binom{n}{[n^\theta]}} = 0.$$

Hint: Use Stirling's formula.

15. Determine the structure of all sets A such that $|A| = k$ and $|2A| = 2k$.

16. Determine the structure of all sets A such that $|A| = k$ and $|2A| = 2k + 1$.

17. Let $h \geq 2$ and $k \geq 3$. Let $A = \{a_0, a_1, \ldots, a_{k-1}\}$ be a set of integers such that

$$0 = a_0 < a_1 < \cdots < a_{k-1} \leq 2k - 3.$$

Define r by $a_{k-1} = k - 1 + r$. Prove that

$$|hA| \geq hk - (h-1) + (h-1)r.$$

18. Prove that $\{x, y\} = \{u, v\}$ if and only if $x + y = u + v$ and $xy = uv$.

19. The following construction, due to Alon (see [79]), shows that Theorem 1.18 is best possible. For $r \geq 2$, define the integer $\ell \geq 6$ by
$$4^r = 3\ell - 2.$$
Let $n = 3\ell + 1$, and let
$$B = \{3, 6, 9, \ldots, 3\ell, 3\ell + 1\} \subseteq [1, n].$$
Then
$$|B| = \ell + 1 > n/3.$$
Show that if $t \leq 2r$, then 2^t is not the sum of any number of elements of B. Show that if $t \geq 2r + 2$, then 2^t is not the sum of three elements of B. Use the congruence
$$2^{2r+1} \equiv 2 \pmod 3$$
to show that 2^{2r+1} is not the sum of three elements of B.

20. A 2-dimensional arithmetic progression of integers is a set Q of the form
$$\begin{aligned} Q &= Q(q_0; q_1, q_2; l_1, l_2) \\ &= \{q_0 + x_1 q_1 + x_2 q_2 : 0 \leq x_1 < l_1, 0 \leq x_2 < l_2\}, \end{aligned}$$
where q_1, q_2, l_1, l_2 are positive integers and $q_0 \in \mathbf{Z}$. Prove that
$$|Q| \leq l_1 l_2$$
and
$$|2Q| \leq (2l_1 - 1)(2l_2 - 1).$$

21. Construct a 2-dimensional arithmetic progression $Q = Q(q_0; q_1, q_2; l_1, l_2)$ such that
$$|Q| = l_1 l_2$$
and
$$|2Q| = (2l_1 - 1)(2l_2 - 1).$$

22. Construct a 2-dimensional arithmetic progression $Q = Q(q_0; q_1, q_2; l_1, l_2)$ such that
$$|Q| = l_1 l_2$$
and
$$|2Q| < (2l_1 - 1)(2l_2 - 1).$$

23. Let k_1, k_2 be positive integers, and let $k = k_1 + k_2$. For the nonnegative integer r, consider the set
$$A_r = [0, k_1 - 1] \cup [r + k_1, r + k_1 + k_2 - 1].$$
Prove that $|2A| = 3k - 3$ if and only if $r \geq \max(k_1 - 1, k_2 - 1)$.

24. Let A be a finite subset of the abelian group G, and let B be a finite subset of the abelian group H. The map $\phi : A \to B$ is a *Freiman isomorphism* if ϕ is a one-to-one correspondence between A and B and if the map $\Phi : 2A \to 2B$ defined by

$$\Phi(a_1 + a_2) = \phi(a_1) + \phi(a_2)$$

is well-defined and a one-to-one correspondence. Let k_1, k_2 be positive integers. For $r \geq 0$, let A_r be the set of integers defined in the preceding exercise. Let B be the subset of the group $\mathbf{Z}^2$ defined by

$$B = \{(i, 0) : 0 \leq i < k_1\} \cup \{(j, 1) : 0 \leq j < k_2\}.$$

Prove that there exists a Freiman isomorphism between A_r and B if and only if $r \geq \max(k_1 - 1, k_2 - 1)$.

25. Fix $r \geq 5$, and let

$$A = \{0, 1, 2, r, r + 1, 2r\} \subseteq \mathbf{Z}.$$

Show that $|2A| = 3|A| - 3 = 15$. Let

$$B = \{(0, 0), (1, 0), (2, 0), (0, 1), (1, 1), (2, 0)\} \subseteq \mathbf{Z}^2.$$

Show that $|2B| = 3|B| - 3 = 15$. Construct a Freiman isomorphism between A and B.

26. Let $A = \{(i, j) \in \mathbf{Z}^2 \mid 0 \leq i < l_1, 0 \leq j < l_2\}$. Then A is the set of lattice points inside the 2-dimensional parallelepiped $\{(x, y) \in \mathbf{R}^2 \mid 0 \leq x < l_1, 0 \leq y < l_2\}$. Contruct a 2-dimensional arithmetic progression Q in $\mathbf{Z}$ such that A and Q are Freiman isomorphic.

2

Sums of congruence classes

2.1 Addition in groups

Let G be an abelian group, and let A and B be finite subsets of G. The sumset $A + B$ is the set of all elements of G that can be written in the form $a + b$, where $a \in A$ and $b \in B$. For $g \in G$, let $r_{A,B}(g)$ denote the number of representations of g as the sum of an element of A and an element of B, that is, $r_{A,B}(g)$ is the number of ordered pairs $(a, b) \in A \times B$ such that $g = a + b$.

The direct problem for addition in groups is to find a lower bound for $|A + B|$ in terms of $|A|$ and $|B|$. This is easy for finite groups if $|A| + |B|$ is large.

Lemma 2.1 *Let G be a finite abelian group, and let A and B be subsets of G such that*

$$|A| + |B| \geq |G| + t.$$

Then

$$r_{A,B}(g) \geq t$$

for all $g \in G$.

Proof. For $g \in G$, let $g - B = \{g - b : b \in B\}$. Since

$$\begin{aligned} |G| &\geq |A \cup (g - B)| \\ &= |A| + |g - B| - |A \cap (g - B)| \\ &= |A| + |B| - |A \cap (g - B)|, \end{aligned}$$

it follows that

$$|A \cap (b - G)| \geq |A| + |B| - |G| \geq t,$$

and so there exist t distinct elements $a_1, \ldots, a_t \in A$ and t distinct elements $b_1, \ldots, b_t \in B$ such that

$$a_i = g - b_i,$$

that is,

$$g = a_i + b_i$$

for $i = 1, \ldots, t$. Therefore, $r_{A,B}(g) \geq t$.

Lemma 2.2 *Let G be a finite abelian group, and let A and B be subsets of G such that $|A| + |B| > |G|$. Then $A + B = G$.*

Proof. Applying Lemma 2.1 with $t = 1$, we see that $r_{A,B}(g) \geq 1$ for all $g \in G$, and so $A + B = G$. This completes the proof.

It follows from Lemma 2.2 that to study the direct problem for addition in groups, it is enough to examine only subsets $A, B \subseteq G$ such that $|A| + |B| \leq |G|$. In this chapter, we shall consider addition in the group $\mathbf{Z}/m\mathbf{Z}$ of congruence classes modulo m.

2.2 The e-transform

A fundamental tool to prove many results in additive number theory is the *e-transform* of an ordered pair (A, B) of nonempty subsets of an abelian group G. Let $e \in G$. The e-transform of (A, B) is the pair $(A(e), B(e))$ of subsets of G defined by

$$\begin{aligned} A(e) &= A \cup (B + e), \\ B(e) &= B \cap (A - e). \end{aligned}$$

The e-transform has the following simple properties.

Lemma 2.3 *Let A and B be nonempty subsets of the abelian group G, and let e be any element of G. Let $(A(e), B(e))$ be the e-transform of the pair (A, B). Then*

$$A(e) + B(e) \subseteq A + B \tag{2.1}$$

and

$$A(e) \setminus A = e + (B \setminus B(e)). \tag{2.2}$$

If A and B are finite sets, then

$$|A(e)| + |B(e)| = |A| + |B|. \tag{2.3}$$

If $e \in A$ and $0 \in B$, then $e \in A(e)$ and $0 \in B(e)$.

Proof. The set inclusion (2.1) follows immediately from the definition of the e-transform of the pair (A, B). To prove (2.2), we observe that

$$\begin{aligned} A(e) \setminus A &= (B + e) \setminus A \\ &= \{b + e : b \in B, b + e \notin A\} \\ &= e + \{b \in B : b \notin A - e\} \\ &= e + \{b \in B : b \notin B(e)\} \\ &= e + (B \setminus B(e)). \end{aligned}$$

Clearly, $A \subseteq A(e)$ and $B(e) \subseteq B$. If A and B are finite sets, then

$$\begin{aligned} |A(e)| - |A| &= |A(e) \setminus A| \\ &= |e + (B \setminus B(e))| \\ &= |B \setminus B(e)| \\ &= |B| - |B(e)|. \end{aligned}$$

This proves (2.3). If $e \in A \subseteq A(e)$ and $0 \in B$, then $0 \in A - e$ and so $0 \in B \cap (A - e) = B(e)$. This completes the proof.

2.3 The Cauchy–Davenport theorem

In this section we study the direct problem for addition in the group $\mathbf{Z}/m\mathbf{Z}$ of congruence classes modulo m. A basic result is the Cauchy–Davenport theorem, which gives a lower bound for the cardinality of the sum of two sets of congruence classes modulo a prime p. This is a consequence of the following result for composite moduli.

Theorem 2.1 (I. Chowla) *Let $m \geq 2$, and let A and B be nonempty subsets of $\mathbf{Z}/m\mathbf{Z}$. If $0 \in B$ and $(b, m) = 1$ for all $b \in B \setminus \{0\}$, then*

$$|A + B| \geq \min(m, |A| + |B| - 1).$$

Proof. By Lemma 2.2, the result is true if $|A| + |B| > m$. Therefore, we can assume that $|A| + |B| \leq m$, and so

$$\min(m, |A| + |B| - 1) = |A| + |B| - 1 \leq m - 1.$$

The theorem also holds if $|A| = 1$ or $|B| = 1$, since in these cases $|A + B| = |A| + |B| - 1$. If the theorem is false, then there exist sets $A, B \subseteq \mathbf{Z}/m\mathbf{Z}$ such that $|A| \geq 2$, $|B| \geq 2$, and

$$|A + B| < |A| + |B| - 1.$$

In particular, $A \neq \mathbf{Z}/m\mathbf{Z}$. Choose the pair (A, B) so that the cardinality of B is minimal. Since $|B| \geq 2$, there exists an element $b^* \in B$, $b^* \neq 0$. If $a + b^* \in A$ for all $a \in A$, then $a + jb^* \in A$ for all $j = 0, 1, 2, \ldots$. Since $(b^*, m) = 1$, this would imply that

$$\mathbf{Z}/m\mathbf{Z} = \{a + jb^* : j = 0, 1, \ldots, m-1\} \subseteq A \subseteq \mathbf{Z}/m\mathbf{Z},$$

and so $A = \mathbf{Z}/m\mathbf{Z}$, which is false. Therefore, there exists an element $e \in A$ such that $e + b^* \notin A$. Apply the e-transform to the pair (A, B). By Lemma 2.3, we have $A(e) + B(e) \subseteq A + B$, and so

$$|A(e) + B(e)| \leq |A + B| < |A| + |B| - 1 = |A(e)| + |B(e)| - 1.$$

Since $e \in A$ and $0 \in B$, it follows that $0 \in B(e) \subseteq B$, and $(b, m) = 1$ for all $b \in B(e) \setminus \{0\}$. Since $e + b^* \notin A$, we have $b^* \notin A - e$, and so

$$b^* \notin B \cap (A - e) = B(e).$$

Therefore, $|B(e)| < |B|$, which contradicts the minimality of $|B|$. This completes the proof.

Theorem 2.2 (Cauchy–Davenport) *Let p be a prime number, and let A and B be nonempty subsets of $\mathbf{Z}/p\mathbf{Z}$. Then*

$$|A + B| \geq \min(p, |A| + |B| - 1).$$

Proof. Let $b_0 \in B$ and $B' = B - b_0$. Then $|B'| = |B|$ and

$$|A + B'| = |A + B - b_0| = |A + B|.$$

Since $0 \in B'$ and $(b, p) = 1$ for all $b \in B' \setminus \{0\}$, we apply Theorem 2.1 to the pair (A, B') and obtain

$$\begin{aligned} |A + B| &= |A + B'| \\ &\geq \min(p, |A| + |B'| - 1) \\ &= \min(p, |A| + |B| - 1). \end{aligned}$$

This completes the proof.

Theorem 2.3 *Let $h \geq 2$. Let p be a prime number, and let $A_1, A_2, \ldots, A_h$ be nonempty subsets of $\mathbf{Z}/p\mathbf{Z}$. Then*

$$|A_1 + A_2 + \cdots + A_h| \geq \min\left(p, \sum_{i=1}^{h} |A_i| - h + 1\right).$$

Proof. By induction on h. The case $h = 2$ is the Cauchy–Davenport theorem. Let $h \geq 3$, and suppose that the result holds for any $h - 1$ subsets of $\mathbf{Z}/p\mathbf{Z}$. Let $A_1, A_2, \ldots, A_h$ be nonempty subsets of $\mathbf{Z}/p\mathbf{Z}$, and let $B = A_1 + \cdots + A_{h-1}$. By the induction hypothesis,

$$|B| = |A_1 + \cdots + A_{h-1}| \geq \min\left(p, \sum_{i=1}^{h-1} |A_i| - h + 2\right),$$

and so

$$\begin{aligned}
|A_1 + A_2 + \cdots + A_h| &= |(A_1 + \cdots + A_{h-1}) + A_h| \\
&= |B + A_h| \\
&\geq \min(p, |B| + |A_h| - 1) \\
&\geq \min\left(p, \left(\sum_{i=1}^{h-1} |A_i| - h + 2\right) + |A_h| - 1\right) \\
&\geq \min\left(p, \sum_{i=1}^{h} |A_i| - h + 1\right).
\end{aligned}$$

This completes the proof.

It is easy to see that this result is best possible. Let $h \geq 2$, and let $k_1, \ldots, k_h$ be positive integers such that

$$k_1 + \cdots + k_h \leq p + h - 1.$$

Let $A_i = \{0, 1, \ldots, k_i - 1\} \subseteq \mathbf{Z}/p\mathbf{Z}$. Then $|A_i| = k_i$ and

$$A_1 + \cdots + A_h = \{0, 1, \ldots, k_1 + \cdots + k_h - h\} \subseteq \mathbf{Z}/p\mathbf{Z}$$

and

$$|A_1 + A_2 + \cdots + A_h| = \sum_{i=1}^{h} |A_i| - h + 1.$$

Theorem 2.4 (Pollard) *Let p be a prime number, and let A and B be nonempty subsets of $\mathbf{Z}/p\mathbf{Z}$. Let*

$$\ell = |B| \leq |A| = k.$$

For $t = 1, \ldots, \ell$, let N_t denote the number of congruence classes in $\mathbf{Z}/p\mathbf{Z}$ that have at least t representations in the form $a + b$, where $a \in A$ and $b \in B$. Then

$$N_1 + N_2 + \cdots + N_t \geq \min(tp, t(k + \ell - t)).$$

Note that the Cauchy–Davenport theorem is the case $t = 1$ of this result.

Proof. For $x \in \mathbf{Z}/p\mathbf{Z}$, let $r_{A,B}(x)$ denote the number of solutions of $x = a + b$ with $a \in A$ and $b \in B$. Then $r_{A,B}(x) \leq \ell$ for all $x \in \mathbf{Z}/p\mathbf{Z}$, and

$$S(A, B, t) = N_1 + N_2 + \cdots + N_t = \sum_{x \in \mathbf{Z}/p\mathbf{Z}} \min(t, r_{A,B}(x)).$$

For $t = \ell$, we have

$$\begin{aligned}
S(A, B, \ell) &= \sum_{x \in \mathbf{Z}/p\mathbf{Z}} \min(\ell, r_{A,B}(x)) \\
&= \sum_{x \in \mathbf{Z}/p\mathbf{Z}} r_{A,B}(x) \\
&= k\ell.
\end{aligned}$$

Therefore, we can assume that

$$1 \leq t < \ell. \tag{2.4}$$

The proof will be by induction on ℓ.

If $\ell = 1$, then $t = 1$ and

$$N_1 = |A + B| = |A| = |A| + |B| - 1 = \min(p, k + \ell - 1).$$

Let $\ell \geq 2$, and suppose that the theorem holds whenever $|B| < \ell$. If $k+\ell-t > p$, then

$$1 \leq t \leq p - k + t < \ell \leq p.$$

Let $\ell' = p-k+t$. Choose $B' \subseteq B$ such that $|B'| = \ell'$. By the induction hypothesis, the theorem holds for the sets A and B', so

$$\begin{aligned} S(A, B, t) &\geq S(A, B', t) \\ &\geq \min(tp, t(k + \ell' - t)) \\ &= tp \\ &= \min(tp, t(k + \ell - t)). \end{aligned}$$

Therefore, it suffices to prove the theorem in the case

$$k + \ell - t \leq p. \tag{2.5}$$

Let A and B be subsets of $\mathbf{Z}/p\mathbf{Z}$ such that

$$\ell = |B| \leq |A| = k,$$

where k, ℓ, and t satisfy inequalities (2.4) and (2.5). These inequalities imply that $k < p$, and so $A \neq \mathbf{Z}/p\mathbf{Z}$.

Let $b \in B$. Replacing the set B with the difference set $B-b$, we can assume that $0 \in B$. Since $\ell \geq 2$, there exists $b^* \in B$ with $b^* \neq 0$. If $a+b^* \in A$ for every $a \in A$, then $a + jb^* \in A$ for all $j \geq 0$, and so $A = \mathbf{Z}/p\mathbf{Z}$, which is impossible. Therefore, there exists $a^* \in A$ such that $a^*+b^* \notin A$, or, equivalently, $b^* \notin A-a^*$. Replacing A with $A-a^*$, we can assume that $0 \in A$ and $B \setminus A \neq \emptyset$. Then $1 \leq |A \cap B| < |B|$. Let

$$U = A \cup B \qquad \text{and} \qquad I = A \cap B$$

Then

$$|U| + |I| = k + \ell$$

and

$$1 \leq |I| < \ell.$$

Let

$$A' = A \setminus I \qquad \text{and} \qquad B' = B \setminus I.$$

Then U is the disjoint union of A', B', and I.

Let $x \in \mathbf{Z}/p\mathbf{Z}$. Every representation of x in the form $x = a + b$ with $a \in A$, $b \in B$ is of one of the following four types:

(i) $x = a' + b'$ with $a' \in A', b' \in B'$

(ii) $x = a' + v$ with $a' \in A', v \in I$

(iii) $x = v + b'$ with $v \in I, b' \in B'$

(iv) $x = v + v'$ with $v, v' \in I$.

The number of representations of the first type is $r_{A',B'}(x)$. The total number of representations of the three other types is $r_{U,I}(x)$. It follows that

$$r_{A,B}(x) = r_{U,I}(x) + r_{A',B'}(x) \geq r_{U,I}(x)$$

for all $x \in \mathbf{Z}/p\mathbf{Z}$.

Let $1 \leq t \leq |I|$. It follows from the induction hypothesis that the theorem holds for the pair of sets U, I. Therefore,

$$\begin{aligned} S(A, B, t) &= \sum_{x \in \mathbf{Z}/p\mathbf{Z}} \min(t, r_{A,B}(x)) \\ &\geq \sum_{x \in \mathbf{Z}/p\mathbf{Z}} \min(t, r_{U,I}(x)) \\ &= S(U, I, t) \\ &\geq \min(tp, t(|U| + |I| - t)) \\ &= \min(tp, t(k + \ell - t)). \end{aligned}$$

Let $|I| < t < \ell$ and $t' = t - |I|$. Since

$$r_{A,B}(x) = r_{U,I}(x) + r_{A',B'}(x),$$

it follows that

$$\begin{aligned} \min(t, r_{A,B}(x)) &\geq \min(|I|, r_{U,I}(x)) + \min(t', r_{A',B'}(x)) \\ &= r_{U,I}(x) + \min(t', r_{A',B'}(x)). \end{aligned}$$

Let $k' = |A'|$ and $\ell' = |B'|$. Then

$$1 \leq t' = t - |I| < \ell - |I| = |B| - |I| = |B'| = \ell'$$

and

$$\begin{aligned} k' + \ell' - t' &= (k - |I|) + (\ell - |I|) - (t - |I|) \\ &= |U| - t \\ &< |U| \\ &\leq p. \end{aligned}$$

It follows from the induction hypothesis that the theorem holds for the pair of sets A', B'. Therefore,

$$\begin{aligned}\sum_{x\in\mathbf{Z}/p\mathbf{Z}} \min(t, r_{A,B}(x)) &\geq \sum_{x\in\mathbf{Z}/p\mathbf{Z}} r_{U,I}(x) + \sum_{x\in\mathbf{Z}/p\mathbf{Z}} \min(t', r_{A',B'}(x)) \\ &\geq |U||I| + t'(k' + \ell' - t') \\ &= |U||I| + (t - |I|)(|U| - t) \\ &= t(|U| + |I| - t) \\ &= t(k + \ell - t).\end{aligned}$$

This completes the proof.

2.4 The Erdős–Ginzburg–Ziv theorem

We shall give two proofs of a simple but important theorem about addition of congruence classes. The first proof uses the Cauchy–Davenport theorem, and the second uses the Chevalley–Warning theorem on the number of solutions of systems of polynomials over a finite field.

Theorem 2.5 (Erdős–Ginzburg–Ziv) *Let $n \geq 1$. If $a_0, a_1, \ldots, a_{2n-2}$ is a sequence of $2n-1$ not necessarily distinct integers, then there exists a subsequence $a_{i_1}, a_{i_2}, \ldots, a_{i_n}$ such that*

$$a_{i_1} + a_{i_2} + \cdots + a_{i_n} \equiv 0 \pmod{n}.$$

Proof. We shall prove the theorem first in the case that $n = p$ is a prime number. Choose $a_i' \in \mathbf{Z}$ such that $a_i' \equiv a_i \pmod{p}$ and $0 \leq a_i' < p$. Renumber the integers a_i so that

$$0 \leq a_0' \leq a_1' \leq \cdots \leq a_{2p-2}' \leq p - 1.$$

If $a_i' = a_{i+p-1}'$ for some $i \in [1, p-1]$, then

$$a_i \equiv a_{i+1} \equiv \cdots \equiv a_{i+p-1} \pmod{p}$$

and

$$a_i + a_{i+1} + \cdots + a_{i+p-1} \equiv pa_i \equiv 0 \pmod{p}.$$

If $a_i' \neq a_{i+p-1}'$ for all $i \in [1, p-1]$, let

$$A_i = \{a_i + p\mathbf{Z}, a_{i+p-1} + p\mathbf{Z}\} \subseteq \mathbf{Z}/p\mathbf{Z}.$$

Then $|A_i| = 2$ for $i = 1, \ldots, p-1$. Applying the Cauchy–Davenport theorem in the form of Theorem 2.3, we see that

$$|A_1 + \cdots + A_{p-1}| \geq \min(p, 2(p-1) - (p-1) + 1) = p,$$

and so

$$A_1 + \cdots + A_{p-1} = \mathbf{Z}/p\mathbf{Z}.$$

It follows that there exist congruence classes $a_{j_i} + p\mathbf{Z} \in A_i$ for $i = 1, \ldots, p-1$ such that $j_i \in \{i, i+p-1\}$ and

$$-a_0 \equiv a_{j_1} + a_{j_2} + \cdots a_{j_{p-1}} \pmod{p},$$

that is,

$$a_0 + a_{j_1} + a_{j_2} + \cdots a_{j_{p-1}} \equiv 0 \pmod{p}.$$

Thus, the theorem holds when $n = p$ is prime.

We shall prove the theorem by induction on n. If $n = 1$, there is nothing to prove. Suppose that $n > 1$ and that the theorem holds for all positive integers less than n. If n is prime, we are done. If n is composite, then

$$n = uv, \text{ where } 1 < u \leq v < n,$$

and so the result holds for both u and v. From the sequence $a_0, \ldots, a_{2n-2}$ of length $2n - 1 = 2uv - 1$ there exists a subsequence $a_{1,i_1}, \ldots, a_{1,i_v}$ such that

$$a_{1,i_1} + \cdots + a_{1,i_v} \equiv 0 \pmod{v}.$$

There are $2n - 1 - v = (2u - 1)v - 1$ integers in the original sequence that are not in this subsequence. Since $2u - 1 \geq 2$, we can find a disjoint subsequence $a_{2,i_1}, \ldots, a_{2,i_v}$ of length v such that

$$a_{2,i_1} + \cdots + a_{2,i_v} \equiv 0 \pmod{v}.$$

There are $2n - 1 - 2v = (2u - 2)v - 1$ terms not in either of the two subsequences already determined. Continuing inductively for $j = 1, \ldots, 2u-1$, we obtain $2u-1$ pairwise disjoint subsequences $a_{j,i_1}, \ldots, a_{j,i_v}$ of length v such that

$$a_{j,i_1} + \cdots + a_{j,i_v} \equiv 0 \pmod{v}.$$

Then

$$a_{j,i_1} + \cdots + a_{j,i_v} = b_j v,$$

where $b_j \in \mathbf{Z}$. Since the theorem holds for u, there is a subsequence $b_{j_1}, \ldots, b_{j_u}$ of the sequence $b_1, \ldots, b_{2u-1}$ such that

$$b_{j_1} + \cdots + b_{j_u} \equiv 0 \pmod{u},$$

that is,

$$b_{j_1} + \cdots + b_{j_u} = cu$$

for some $c \in \mathbf{Z}$. Then

$$\sum_{r=1}^{u} \sum_{s=1}^{v} a_{j_r,i_s} = \sum_{r=1}^{u} b_{j_r} v = cuv = cn \equiv 0 \pmod{n}.$$

This completes the proof.

Theorem 2.6 (Chevalley–Warning) *Let p be a prime number, and let $\mathbf{F}_q$ be the finite field with $q = p^t$ elements. For $i = 1, \ldots, m$, let $f_i(x_1, \ldots, x_n)$ be a polynomial of degree d_i in n variables with coefficients in $\mathbf{F}_q$. Let N denote the number of n-tuples $(x_1, \ldots, x_n)$ of elements of $\mathbf{F}_q$ such that*

$$f_i(x_1, \ldots, x_n) = 0$$

for all $i = 1, \ldots, m$. If

$$\sum_{i=1}^{m} d_i < n,$$

then

$$N \equiv 0 \pmod{p}.$$

Proof. The multiplicative group of the nonzero elements of a finite field is cyclic, and so, for any $x \in \mathbf{F}_q$,

$$x^{q-1} = \begin{cases} 1 & \text{if } x \neq 0 \\ 0 & \text{if } x = 0. \end{cases} \tag{2.6}$$

Moreover, with the convention that $0^0 = 1$, we have

$$\sum_{x \in \mathbf{F}_q} x^r = 0 \qquad \text{if } 0 \leq r < q-1.$$

Let $x_1, \ldots, x_n \in \mathbf{F}_q$. Then

$$\prod_{i=1}^{m} \left(1 - f_i(x_1, \ldots, x_n)^{q-1}\right) = \begin{cases} 1 & \text{if } f_i(x_1, \ldots, x_n) = 0 \quad \text{for all } i \\ 0 & \text{otherwise,} \end{cases}$$

and so

$$N = \sum_{x_1, \ldots, x_n \in \mathbf{F}_q} \prod_{i=1}^{m} \left(1 - f_i(x_1, \ldots, x_n)^{q-1}\right).$$

Since the degree of $f_i(x_1, \ldots, x_n)$ is d_i, it follows that

$$\prod_{i=1}^{m} \left(1 - f_i(x_1, \ldots, x_n)^{q-1}\right) = \sum_{r_1, \ldots, r_n} a_{r_1, \ldots, r_n} x_1^{r_n} \cdots x_n^{r_n}$$

is a polynomial of degree at most $(q-1)\sum_{i=1}^{m} d_i$ with coefficients $a_{r_1, \ldots, r_n} \in \mathbf{F}_q$. Then

$$\begin{aligned} N &\equiv \sum_{x_1, \ldots, x_n \in \mathbf{F}_q} \prod_{i=1}^{m} \left(1 - f_i(x_1, \ldots, x_n)^{q-1}\right) \pmod{p} \\ &\equiv \sum_{x_1, \ldots, x_n \in \mathbf{F}_q} \sum_{r_1, \ldots, r_n} a_{r_1, \ldots, r_n} x_1^{r_n} \cdots x_n^{r_n} \pmod{p} \\ &\equiv \sum_{r_1, \ldots, r_n} a_{r_1, \ldots, r_n} \sum_{x_1, \ldots, x_n \in \mathbf{F}_q} x_1^{r_n} \cdots x_n^{r_n} \pmod{p} \\ &\equiv \sum_{r_1, \ldots, r_n} a_{r_1, \ldots, r_n} \prod_{j=1}^{n} \sum_{x_j \in \mathbf{F}_q} x_j^{r_j} \pmod{p}, \end{aligned}$$

where the summation runs over all n-tuples $r_1, \ldots, r_n$ of nonnegative integers such that

$$\sum_{j=1}^{n} r_j \leq (q-1)\sum_{j=1}^{n} d_j < n(q-1).$$

This implies that $0 \leq r_j < q-1$ for some j, and so

$$\prod_{j=1}^{n} \sum_{x_j \in \mathbf{F}_q} x_j^{r_j} \equiv 0 \pmod{p}.$$

Therefore,

$$N \equiv 0 \pmod{p}.$$

This completes the proof.

In the case when n is a prime number p, the Erdős–Ginzburg–Ziv theorem (Theorem 2.5) is a corollary of the Chevalley–Warning theorem. Let $a_1, \ldots, a_{2p-1}$ be a sequence of elements in the finite field $\mathbf{F}_p = \mathbf{Z}/p\mathbf{Z}$. Consider the polynomials $f_1, f_2 \in \mathbf{F}_p[x_1, \ldots, x_{2p-1}]$ defined by

$$f_1(x_1, \ldots, x_{2p-1}) = \sum_{j=1}^{2p-1} x_j^{p-1}$$

and

$$f_2(x_1, \ldots, x_{2p-1}) = \sum_{j=1}^{2p-1} a_j x_j^{p-1}.$$

Let d_i be the degree of the polynomial f_i. Then $d_1 = d_2 = p-1$. Let N denote the number of simultaneous solutions of these two polynomials. Since

$$d_1 + d_2 = 2p - 2 < 2p - 1,$$

it follows from Theorem 2.6 that $N \equiv 0 \pmod{p}$. Since

$$f_1(0, \ldots, 0) = f_2(0, \ldots, 0) = 0,$$

it follows that $N \geq 1$ and so $N \geq p \geq 2$. Therefore, the polynomials f_1 and f_2 have a nontrivial solution, that is, there exist $x_1, \ldots, x_{2p-1} \in \mathbf{Z}/p\mathbf{Z}$ not all zero such that

$$f_1(x_1, \ldots, x_{2p-1}) = \sum_{j=1}^{2p-1} x_j^{p-1} = 0$$

and

$$f_2(x_1, \ldots, x_{2p-1}) = \sum_{j=1}^{2p-1} a_j x_j^{p-1} = 0.$$

For $x \in \mathbf{Z}/p\mathbf{Z}$, we have $x^{p-1} = 1$ if and only if $x \not\equiv 0$. It follows from the first equation that $x_j \not\equiv 0$ for exactly p variables $x_{j_1}, \ldots, x_{j_p}$. Then the second equation implies that

$$a_{j_1} + \cdots a_{j_p} \equiv 0 \pmod{p}.$$

This completes the proof.

2.5 Vosper's theorem

The inverse problem for addition in groups is to describe the structure of the pairs of subsets (A, B) for which the cardinality of the sumset $A + B$ is small. For most pairs (A, B) in an abelian group G, the sumset $A + B$ will contain at least $|A| + |B|$ elements. The simplest inverse problem is to classify the pairs (A, B) of finite subsets of G such that $A + B \neq G$ and $|A + B| < |A| + |B|$. Such pairs are called *critical.* This is an open problem for arbitrary groups. However, Vosper completely solved the problem of classifying the critical pairs for the groups $\mathbf{Z}/p\mathbf{Z}$, where p is a prime number. He proved that if $A, B \subseteq \mathbf{Z}/p\mathbf{Z}$ and $|A + B| = |A| + |B| - 1$, then, except for two special cases, the sets A and B are arithmetic progressions with the same common difference, where an *arithmetic progression* in an abelian group G is a set of the form

$$\{a + id : i = 0, 1, \ldots, k - 1\}.$$

The group element d is called the *common difference* of the progression, and k is called the *length* of the progression. The order of the group element d in G must be at least k.

Using exponential sums and analytic methods, Freiman generalized Vosper's theorem for sumsets of the form $2A = A + A$ in $\mathbf{Z}/p\mathbf{Z}$. In particular, he proved that if $A \subseteq \mathbf{Z}/p\mathbf{Z}$, $|A| = k \leq p/35$, and $|2A| = 2k - 1 + r \leq 12k/5 - 3$, then A is contained in an arithmetic progression of length $k + r$.

In this section, we shall denote the complement of the set S in G by $\overline{S}$.

Theorem 2.7 (Vosper) *Let p be a prime number, and let A and B be nonempty subsets of the group $G = \mathbf{Z}/p\mathbf{Z}$ such that $A + B \neq G$. Then*

$$|A + B| = |A| + |B| - 1$$

if and only if at least one of the following three conditions holds:

(i) $\min(|A|, |B|) = 1$,

(ii) $|A + B| = p - 1$ *and* $B = \overline{c - A}$, *where* $\{c\} = G \setminus (A + B)$

(iii) *A and B are arithmetic progressions with the same common difference.*

Proof. By Lemma 2.2, if $A + B \neq G$, then $|A| + |B| \leq p$. If $\min(|A|, |B|) = |B| = 1$, then $|A + B| = |A| = |A| + |B| - 1$, and so (A, B) is a critical pair.

Let $c \in G$, and let A be any subset of G such that $1 \leq |A| \leq p - 1$. Let $B = \overline{c - A}$. Then $c \notin A + B$, and so $|A + B| \leq p - 1$. Since

$$|B| = |\overline{c - A}| = p - |c - A| = p - |A|,$$

the Cauchy–Davenport theorem implies that

$$p - 1 = |A| + |B| - 1 \leq |A + B| \leq p - 1,$$

and so $|A + B| = |A| + |B| - 1$.

If A and B are arithmetic progressions in G with the same common difference d, then there exist group elements $a, b \in G$ and positive integers k, l with $k+l \leq p$ such that

$$A = \{a + id : i = 0, 1, \ldots, k - 1\}$$

and

$$B = \{b + id : i = 0, 1, \ldots, l - 1\}.$$

Since $d \in G \setminus \{0\}$, the order of d is p. Then

$$A + B = \{a + b + id : i = 0, 1, \ldots, k + l - 2\},$$

and $|A+B| = k+l-1 = |A|+|B|-1$. Therefore, if the sets A, B satisfy conditions (i), (ii), or (iii), then the pair (A, B) is critical.

Conversely, let (A, B) be a critical pair, that is,

$$|A + B| = |A| + |B| - 1.$$

If $|A| = 1$ or $|B| = 1$, the pair is of the form (i).

If $|A + B| = p - 1$, then $\overline{A + B} = \{c\}$ for some $c \in G$. Since $c \notin A + B$, it follows that $B \cap (c - A) = \emptyset$ and so

$$B \subseteq \overline{c - A}.$$

Then

$$|B| \leq |\overline{c - A}| = p - |c - A| = p - |A|.$$

Since

$$p - 1 = |A + B| = |A| + |B| - 1 \leq p - 1,$$

it follows that $|B| = p - |A|$ and so $B = \overline{c - A}$. Thus, the pair (A, B) is of the form (ii).

For the remainder of the proof, we can assume that (A, B) is a critical pair such that

$$\min(|A|, |B|) \geq 2$$

and

$$|A + B| < p - 1.$$

We shall show that A and B are arithmetic progressions with the same common difference. This will require several lemmas.

Lemma 2.4 *Let A and B be subsets of $\mathbf{Z}/p\mathbf{Z}$ such that*

$$\min(|A|, |B|) \geq 2$$

and

$$|A + B| = |A| + |B| - 1 < p - 1.$$

If A is an arithmetic progression, then B is an arithmetic progression with the same common difference.

Proof. Let $|A| = k$ and $|B| = \ell$. Since A is an arithmetic progression, there exists $d \in \mathbf{Z}/p\mathbf{Z}$ such that $d \neq 0$ and

$$A = \{a_0 + id : i = 0, 1, \ldots, k-1\}.$$

Then

$$A' = \{(a - a_0)d^{-1} \mid a \in A\} = \{i + p\mathbf{Z} : i = 0, 1, \ldots, k-1\} \subseteq \mathbf{Z}/p\mathbf{Z}.$$

Choose $b_0 \in B$, and let

$$B' = \{(b - b_0)d^{-1} \mid b \in B\}.$$

Then

$$\begin{aligned} 0 &\in B', \\ |A'| &= |A| = k \geq 2, \\ |B'| &= |B| = \ell \geq 2, \\ A' + B' &= \{(c - a_0 - b_0)d^{-1} \mid c \in A + B\}, \end{aligned}$$

and so

$$|A' + B'| = |A + B| = |A'| + |B'| - 1 < p - 1.$$

Therefore, we can assume without loss of generality that $A = A'$ and $B = B'$. We shall prove that $B = \{b, b+1, b+2, \ldots, b+\ell-1\}$ for some $b \in B$.

Let $B = \{b_0, b_1, \ldots, b_{\ell-1}\}$. For $j = 0, 1, \ldots, \ell-1$, choose $r_j \in [0, p-1]$ such that $b_j = r_j + p\mathbf{Z}$. By appropriately renumbering the congruence classes b_j, we can assume that

$$0 = r_0 < r_1 < \cdots < r_{\ell-1} < p.$$

Let $r_\ell = p$. Since every element of $A + B$ is of the form $b_j + i = r_j + i + p\mathbf{Z}$ for some $i \in [0, k-1]$ and $j \in [0, \ell-1]$, it follows that

$$A + B = \bigcup_{j=0}^{\ell-1} [r_j, r_j + \min(k-1, r_{j+1} - r_j - 1)] + p\mathbf{Z}.$$

Since the ℓ sets in this union are pairwise disjoint, we have

$$\begin{aligned} k + \ell - 1 &= |A + B| \\ &= \sum_{j=0}^{\ell-1} \big(1 + \min(k-1, r_{j+1} - r_j - 1)\big) \\ &= \ell + \sum_{j=0}^{\ell-1} \min(k-1, r_{j+1} - r_j - 1). \end{aligned}$$

If $r_{j+1} - r_j - 1 \leq k - 1$ for all $j = 0, 1, \ldots, \ell-1$, then

$$k + \ell - 1 = \ell + \sum_{j=0}^{\ell-1} (r_{j+1} - r_j - 1) = r_\ell - r_0 = p,$$

which is false. Therefore, $r_{j_0+1} - r_{j_0} - 1 > k - 1$ for some $j_0 \in [0, \ell - 1]$, and so

$$\begin{aligned} k + \ell - 1 &= |A + B| \\ &= k + \ell - 1 + \sum_{\substack{j=0 \\ j \neq j_0}}^{\ell-1} \min(k - 1, r_{j+1} - r_j - 1). \end{aligned}$$

It follows that $r_{j+1} - r_j = 1$ for all $j \in [0, \ell - 1]$, $j \neq j_0$ and so B is the arithmetic progression

$$[r_{j_0+1}, r_{j_0+1} + \ell - 1] + p\mathbf{Z}.$$

Lemma 2.5 *Let A and B be subsets of $\mathbf{Z}/p\mathbf{Z}$ such that*

$$\min(|A|, |B|) = 2$$

and

$$|A + B| = |A| + |B| - 1 < p - 1.$$

Then A and B are arithmetic progressions with the same common difference.

Proof. This follows immediately from Lemma 2.4 since a set with two elements is an arithmetic progression.

Lemma 2.6 *Let (A, B) be a critical pair in $\mathbf{Z}/p\mathbf{Z}$ such that*

$$\min(|A|, |B|) \geq 2$$

and

$$|A + B| = |A| + |B| - 1 < p - 1.$$

Let $D = \overline{A + B}$. Then $(D, -A)$ is a critical pair.

Proof. Let $|A| = k$ and $|B| = \ell$. Since $k + \ell - 1 \leq p - 2$, it follows that

$$|D| = |\overline{A + B}| = p - (k + \ell - 1) \geq 2.$$

We must show that $|D - A| = |D| + |-A| - 1 = p - \ell$. By the Cauchy–Davenport theorem,

$$\begin{aligned} |D - A| &\geq \min(p, |D| + |-A| - 1) \\ &= \min(p, (p - k - \ell + 1) + k - 1) \\ &= p - \ell. \end{aligned}$$

Since $(A + B) \cap D = \emptyset$, it follows that $B \cap (D - A) = \emptyset$ and so $D - A \subseteq \overline{B}$. Therefore,

$$|D - A| \leq |\overline{B}| = p - |B| = p - \ell.$$

This completes the proof.

Lemma 2.7 *Let (A, B) be a critical pair in $\mathbf{Z}/p\mathbf{Z}$ such that*

$$\min(|A|, |B|) \geq 2$$

and

$$|A + B| = |A| + |B| - 1 < p - 1.$$

If $A + B$ is an arithmetic progression, then A and B are arithmetic progressions with the same common difference.

Proof. If $A+B$ is an arithmetic progression, then $D = \overline{A + B}$ is also an arithmetic progression. By Lemma 2.6, the pair $(D, -A)$ is critical, and so, by Lemma 2.4, the set $-A$ is an arithmetic progression. It follows that A is an arithmetic progression, and, since the pair (A, B) is critical, the sets A and B are arithmetic progressions with the same common difference.

Lemma 2.8 *Let (A, B) be a critical pair in $\mathbf{Z}/p\mathbf{Z}$ such that*

$$|A| = k \geq 2,$$

$$|B| = \ell \geq 3,$$

$$0 \in B,$$

and

$$|A + B| = |A| + |B| - 1 < p - 1.$$

Then there exists a congruence class $e \in A$ with the property that the e-transform $(A(e), B(e))$ is a critical pair such that $A(e) + B(e) = A + B$ and $2 \leq |B(e)| < |B|$.

Proof. If $(A(e), B(e))$ is any e-transform of the critical pair (A, B), then it follows from Lemma 2.3 and the Cauchy–Davenport theorem that

$$\begin{aligned}
|A| + |B| - 1 &= |A(e)| + |B(e)| - 1 \\
&\leq |A(e) + B(e)| \\
&\leq |A + B| \\
&= |A| + |B| - 1.
\end{aligned}$$

Therefore,

$$|A(e)| + |B(e)| - 1 = |A(e) + B(e)| = |A + B|,$$

and so $(A(e), B(e))$ is also a critical pair. Since $A(e) + B(e) \subseteq A + B$, it follows that $A(e) + B(e) = A + B$.

Let

$$X = \{e \in A : B(e) \neq B\}.$$

Since $B(e) \subseteq B$ for all $e \in G$, it follows that $|B(e)| < |B|$ for all $e \in X$.

We shall show that $|X| \geq 2$. Let

$$Y = A \setminus X = \{e \in A : B(e) = B\}.$$

If $Y = \emptyset$, then $X = A$ and $|X| = |A| \geq 2$. If $Y \neq \emptyset$, choose $e \in Y$. Then $B = B(e) = B \cap (A - e)$ and so $B \subseteq A - e$. It follows that $e + B \subseteq A$ for all $e \in Y$, and so $Y + B \subseteq A$. By the Cauchy–Davenport theorem,

$$\begin{aligned} k &= |A| \\ &\geq |Y + B| \\ &\geq \min(p, |Y| + \ell - 1) \\ &= |Y| + \ell - 1 \\ &= k - |X| + \ell - 1, \end{aligned}$$

and so $|X| \geq \ell - 1 \geq 2$.

We shall show that $|B(e)| \geq 2$ for some $e \in X$. Since $e \in X \subseteq A$ and $0 \in B$, we have $0 \in B(e)$. Suppose that $B(e) = B \cap (A - e) = \{0\}$ for all $e \in X$. Let $B' = B \setminus \{0\}$. Then $B' \cap (A - e) = \emptyset$ and so $(e + B') \cap A = \emptyset$ for all $e \in X$. Therefore, $(X + B') \cap A = \emptyset$. Since $X + B' \subseteq A + B$, it follows that

$$X + B' \subseteq (A + B) \setminus A$$

and so, again by the Cauchy–Davenport theorem,

$$|X| + \ell - 2 = |X| + (\ell - 1) - 1 \leq |X + B'| \leq |A + B| - |A| = \ell - 1,$$

which is impossible, because $|X| \geq 2$. This completes the proof.

Proof of Vosper's theorem, concluded. Let (A, B) be a critical pair with $|B| = l \geq 2$. The proof is by induction on l. If $l = 2$, the result follows from Lemma 2.5. Let $l \geq 3$, and assume that the theorem holds for all critical pairs (A, B) with $|B| < l$. By Lemma 2.8, there exists $e \in A$ such that $(A(e), B(e))$ is a critical pair with $A(e) + B(e) = A + B$ and $2 \leq |B(e)| < l$. The induction assumption implies that $A(e)$ and $B(e)$ are arithmetic progressions with the same common difference. Therefore, $A(e)+B(e) = A+B$ is an arithmetic progression, and Lemma 2.7 implies that A and B are arithmetic progressions with the same common difference. This completes the proof of Vosper's theorem.

2.6 Application: The range of a diagonal form

Let $k \geq 1$ and p be a prime number. The polynomial

$$f(x_1, \ldots, x_n) = c_1 x_1^k + \cdots + c_n x_n^k$$

with coefficients in the field $\mathbf{Z}/p\mathbf{Z}$ is called a *diagonal form* of degree k. We shall assume that $c_i \neq 0$ for all i. The *range* of f is the set

$$R(f) = \{f(x_1, \ldots, x_n) : x_1, \ldots, x_n \in \mathbf{Z}/p\mathbf{Z}\}.$$

Lemma 2.9 *Let $p \equiv 1 \pmod{k}$, and let f be a diagonal form of degree k over the field $\mathbf{Z}/p\mathbf{Z}$. If $p = ks + 1$, then*

$$|R(f)| \equiv 1 \pmod{s}.$$

Proof. Let $(\mathbf{Z}/p\mathbf{Z})^*$ denote the multiplicative group of the nonzero elements of $\mathbf{Z}/p\mathbf{Z}$. This is a cyclic group of order $p - 1$. Since k divides $p - 1$, it follows that $A_k^* = \{x^k : x \in (\mathbf{Z}/p\mathbf{Z})^*\}$ is a subgroup of $\mathbf{Z}/p\mathbf{Z}$ of order s. Note that $0 = f(0, \ldots, 0) \in R(f)$. Let $R(f)^* = R(f) \setminus \{0\}$. If $z \in R(f)^*$, then

$$z = \sum_{i=1}^{n} a_i x_i^k$$

for some $x_1, \ldots, x_n \in \mathbf{Z}/p\mathbf{Z}$, and so, for any $y^k \in A_k^*$,

$$zy^k = \sum_{i=1}^{n} a_i (x_i y)^k \in R(f)^*.$$

Therefore,

$$zA_k^* \subseteq R(f)^*.$$

This implies that $R(f)^*$ is a union of cosets of A_k^*, hence

$$|R(f)^*| \equiv 0 \pmod{s}$$

and

$$|R(f)| = |R(f)^*| + 1 \equiv 1 \pmod{s}.$$

This completes the proof.

Lemma 2.10 *Let $p > 3$ be a prime number, let $1 < s < p - 1$, and let A be a set of s distinct elements from the field $\mathbf{Z}/p\mathbf{Z}$. If*

$$\sum_{a \in A} a = \sum_{a \in A} a^2 = 0,$$

then the set A is not an arithmetic progression.

Proof. If A is an arithmetic progression, then there exists $d \in (\mathbf{Z}/p\mathbf{Z})^*$ such that

$$A = \{a_0 + id : i = 0, 1, \ldots, s - 1\}.$$

Then

$$\begin{aligned}
\sum_{a \in A} a &= \sum_{i=0}^{s-1} (a_0 + id) \\
&= sa_0 + \frac{s(s-1)d}{2} \\
&= 0,
\end{aligned}$$

and so

$$a_0 = -\frac{(s-1)d}{2}.$$

It follows that

$$\begin{aligned}
\sum_{a \in A} a^2 &= \sum_{i=0}^{s-1} (a_0 + id)^2 \\
&= \sum_{i=0}^{s-1} (a_0^2 + 2a_0 id + i^2 d^2) \\
&= sa_0^2 + s(s-1)a_0 d + \frac{s(s-1)(2s-1)d^2}{6} \\
&= -\frac{s(s-1)^2 d^2}{4} + \frac{s(s-1)(2s-1)d^2}{6} \\
&= \frac{(s-1)s(s+1)d^2}{12} \\
&= 0,
\end{aligned}$$

which is impossible. This completes the proof.

Lemma 2.11 *Let $p > 3$ be a prime number, and let $p \equiv 1 \pmod{k}$, where*

$$1 < k < \frac{p-1}{2}.$$

Let $A_k = \{x^k : x \in \mathbf{Z}/p\mathbf{Z}\}$. Then A_k is not an arithmetic progression in $\mathbf{Z}/p\mathbf{Z}$.

Proof. Let $p = ks + 1$. Since $2k < p - 1$, it follows that $s \geq 3$. Let g be a primitive root modulo p, that is, a generator of the cyclic group $(\mathbf{Z}/p\mathbf{Z})^*$. The kth powers in this group are the s distinct elements $1, g^k, g^{2k}, \ldots, g^{(s-1)k}$. Since $s \geq 3$, we have

$$\begin{aligned}
\sum_{a \in A_k} a &= \sum_{i=0}^{s-1} g^{ik} \\
&= \frac{g^{sk} - 1}{g^k - 1} \\
&= \frac{g^{p-1} - 1}{g^k - 1} \\
&= 0
\end{aligned}$$

and

$$\sum_{a \in A_k} a^2 = \sum_{i=0}^{s-1} g^{2ik}$$

$$\begin{aligned} &= \frac{g^{2sk}-1}{g^{2k}-1} \\ &= \frac{g^{2(p-1)}-1}{g^k-1} \\ &= 0. \end{aligned}$$

By Lemma 2.10, the set A_k is not an arithmetic progression.

Lemma 2.12 *For $k \geq 1$, let $A_k = \{x^k : x \in \mathbf{Z}/p\mathbf{Z}\}$. If $d = (k, p-1)$, then $A_k = A_d$.*

Proof. There exist integers u and v such that $d = uk + v(p-1)$. Let $x \in \mathbf{Z}/p\mathbf{Z}$. If $x \neq 0$, then

$$x^d = x^{uk+v(p-1)} = (x^u)^k (x^{p-1})^v = (x^u)^k \in A_k,$$

and so $A_d \subseteq A_k$.

Similarly, since d divides k, we have $k = rd$ for some integer r, and

$$x^k = x^{rd} = (x^r)^d \in A_d.$$

Therefore, $A_k \subseteq A_d$.

Theorem 2.8 *Let $p > 3$ be a prime number, and let k be a positive integer such that*

$$1 < (k, p-1) < \frac{p-1}{2}.$$

Let $c_1, \ldots, c_n$ be nonzero elements of the field $\mathbf{Z}/p\mathbf{Z}$, and let

$$f(x_1, \ldots, x_n) = c_1 x_1^k + \cdots + c_n x_n^k.$$

Let $R(f)$ be the range of the diagonal form f. Then

$$|R(f)| \geq \min\left(p, \frac{(2n-1)(p-1)}{(k, p-1)} + 1\right).$$

Proof. Let $d = (k, p-1)$, and let

$$g(x_1, \ldots, x_n) = c_1 x_1^d + \cdots + c_n x_n^d.$$

Let $R(g)$ be the range of the diagonal form g. Let $A_k = \{x^k : x \in \mathbf{Z}/p\mathbf{Z}\}$ and $A_d = \{x^d : x \in \mathbf{Z}/p\mathbf{Z}\}$. Since $A_k = A_d$ by Lemma 2.12, it follows that $R(f) = R(g)$, and so we can assume that $k = (k, p-1)$. Then

$$p = ks + 1,$$

where $s \geq 3$, and $|A_k| = s + 1$. We must prove that

$$|R(f)| \geq \min(p, (2n-1)s + 1).$$

The proof is by induction on n. If $n = 1$, then $f(x_1) = c_1 x_1^k$, where $c_1 \not\equiv 0$, and so

$$|R(f)| = |A_k| = s + 1 = \min(p, s + 1).$$

Let $n \geq 2$, and assume that the theorem holds for $n - 1$. Let

$$A = \left\{ \sum_{i=1}^{n-1} c_i x_i^k : x_1, \ldots, x_{n-1} \in \mathbf{Z}/p\mathbf{Z} \right\}$$

and

$$B = \left\{ c_n x_n^k : x_n \in \mathbf{Z}/p\mathbf{Z} \right\}.$$

Then

$$|B| = |A_k| = s + 1.$$

Since the set A is the range of a diagonal form in $n - 1$ variables, the induction hypothesis implies that

$$|A| \geq \min(p, (2n - 3)s + 1).$$

Since

$$R(f) = A + B,$$

it follows from the Cauchy–Davenport theorem that

$$\begin{aligned} |R(f)| &= |A + B| \\ &\geq \min(p, |A| + |B| - 1) \\ &\geq \min(p, (2n - 2)s + 1). \end{aligned}$$

If $|R(f)| = p$, we are done. If $|R(f)| \leq p - 1$, then

$$|R(f)| = |A + B| \geq |A| + |B| - 1 \geq (2n - 2)s + 1.$$

If $|R(f)| = (2n - 2)s + 1$, then it follows from Vosper's inverse theorem (Theorem 2.7) that

(i) either $\min(|A|, |B|) = 1$, which is false,

(ii) or

$$|R(f)| = |A + B| = p - 1 = ks,$$

which is also false since $R(f) \equiv 1 \pmod{s}$ by Lemma 2.9,

(iii) or, finally, A and B are arithmetic progressions with the same common difference, which is false, since by Lemma 2.11 the sets A_k and B are not arithmetic progressions.

Therefore,

$$|R(f)| \geq (2n - 2)s + 2.$$

Since

$$|R(f)| \equiv 1 \pmod{s},$$

it follows that

$$|R(f)| \geq (2n - 1)s + 1.$$

This completes the proof.

2.7 Exponential sums

Let m and x be integers such that $m \geq 2$, and let $a = r + m\mathbf{Z}$ be an element of the group $\mathbf{Z}/m\mathbf{Z}$ of congruence classes modulo m. We define

$$e^{2\pi iax/m} = e^{2\pi irx/m}.$$

This function on $\mathbf{Z}/m\mathbf{Z}$ is well defined, since if $r, r' \in \mathbf{Z}$ and $r \equiv r' \pmod{m}$, then

$$e^{2\pi irx/m} = e^{2\pi ir'x/m}$$

for every $x \in \mathbf{Z}$.

Let $A = \{a_0, a_1, \ldots, a_{k-1}\}$ be a sequence of k not necessarily distinct congruence classes in the group $\mathbf{Z}/m\mathbf{Z}$. We define the exponential sum

$$S_A(x) = \sum_{j=0}^{k-1} e^{2\pi ia_jx/m}. \tag{2.7}$$

For all $x \in \mathbf{Z}$,

$$|S_A(x)| \leq |S_A(0)| = k.$$

The basic identity for exponential sums is the following.

Lemma 2.13 *Let $m \geq 2$ and $a \in \mathbf{Z}/m\mathbf{Z}$. Then*

$$\sum_{x=0}^{m-1} e^{2\pi iax/m} = \begin{cases} m & \text{if } a = 0 \\ 0 & \text{if } a \neq 0. \end{cases} \tag{2.8}$$

Proof. Let $a = r + m\mathbf{Z}$. If $r \equiv 0 \pmod{m}$, then

$$\sum_{x=0}^{m-1} e^{2\pi irx/m} = \sum_{x=0}^{m-1} 1 = m.$$

If $r \not\equiv 0 \pmod{m}$, then the series is a finite geometric progression and

$$\sum_{x=0}^{m-1} e^{2\pi irx/m} = \frac{e^{2\pi ir} - 1}{e^{2\pi ir/m} - 1} = \frac{1 - 1}{e^{2\pi ir/m} - 1} = 0.$$

This completes the proof.

Let $\overline{z}$ denote the complex conjugate of the complex number z, and let $-A = \{-a \mid a \in A\}$. Then

$$\overline{S_A(x)} = \overline{\sum_{j=0}^{k-1} e^{2\pi ir_jx/p}} = \sum_{j=0}^{k-1} e^{-2\pi ir_jx/p} = S_{-A}(x).$$

Lemma 2.14 *Let $A_1, \ldots, A_{n_1}, B_1, \ldots, B_{n_2}$ be nonempty subsets of $\mathbf{Z}/m\mathbf{Z}$, and let N be the number of solutions of the equation*

$$a_1 + \cdots + a_{n_1} = b_1 + \cdots + b_{n_2}$$

in $\mathbf{Z}/m\mathbf{Z}$ with $a_i \in A_i$ for $i = 1, \ldots, n_1$, and $b_j \in B_j$ for $j = 1, \ldots, n_2$. Then

$$N = \frac{1}{m} \sum_{x=0}^{m-1} S_{A_1}(x) \cdots S_{A_{n_1}}(x) \overline{S_{B_1}(x)} \cdots \overline{S_{B_{n_2}}(x)}.$$

Proof. This follows immediately from equation (2.8), since

$$\begin{aligned}
&\sum_{x=0}^{m-1} e^{2\pi i (a_1 + \cdots + a_{n_1} - b_1 - \cdots - b_{n_2})x/m} \\
&\quad = \begin{cases} m & \text{if } a_1 + \cdots + a_{n_1} = b_1 + \cdots b_{n_2} \\ 0 & \text{if } a_1 + \cdots + a_{n_1} \neq b_1 + \cdots b_{n_2} \end{cases}
\end{aligned}$$

and so

$$\begin{aligned}
&\sum_{x=0}^{m-1} S_{A_1}(x) \cdots S_{A_{n_1}}(x) \overline{S_{B_1}(x)} \cdots \overline{S_{B_{n_2}}(x)} \\
&= \sum_{x=0}^{m-1} \sum_{a_1 \in A_1} \cdots \sum_{a_{n_1} \in A_{n_1}} \sum_{b_1 \in B_1} \cdots \sum_{b_{n_2} \in B_{n_2}} e^{2\pi i (a_1 + \cdots + a_{n_1} - b_1 - \cdots - b_{n_2})x/m} \\
&= \sum_{a_1 \in A_1} \cdots \sum_{a_{n_1} \in A_{n_1}} \sum_{b_1 \in B_1} \cdots \sum_{b_{n_2} \in B_{n_2}} \sum_{x=0}^{m-1} e^{2\pi i (a_1 + \cdots + a_{n_1} - b_1 - \cdots - b_{n_2})x/m} \\
&= Nm.
\end{aligned}$$

Lemma 2.15 *Let A be a nonempty subset of $\mathbf{Z}/m\mathbf{Z}$ with $|A| = k$. Then*

$$\sum_{x=0}^{m-1} |S_A(x)|^2 = km$$

and

$$\sum_{x=0}^{m-1} S_A(x)^2 \overline{S_{2A}(x)} = k^2 m.$$

Proof. These identities follow from Theorem 2.14. The first comes from the fact that

$$\sum_{x=0}^{m-1} |S_A(x)|^2 = \sum_{x=0}^{m-1} S_A(x) \overline{S_A(x)}$$

and the number of solutions of the equation $a_1 = a_2$ with $a_1, a_2 \in A$ is $|A| = k$. The second follows from the observation that the number of solutions of the equation $a_1 + a_2 = b$ with $a_1, a_2 \in A$ and $b \in 2A$ is $|A|^2 = k^2$. This completes the proof.

For $\alpha, \alpha' \in \mathbf{R}$, we write

$$\alpha \equiv \alpha' \pmod{1}$$

if $\alpha - \alpha' \in \mathbf{Z}$. If $r \equiv r' \pmod{m}$, then

$$\frac{rn}{m} \equiv \frac{r'n}{m} \pmod{1}$$

for all $n \in \mathbf{Z}$. Let U be a set of real numbers. We write

$$\alpha \in U \pmod{1}$$

if there exists $\alpha' \in U$ such that $\alpha \equiv \alpha' \pmod{1}$. This means that there exists an integer $n \in \mathbf{Z}$ such that $\alpha - n \in U$. For example, let $U = [\beta, \beta + 1/2)$ be the interval of all real numbers t such that $\beta \leq t < \beta + 1/2$. Then

$$\alpha \in [\beta, \beta + 1/2) \pmod{1}$$

if and only if there exists an integer $n \in \mathbf{Z}$ such that

$$\beta \leq \alpha - n < \beta + 1/2.$$

Lemma 2.16 *Let $\alpha_0, \alpha_1, \ldots, \alpha_{k-1}$ be real numbers such that*

$$a \leq \alpha_0 \leq \alpha_1 \leq \cdots \leq \alpha_{k-1} < b.$$

Let $\alpha_k = b$, and let $n(t) : [a, b] \to \mathbf{R}$ be any function such that

$$n(t) = \begin{cases} 0 & \text{for } a < t < \alpha_0 \\ l + 1 & \text{for } \alpha_l < t < \alpha_{l+1} \text{ and } l = 0, 1, \ldots, k-1. \end{cases}$$

Let f be a Riemann-integrable function on the interval $a \leq t \leq b$. Then

$$\sum_{a \leq \alpha_j < b} \int_{\alpha_j}^{b} f(t)dt = \int_a^b n(t) f(t)dt.$$

The function $n(t)$ is a *counting function* for the sequence $\alpha_0, \alpha_1, \ldots, \alpha_{k-1}$.

Proof. Since $n(t) = 0$ for $a \leq t < \alpha_0$, and $n(t) = l + 1$ for $\alpha_l < t < \alpha_{l+1}$ and $l = 0, 1, \ldots, k-1$, it follows by a simple interchange of summation that

$$\begin{aligned} \sum_{a \leq \alpha_j < b} \int_{\alpha_j}^{b} f(t)dt &= \sum_{j=0}^{k-1} \int_{\alpha_j}^{b} f(t)dt \\ &= \sum_{j=0}^{k-1} \sum_{l=j}^{k-1} \int_{\alpha_l}^{\alpha_{l+1}} f(t)dt \\ &= \sum_{l=0}^{k-1} \sum_{j=0}^{l} \int_{\alpha_l}^{\alpha_{l+1}} f(t)dt \end{aligned}$$

$$\begin{aligned}
&= \sum_{l=0}^{k-1} \int_{\alpha_l}^{\alpha_{l+1}} (l+1) f(t) dt \\
&= \sum_{l=0}^{k-1} \int_{\alpha_l}^{\alpha_{l+1}} n(t) f(t) dt \\
&= \int_{\alpha_0}^{\alpha_k} n(t) f(t) dt \\
&= \int_a^b n(t) f(t) dt.
\end{aligned}$$

Theorem 2.9 *Let $\alpha_0, \alpha_1, \ldots, \alpha_{k-1} \in \mathbf{R}$, and let $N(\beta)$ denote the number of α_j such that*

$$\alpha_j \in [\beta, \beta + 1/2) \pmod 1.$$

If

$$\left| \sum_{j=0}^{k-1} e^{2\pi i \alpha_j} \right| \geq \theta k \tag{2.9}$$

for some $\theta \in [0, 1)$, then there exists $\beta \in \mathbf{R}$ such that

$$N(\beta) \geq \frac{(1+\theta)k}{2}.$$

Proof. Choose $\gamma \in \mathbf{R}$ such that

$$S = \sum_{j=0}^{k-1} e^{2\pi i \alpha_j} = |S| e^{2\pi i \gamma}.$$

Then

$$|S| = \sum_{j=0}^{k-1} e^{2\pi i (\alpha_j - \gamma)}.$$

Let $\alpha'_j = \alpha_j - \gamma$ for $j = 0, 1, \ldots, k-1$, and let $N'(\beta)$ denote the number of α'_j such that

$$\alpha'_j \in [\beta, \beta + 1/2) \pmod 1.$$

Then $N'(\beta - \gamma) = N(\beta)$ for all $\beta \in \mathbf{R}$, and so $N(\beta) \geq (1+\theta)k/2$ for $\beta \in \mathbf{R}$ if and only if $N'(\beta') \geq (1+\theta)k/2$ for $\beta' = \beta - \gamma \in \mathbf{R}$. Therefore, without loss of generality, we can replace α_j with $\alpha_j - \gamma$ and assume that $|S| = S$. Since the exponential function $e^{2\pi i t}$ has period 1, we can also replace each real number α_j by its fractional part and assume that $0 \leq \alpha_j < 1$ for $j = 0, 1, \ldots, k-1$.

Suppose that $N(\beta) < (1+\theta)k/2$ for all $\beta \in \mathbf{R}$. Let $|X|$ denote the cardinality of the set X. For $0 \leq t < 1/4$, we define the counting functions $n_i(t)$ by

$$\begin{aligned}
n_1(t) &= |\{j \in [0, k-1] \mid 0 \leq \alpha_j < t\}| \\
n_2(t) &= |\{j \in [0, k-1] \mid 1/2 - t \leq \alpha_j < 1/2\}|
\end{aligned}$$

$$\begin{aligned}
&= |\{j \in [0, k-1] \mid 0 < 1/2 - \alpha_j \leq t\}| \\
n_3(t) &= |\{j \in [0, k-1] \mid 1/2 \leq \alpha_j < 1/2 + t\}| \\
&= |\{j \in [0, k-1] \mid 0 \leq \alpha_j - 1/2 < t\}| \\
n_4(t) &= |\{j \in [0, k-1] \mid 1 - t \leq \alpha_j < 1\}| \\
&= |\{j \in [0, k-1] \mid 0 < 1 - \alpha_j \leq t\}|.
\end{aligned}$$

Then

$$\begin{aligned}
n_1(t) + n_4(t) &= |\{j \in [0, k-1] \mid \alpha_j \in [0, t) \cup [1-t, 1)\}| \\
&= |\{j \in [0, k-1] \mid \alpha_j \in [1, 1+t) \cup [1-t, 1) \pmod 1\}|
\end{aligned}$$

and

$$\begin{aligned}
n_2(t) + n_3(t) &= |\{j \in [0, k-1] \mid \alpha_j \in [1/2 - t, 1/2) \cup [1/2, 1/2 + t)\}| \\
&= |\{j \in [0, k-1] \mid \alpha_j \in [1/2 - t, 1/2 + t)\}|.
\end{aligned}$$

Therefore,

$$\begin{aligned}
k - n_2(t) - n_3(t) &= |\{j \in [0, k-1] \mid \alpha_j \notin [1/2 - t, 1/2 + t)\}| \\
&= |\{j \in [0, k-1] \mid \alpha_j \in [0, 1/2 - t) \cup [1/2 + t, 1)\}| \\
&= |\{j \in [0, k-1] \mid \alpha_j \in [1, 3/2 - t) \cup [1/2 + t, 1) \pmod 1\}|
\end{aligned}$$

and so

$$\begin{aligned}
&k + n_1(t) - n_2(t) - n_3(t) + n_4(t) = \\
&= |\{j \in [0, k-1] \mid \alpha_j \in [1, 3/2 - t) \cup [1/2 + t, 1) \pmod 1\}| \\
&\quad + |\{j \in [0, k-1] \mid \alpha_j \in [1 - t, 1) \cup [1, 1+t) \pmod 1\}| \\
&= |\{j \in [0, k-1] \mid \alpha_j \in [1 - t, 3/2 - t) \pmod 1\}| \\
&\quad + |\{j \in [0, k-1] \mid \alpha_j \in [1/2 + t, 1 + t) \pmod 1\}| \\
&= N(1 - t) + N(1/2 + t) \\
&< (1 + \theta)k.
\end{aligned}$$

Thus,

$$n_1(t) - n_2(t) - n_3(t) + n_4(t) < \theta k$$

for $0 \leq t < 1/4$. Applying Lemma 2.16 to the counting functions $n_i(t)$ for $i = 1, \ldots, 4$, and using the fact that $S = |S|$ is a real number, we obtain

$$\begin{aligned}
S &= \sum_{j=0}^{k-1} e^{2\pi i \alpha_j} \\
&= \sum_{j=0}^{k-1} \cos(2\pi \alpha_j) + i \sum_{j=0}^{k-1} \sin(2\pi \alpha_j)
\end{aligned}$$

$$
\begin{aligned}
&= \sum_{j=0}^{k-1} \cos(2\pi\alpha_j) \\
&= \sum_{0\leq\alpha_j<1/4} \cos 2\pi\alpha_j + \sum_{1/4\leq\alpha_j<1/2} \cos 2\pi\alpha_j \\
&\quad + \sum_{1/2\leq\alpha_j<3/4} \cos 2\pi\alpha_j + \sum_{3/4\leq\alpha_j<1} \cos 2\pi\alpha_j \\
&= \sum_{0\leq\alpha_j<1/4} \cos 2\pi\alpha_j - \sum_{1/4\leq\alpha_j<1/2} \cos 2\pi(1/2-\alpha_j) \\
&\quad - \sum_{1/2\leq\alpha_j<3/4} \cos 2\pi(\alpha_j-1/2) + \sum_{3/4\leq\alpha_j<1} \cos 2\pi(1-\alpha_j) \\
&= 2\pi \sum_{0\leq\alpha_j<1/4} \int_{\alpha_j}^{1/4} \sin 2\pi t\,dt - 2\pi \sum_{1/4\leq\alpha_j<1/2} \int_{1/2-\alpha_j}^{1/4} \sin 2\pi t\,dt \\
&\quad -2\pi \sum_{1/2\leq\alpha_j<3/4} \int_{\alpha_j-1/2}^{1/4} \sin 2\pi t\,dt + 2\pi \sum_{3/4\leq\alpha_j<1} \int_{1-\alpha_j}^{1/4} \sin 2\pi t\,dt \\
&= 2\pi \int_0^{1/4} (n_1(t)-n_2(t)-n_3(t)+n_4(t)) \sin 2\pi t\,dt \\
&< 2\pi \int_0^{1/4} \theta k \sin 2\pi t\,dt \\
&= \theta k,
\end{aligned}
$$

which contradicts condition (2.9). Therefore, $N(\beta) \geq (1+\theta)k/2$ for some $\beta \in \mathbf{R}$. This completes the proof.

2.8 The Freiman–Vosper theorem

In this section we prove Freiman's generalization of Vosper's inverse theorem for the group of congruence classes modulo a prime number p. The proof uses two fundamental methods in additive number theory. The first is the estimation of exponential sums to construct a "large" subset of a set $A \subseteq \mathbf{Z}/p\mathbf{Z}$. The second is the use of arithmetic arguments to replace the set A of congruence classes with a set T of integers such that there is a one-to-one correspondence between the elements of the sumsets $2A$ and $2T$. We can then apply an inverse theorem for sums of sets of integers.

Theorem 2.10 *Let c_0 and c_1 be real numbers such that*

$$
0 < c_0 \leq \frac{1}{12}, \tag{2.10}
$$

$$
c_1 > 2, \tag{2.11}
$$

and

$$\frac{2c_1 - 3}{3} < \frac{1 - c_0 c_1}{c_1^{1/2}}. \tag{2.12}$$

Let p be an odd prime number, and let A be a nonempty set of congruence classes modulo p such that

$$3 \leq k = |A| \leq c_0 p \tag{2.13}$$

and

$$|2A| \leq c_1 k - 3. \tag{2.14}$$

Define the integer b by $|2A| = 2k - 1 + b$. Then A is contained in an arithmetic progression in $\mathbf{Z}/p\mathbf{Z}$ of length $k + b$.

Theorem 2.11 *Let A be a nonempty set of congruence classes modulo p such that*

$$|A| = k \leq \frac{p}{35}$$

and

$$|2A| \leq \frac{12k}{5} - 3.$$

Define the integer r by $|2A| = 2k - 1 + r$. Then A is contained in an arithmetic progression in $\mathbf{Z}/p\mathbf{Z}$ of length $k + r$.

Proof. If $c_0 = 1/35$ and $c_1 = 12/5$, then

$$\frac{2c_1 - 3}{3} = 0.6 < 0.601 < \frac{1 - c_0 c_1}{c_1^{1/2}}.$$

The result follows from Theorem 2.10.

Proof of Theorem 2.10. Inequalities (2.10), (2.11), and (2.12) imply that

$$c_1(2c_1 - 3)^2 < 9.$$

Since the polynomial $x(2x - 3)^2$ is strictly increasing for $x \geq 3/2$, it follows that

$$c_1 < 2.5. \tag{2.15}$$

Let $|2A| = \ell$. By inequalities (2.13) and (2.10), we have

$$2k - 1 < 2c_0 p \leq \frac{p}{6},$$

and so, by the Cauchy–Davenport theorem,

$$\ell = |2A| = 2k - 1 + b \geq \mathrm{mir}(p, 2k - 1) = 2k - 1$$

and $b \geq 0$. Moreover, by inequality (2.14),

$$\ell = |2A| < c_1 k \leq c_0 c_1 p. \tag{2.16}$$

By inequality (2.12), we can choose a positive real number θ such that

$$\frac{1}{3} < \frac{2c_1 - 3}{3} \leq \theta < \frac{1 - c_0 c_1}{c_1^{1/2}}. \tag{2.17}$$

Then

$$c_1 \leq \frac{3(1+\theta)}{2} \tag{2.18}$$

and

$$c_0 c_1 + \theta c_1^{1/2} < 1. \tag{2.19}$$

Let $A = \{a_0, a_1, \ldots, a_{k-1}\} \subseteq \mathbf{Z}/p\mathbf{Z}$. Choose $r_j \in \{0, 1, \ldots, p-1\}$ such that $a_j = r_j + p\mathbf{Z}$ for $j = 0, 1, \ldots, k-1$, and let $R = \{r_0, r_1, \ldots, r_{k-1}\} \subseteq \mathbf{Z}$. We consider the exponential sums $S_A(x)$ and $S_{2A}(x)$ defined by

$$S_A(x) = \sum_{a \in A} e^{2\pi i a x/p} = \sum_{j=0}^{k-1} e^{2\pi i r_j x/p}$$

and

$$S_{2A}(x) = \sum_{b \in 2A} e^{2\pi i b x/p}.$$

We shall prove that there exists an integer $z \not\equiv 0 \pmod{p}$ such that $|S_A(z)| > \theta k$. If not, then $|S_A(x)| \leq \theta k$ for all $x \not\equiv 0 \pmod{p}$. Using Lemma 2.15, the Cauchy–Schwartz inequality, and inequalities (2.17) and (2.16), we obtain

$$\begin{aligned}
k^2 p &= \sum_{x=0}^{p-1} S_A(x)^2 \overline{S_{2A}(x)} \\
&= S_A(0)^2 \overline{S_{2A}(0)} + \sum_{x=1}^{p-1} S_A(x)^2 \overline{S_{2A}(x)} \\
&= k^2 \ell + \sum_{x=1}^{p-1} S_A(x)^2 \overline{S_{2A}(x)} \\
&\leq k^2 \ell + \sum_{x=1}^{p-1} |S_A(x)|^2 |S_{2A}(x)| \\
&\leq k^2 \ell + \theta k \sum_{x=1}^{p-1} |S_A(x)| |S_{2A}(x)| \\
&< k^2 \ell + \theta k \sum_{x=0}^{p-1} |S_A(x)| |S_{2A}(x)| \\
&\leq k^2 \ell + \theta k \left(\sum_{x=0}^{p-1} |S_A(x)|^2 \right)^{1/2} \left(\sum_{x=0}^{p-1} |S_{2A}(x)|^2 \right)^{1/2} \\
&= k^2 \ell + \theta k (kp)^{1/2} (\ell p)^{1/2}
\end{aligned}$$

$$
\begin{aligned}
&= k^2\ell + \theta k^{3/2}\ell^{1/2}p \\
&< c_0c_1k^2p + \theta c_1^{1/2}k^2p \\
&= (c_0c_1 + \theta c_1^{1/2})k^2p \\
&< k^2p,
\end{aligned}
$$

which is absurd. Therefore,

$$|S_A(z)| = \left|\sum_{a\in A} e^{2\pi iaz/p}\right| = \left|\sum_{j=0}^{p-1} e^{2\pi ir_jz/p}\right| > \theta k$$

for some integer $z \not\equiv 0 \pmod p$.

Applying Theorem 2.9 to the real numbers $\alpha_j = r_jz/p$ for $j = 1, \ldots, k$, we obtain a real number β and a subset $R' \subseteq R$ such that

$$k' = |R'| > \frac{(1+\theta)k}{2} \geq 2$$

and

$$\frac{r_jz}{p} \in \left[\beta, \beta + \frac{1}{2}\right) \pmod 1$$

for all $r_j \in R'$. Since p is odd, the interval $[\beta, \beta + 1/2)$ contains $(p\pm 1)/2$ fractions with denominator p, and these fractions are consecutive. This means that there is an integer u_0 such that the fractions in the interval can be written in the form $(u_0 + s)/p$ with

$$s \in \left\{0, 1, \ldots, \frac{p-1}{2}\right\}.$$

Therefore, for each $r_j \in R'$ there exist integers m_j and s_j such that

$$s_j \in \left\{0, 1, \ldots, \frac{p-1}{2}\right\}$$

and

$$\beta \leq \frac{r_jz}{p} - m_j = \frac{u_0 + s_j}{p} < \beta + \frac{1}{2}.$$

Then

$$r_jz \equiv u_0 + s_j \pmod p.$$

Since $z \not\equiv 0 \pmod p$, there exists an integer v_1 such that $v_1z \equiv 1 \pmod p$. Let $u_1 = v_1u_0$. Then

$$r_j \equiv u_1 + v_1s_j \pmod p.$$

Reorder the elements of R so that

$$R' = \{r_0, r_1, \ldots, r_{k'-1}\} \subseteq R = \{r_0, r_1, \ldots, r_{k'-1}, r_{k'}, \ldots, r_{k-1}\}$$

and

$$0 \leq s_0 < s_1 < \cdots < s_{k'-1} \leq \frac{p-1}{2},$$

Let

$$d = (s_1 - s_0, s_2 - s_0, \ldots, s_{k-1} - s_0),$$

and let

$$t_j = \frac{s_j - s_0}{d}$$

for $j = 0, 1, \ldots, k' - 1$. Then

$$0 = t_0 < t_1 < \cdots < t_{k'-1} \leq \frac{p-1}{2}$$

and

$$(t_1, \ldots, t_{k'-1}) = 1.$$

The set

$$T' = \{t_0, t_1, \ldots, t_{k'-1}\}$$

is in normal form. If $r_j \in R'$, then

$$r_j \equiv u_1 + v_1 s_j = u_1 + v_1(s_0 + dt_j) = u_2 + v_2 t_j \pmod{p},$$

where $u_2 = u_1 + v_1 s_0$ and $v_2 = v_1 d \not\equiv 0 \pmod{p}$. Let

$$A' = \{r_j + p\mathbf{Z} : r_j \in R'\} = \{u_2 + v_2 t_j + p\mathbf{Z} : t_j \in T'\} \subseteq A.$$

The following statement, which reduces sums of congruence classes to sums of integers, is the key step in the proof of the theorem. Let $j_1, j_2, j_3, j_4 \in [0, k' - 1]$. Since each integer t_j belongs to the interval $[0, (p-1)/2]$, it follows that

$$r_{j_1} + r_{j_2} \equiv r_{j_3} + r_{j_4} \pmod{p}$$

if and only if

$$t_{j_1} + t_{j_2} \equiv t_{j_3} + t_{j_4} \pmod{p}$$

if and only if

$$t_{j_1} + t_{j_2} = t_{j_3} + t_{j_4}$$

in $\mathbf{Z}$. It follows that

$$|2T'| = |2A'| \leq |2A| \leq c_1 k - 3, \tag{2.20}$$

where $2T'$ is a set of integers and $2A'$ and $2A$ are sets of congruence classes modulo p.

If $t_{k'-1} \geq 2k' - 3$, then Theorem 1.14 and inequality (2.18) imply that

$$|2T'| \geq 3k' - 3 > \frac{3(1+\theta)k}{2} - 3 \geq c_1 k - 3,$$

which contradicts inequality (2.20). Therefore, $t_{k'-1} \leq 2k' - 4$ and so

$$T' \subseteq [0, 2k' - 4]$$

and

$$2T' \subseteq [0, 4k' - 8].$$

Let

$$T = \{t \in [0, p-1] : r_j \equiv u_2 + v_2 t \pmod{p} \text{ for some } r_j \in R\}.$$

Then $T' \subseteq T$. If there exists an integer $t^* \in T$ such that

$$4k' - 7 \leq t^* \leq p - 2k' + 3,$$

then

$$T' + \{t^*\} \subseteq [4k' - 7, p - 1]$$

and

$$2T' \cap (T' + \{t^*\}) = \emptyset.$$

Let $r^* \equiv u_2 + v_2 t^* \pmod{p}$. Since

$$t^* \geq 4k' - 7 > 2k' - 2,$$

it follows that $r^* \in R \setminus R' = \{r_{k'}, r_{k'-1}, \ldots, r_{k-1}\}$ and $a^* = r^* + p\mathbf{Z} \in A \setminus A'$. Let $0 \leq j_1, j_2, j_3 \leq k' - 1$. Since

$$r_{j_1} + r_{j_2} \equiv 2u_2 + v_2(t_{j_1} + t_{j_2}) \pmod{p}$$

and

$$r_{j_3} + r^* \equiv 2u_2 + v_2(t_{j_3} + t) \pmod{p},$$

and since the sets $2T'$ and $T' + \{t\}$ are disjoint subsets of $[0, p-1]$, it follows that no integer in the set

$$2R' = \{r_{j_1} + r_{j_2} : 0 \leq j_1, j_2 \leq k' - 1\}$$

is congruent modulo p to an integer in the set

$$R' + \{r^*\} = \{r_{j_3} + r^* : 0 \leq j_3 \leq k' - 1\}.$$

Since $2R' \cup (R' + \{r^*\})$ is a complete set of representatives of the congruence classes in

$$2A' \cup (A' + \{a^*\}) \subseteq 2A \subseteq \mathbf{Z}/p\mathbf{Z}.$$

it follows from the Cauchy–Davenport theorem and from inequalities (2.18) and (2.20) that

$$\begin{aligned}
|2A| &\geq |2A'| + |A' \cup \{a^*\}| \\
&\geq (2k' - 1) + k' \\
&= 3k' - 1 \\
&> \frac{3(1+\theta)k}{2} - 1 \\
&\geq c_1 k - 1 \\
&> c_1 k - 3 \\
&\geq |2A|,
\end{aligned}$$

which is absurd. Therefore,

$$T \subseteq [0, 4k' - 8] \cup [p - (2k' - 4), p - 1].$$

The set $[0, 4k' - 8] \cup [p - (2k' - 4), p - 1]$ and the interval

$$[-(2k' - 4), 4k' - 8] = -(2k' - 4) + [0, 6k' - 12]$$

represent exactly the same integers modulo p, and so for every $a \in A$ there exist integers $t \in T$ and $w \in [0, 6k' - 12]$ such that

$$a \equiv u_2 + v_2 t \equiv u_2 + v_2(-(2k' - 4) + w) \equiv u_3 + v_2 w \pmod{p},$$

where $u_3 = u_2 - v_2(2k' - 4)$. Let

$$W = \{w \in [0, 6k' - 12] : u_3 + v_2 w \equiv a \pmod{p} \text{ for some } a \in A\}.$$

Since $k \leq c_0 p \leq p/12$ by inequality (2.10), it follows that

$$6k' - 12 < 6k \leq \frac{p}{2}.$$

Since $c_1 < 2.5$ by inequality (2.15), it follows that

$$|2W| = |2A| = 2k - 1 + r \leq c_1 k - 3 < 3k - 3,$$

where $2W$ is a sumset of integers and $2A$ is a sumset of congruence classes modulo p. By Theorem 1.16, the set W is contained in an arithmetic progression of length $k + b$, and so A is contained in an arithmetic progression of length $k + b$ in $\mathbf{Z}/p\mathbf{Z}$. This completes the proof.

2.9 Notes

The Cauchy–Davenport theorem was proved by Cauchy [16] in 1813. Davenport [22] rediscovered the result in 1935. I. Chowla [18] immediately extended the Cauchy–Davenport theorem to composite moduli. Other generalizations have been obtained by Pillai [101], Shatrovskii [117], Brakemaier [13], and Hamidoune [61]. Pollard [106, 107] also extended Theorem 2.4 to the case of sums of h subsets of $\mathbf{Z}/m\mathbf{Z}$ for composite m. Davenport [25] discovered in 1947 that Cauchy had proved the Cauchy–Davenport theorem first.

The Erdős–Ginzburg–Ziv theorem appears in [40]. See Alon and Dubiner [1], Bialostocki and Lotspeich [8], and Hamidoune, Ordaz, and Ortunio [67] for refinements of this important result. There is a different proof of the Erdős–Ginzburg–Ziv theorem in Bailey and Richter [5].

Vosper's inverse theorem [126] was published in 1956. The application to diagonal forms (Theorem 2.8) is due to Chowla, Mann, and Straus [19].

The Freiman–Vosper theorem (Theorem 2.10) and Theorem 2.9 appear in Freiman [50, 51]. The proof of Theorem 2.9 is due to Postnikova [108]. This result has been generalized by Moran and Pollington [88].

2.10 Exercises

1. Let G be a finite group, not necessarily abelian, with a multiplicative operation. For nonempty subsets A, B of G, let

$$AB = \{ab : a \in A, b \in B\}.$$

Show that Lemmas 2.1 and 2.2 are also true in the nonabelian case.

2. Let A be a nonempty subset of $\mathbf{Z}/p\mathbf{Z}$ such that $|A| = k$ and $|2A| = 2k - 1 < p$. Prove that A is an arithmetic progression.

3. Let $h \geq 2$, and let A be a nonempty subset of $\mathbf{Z}/p\mathbf{Z}$ such that $|A| = k$ and $|hA| = hk - h + 1 < p$. Prove that A is an arithmetic progression.

4. Extend I. Chowla's theorem (Theorem 2.1) to sums of $h \geq 3$ subsets of $\mathbf{Z}/m\mathbf{Z}$.

5. Let $m \geq 2$, and let u and v be integers such that $(u - v, m) = 1$. Let $a_1, a_2, \ldots, a_{2m-2}$ be a sequence of $2m - 2$ not necessarily distinct integers such that exactly $m - 1$ integers a_i satisfy

$$a_i \equiv u \pmod{m}$$

and exactly $m - 1$ integers a_i satisfy

$$a_i \equiv v \pmod{m}.$$

Prove that there does not exist a sequence $1 \leq i_1 < \cdots < i_m \leq 2m - 2$ such that

$$a_{i_1} + a_{i_2} + \cdots a_{i_m} \equiv 0 \pmod{m}.$$

This example shows that the Erdős–Ginzburg–Ziv theorem is best possible.

6. Let p be a prime number, and let $(ab, p) = 1$. Let $f(x, y) = ax^2 + by^2$. Use the Cauchy–Davenport Theorem to prove that $f(x, y) \equiv n \pmod{p}$ is solvable for all n.

7. Let p be a prime number, and let $k \geq 3$. Let $c_1, c_2, \ldots, c_k$ be nonzero elements of $G = \mathbf{Z}/p\mathbf{Z}$. Let

$$f(x_1, x_2, \ldots, x_k) = c_1 x_1^k + c_2 x_2^k + \cdots + c_k x_k^k.$$

Prove that the congruence $f(x_1, x_2, \ldots, x_k) \equiv n \pmod{p}$ is solvable for all n.

8. Let p be a prime number, and let $1 \leq k \leq l \leq p$. Let

$$A = \{0, 1, 2, \ldots, k - 1\} \subseteq \mathbf{Z}/p\mathbf{Z}$$

and

$$B = \{0, 1, 2, \ldots, l-1\} \subseteq \mathbf{Z}/p\mathbf{Z}.$$

For $t = 1, \ldots, k$, let N_t denote the number of $x \in \mathbf{Z}/p\mathbf{Z}$ such that x has at least t representations in the form $x = a + b$ with $a \in A, b \in B$. Prove that

$$N_t = \begin{cases} p & \text{if } 1 \leq t \leq k+l-p \\ k+l+1-2t & \text{if } k+l+1-p \leq t \leq k. \end{cases}$$

9. Let $p > 3$ be a prime number, and let k be a positive integer such that $(k, p-1) < (p-1)/2$. Let

$$n \geq \frac{(k, p-1)+1}{2}$$

and

$$f(x_1, \ldots, x_n) = c_1 x_1^k + \cdots + c_n x_n^k,$$

where $c_1, \ldots, c_n \in (\mathbf{Z}/p\mathbf{Z})^*$. Prove that $R(f) = \mathbf{Z}/p\mathbf{Z}$.

10. Let $\alpha_0, \alpha_1, \ldots, \alpha_{k-1} \in \mathbf{R}$. Prove that if

$$\left| \sum_{j=0}^{k-1} e^{2\pi i \alpha_j} \right| = k,$$

then

$$\alpha_j \equiv \alpha_0 \pmod 1$$

for $j = 1, \ldots, k-1$.

11. Let h, m, and t be positive integers such that $h \geq 2$ and $m+1$ is divisible by h. Let $k = mt$, and let A be the sequence of k real numbers $\alpha_0, \alpha_1, \ldots, \alpha_{k-1}$ defined by

$$\alpha_{j-1+(l-1)m} = \frac{j}{m+1}$$

for $j = 1, \ldots, m$ and $l = 1, \ldots, t$. Let $N_h(\beta)$ denote the number of α_j such that

$$\alpha_j \in [\beta, \beta + 1/h) \pmod 1,$$

and let

$$S_A = \sum_{j=0}^{k-1} e^{2\pi i \alpha_j}.$$

Let $\theta = 1/m$. Prove that $|S_A| = \theta k$ and

$$\max_{\beta \in \mathbf{R}} N_h(\beta) = \frac{(1+\theta)k}{h}.$$

In the case $h = 2$, this example shows that Theorem 2.9 is best possible.

3

Sums of distinct congruence classes

3.1 The Erdős–Heilbronn conjecture

Let A be a set of k congruence classes modulo a prime p. It follows from the Cauchy–Davenport theorem that

$$|2A| \geq \min(p, 2k - 1)$$

and, more generally (by Theorem 2.3), that

$$|hA| \geq \min(p, hk - h + 1)$$

for every $h \geq 2$. Denote by $h^\wedge A$ the set consisting of all sums of h distinct elements of A, that is, all sums of the form $a_1 + \cdots + a_h$, where $a_1, \ldots, a_h \in A$ and $a_i \neq a_j$ for $i \neq j$. More than thirty years ago, Erdős and Heilbronn conjectured that

$$|2^\wedge A| \geq \min(p, 2k - 3).$$

The $h$fold generalization of this conjecture is

$$|h^\wedge A| \geq \min(p, hk - h^2 + 1)$$

for all $h \geq 2$. We shall give two proofs of this statement. The first uses the combinatorics of the h-dimensional ballot numbers and some facts from exterior algebra. These prerequisites are developed in the following sections. The second proof uses only the simplest properties of polynomials.

3.2 Vandermonde determinants

A *permutation* of a set X is a map $\sigma : X \to X$ that is one-to-one and onto. The *symmetric group* S_h is the group of all permutations of the set $\{0, 1, 2, \ldots, h-1\}$. Let $F[x_0, x_1, \ldots, x_{h-1}]$ be the ring of polynomials in h variables with coefficients in a field F. The group S_h acts on $F[x_0, \ldots, x_{h-1}]$ as follows. For $\sigma \in S_h$ and $p \in F[x_0, \ldots, x_{h-1}]$, we define $\sigma p \in F[x_0, \ldots, x_{h-1}]$ by

$$(\sigma p)(x_0, x_1, \ldots, x_{h-1}) = p(x_{\sigma(0)}, x_{\sigma(1)}, \ldots, x_{\sigma(h-1)}). \tag{3.1}$$

Then

$$\sigma(\tau p) = (\sigma\tau)p \tag{3.2}$$

for all $\sigma, \tau \in S_h$ (see Exercise 1). The function

$$\Delta(x_0, x_1, \ldots, x_{h-1}) = \prod_{0 \leq i < j \leq h-1} (x_j - x_i)$$

is a homogeneous polynomial of degree $\binom{h}{2}$. We define the *sign* of the permutation $\sigma \in S_h$ as follows:

$$\begin{aligned}
(\sigma\Delta)(x_0, x_1, \ldots, x_{h-1}) &= \prod_{0 \leq i < j \leq h-1} \left(x_{\sigma(j)} - x_{\sigma(i)}\right) \\
&= \operatorname{sign}(\sigma) \prod_{0 \leq i < j \leq h-1} (x_j - x_i) \\
&= \operatorname{sign}(\sigma)\Delta(x_0, \ldots, x_{h-1}),
\end{aligned}$$

and so

$$\operatorname{sign}(\sigma) = \pm 1.$$

It follows from (3.2) that

$$\operatorname{sign}(\sigma\tau) = \operatorname{sign}(\sigma)\operatorname{sign}(\tau)$$

for all $\sigma, \tau \in S_h$, and so

$$\operatorname{sign} : S_h \to \{1, -1\}$$

is a group homomorphism. Thus, $\operatorname{sign}(\sigma^{-1}) = \operatorname{sign}(\sigma)$ for all $\sigma \in S_h$.

A permutation σ is called *even* if $\operatorname{sign}(\sigma) = 1$ and *odd* if $\operatorname{sign}(\sigma) = -1$. Every transposition $\tau = (i, j) \in S_h$ is odd (Exercise 2). Let $\sigma, \tau \in S_h$, where τ is a transposition. Then σ is even if and only if $\tau\sigma$ is odd.

Let

$$A = \begin{pmatrix}
a_{0,0} & a_{0,1} & a_{0,2} & \cdots & a_{0,h-1} \\
a_{1,0} & a_{1,1} & a_{1,3} & \cdots & a_{1,h-1} \\
\vdots & & & & \vdots \\
a_{h-1,0} & a_{h-1,1} & a_{h-1,2} & \cdots & a_{h-1,h-1}
\end{pmatrix}$$

be an $h \times h$ matrix with coefficients in a ring. The *determinant* of A, denoted $|A|$, is defined by

$$\sum_{\sigma \in S_h} \operatorname{sign}(\sigma) \prod_{i=0}^{h-1} a_{i,\sigma(i)}.$$

We require only the basic properties of determinants.

Lemma 3.1 *Let $h \geq 2$, and let $x_0, x_1, \ldots, x_{h-1}$ be variables. Then*

$$\begin{vmatrix} 1 & x_0 & x_0^2 & \cdots & x_0^{h-1} \\ 1 & x_1 & x_1^2 & \cdots & x_1^{h-1} \\ 1 & x_2 & x_2^2 & \cdots & x_2^{h-1} \\ \vdots & & & & \\ 1 & x_{h-1} & x_{h-1}^2 & \cdots & x_{h-1}^{h-1} \end{vmatrix} = \Delta(x_0, x_1, \ldots, x_{h-1}). \tag{3.3}$$

This polynomial identity is called the *Vandermonde determinant*.

Proof. The proof is by induction on h. For $h = 2$ we have

$$\begin{vmatrix} 1 & x_0 \\ 1 & x_1 \end{vmatrix} = x_1 - x_0 = \Delta(x_0, x_1).$$

Assume that the Lemma is true for some $h - 1 \geq 2$. Let A be the $h \times h$ determinant in (3.3). Subtracting x_0 times the first column from the second column, we obtain

$$\begin{vmatrix} 1 & 0 & x_0^2 & \cdots & x_0^{h-1} \\ 1 & x_1 - x_0 & x_1^2 & \cdots & x_1^{h-1} \\ 1 & x_2 - x_0 & x_2^2 & \cdots & x_2^{h-1} \\ \vdots & & & & \\ 1 & x_{h-1} - x_0 & x_{h-1}^2 & \cdots & x_{h-1}^{h-1} \end{vmatrix}.$$

In this new determinant, we subtract x_0^2 times the first column from the third column and obtain

$$\begin{vmatrix} 1 & 0 & 0 & x_0^3 & \cdots & x_0^{h-1} \\ 1 & x_1 - x_0 & x_1^2 - x_0^2 & x_1^3 & \cdots & x_1^{h-1} \\ 1 & x_2 - x_0 & x_2^2 - x_0^2 & x_2^3 & \cdots & x_2^{h-1} \\ \vdots & & & & & \vdots \\ 1 & x_{h-1} - x_0 & x_{h-1}^2 - x_0^2 & x_{h-1}^3 & \cdots & x_{h-1}^{h-1} \end{vmatrix}.$$

After subtracting x_0^{j-1} times the first column from the jth column for $j = 2, 3, \ldots, h$, we obtain the determinant

$$\begin{vmatrix} 1 & 0 & 0 & \cdots & 0 \\ 1 & x_1 - x_0 & x_1^2 - x_0^2 & \cdots & x_1^{h-1} - x_0^{h-1} \\ 1 & x_2 - x_0 & x_2^2 - x_0^2 & \cdots & x_2^{h-1} - x_0^{h-1} \\ \vdots & & & & \\ 1 & x_{h-1} - x_0 & x_{h-1}^2 - x_0^2 & \cdots & x_{h-1}^{h-1} - x_0^{h-1} \end{vmatrix},$$

which is equal to

$$\begin{vmatrix} x_1 - x_0 & x_1^2 - x_0^2 & \cdots & x_1^{h-1} - x_0^{h-1} \\ x_2 - x_0 & x_2^2 - x_0^2 & \cdots & x_2^{h-1} - x_0^{h-1} \\ \vdots & & & \\ x_{h-1} - x_0 & x_{h-1}^2 - x_0^2 & \cdots & x_{h-1}^{h-1} - x_0^{h-1} \end{vmatrix}.$$

For $j = 1, \ldots, h-1$, every polynomial in the jth row is a multiple of $x_j - x_0$, and so this determinant equals

$$\prod_{j=1}^{h-1}(x_j - x_0) \begin{vmatrix} 1 & x_1 + x_0 & x_1^2 + x_1x_0 + x_0^2 & \cdots & x_1^{h-2} + x_1^{h-3}x_0 + \cdots + x_0^{h-2} \\ 1 & x_2 + x_0 & x_2^2 + x_2x_0 + x_0^2 & \cdots & x_2^{h-2} + x_2^{h-3}x_0 + \cdots + x_0^{h-2} \\ \vdots & & & & \\ 1 & x_{h-1} + x_0 & x_{h-1}^2 + x_{h-1}x_0 + x_0^2 & \cdots & x_{h-1}^{h-2} + x_{h-1}^{h-3}x_0 + \cdots + x_0^{h-2} \end{vmatrix}.$$

Continuing to subtract appropriate multiples of one column from another, we find that this is equal to

$$\begin{aligned} \prod_{j=1}^{h-1}(x_j - x_0) \begin{vmatrix} 1 & x_1 & x_1^2 & \cdots & x_1^{h-2} \\ 1 & x_2 & x_2^2 & \cdots & x_2^{h-2} \\ \vdots & & & & \\ 1 & x_{h-1} & x_{h-1}^2 & \cdots & x_{h-1}^{h-2} \end{vmatrix} &= \\ &= \prod_{j=1}^{h-1}(x_j - x_0) \prod_{1 \leq i < j \leq h-1}^{h-1} (x_j - x_i) \\ &= \prod_{0 \leq i < j \leq h-1}^{h-1} (x_j - x_i) \\ &= \Delta(x_0, x_1, \ldots, x_{h-1}). \end{aligned}$$

This completes the proof.

Let $[x]_0 = 1$. For $r \geq 1$, let $[x]_r$ be the polynomial of degree r defined by

$$[x]_r = x(x-1)(x-2)\cdots(x-r+1).$$

Lemma 3.2 *Let $h \geq 2$, and let $x_0, x_1, \ldots, x_{h-1}$ be variables. Then*

$$\begin{vmatrix} 1 & [x_0]_1 & [x_0]_2 & \cdots & [x_0]_{h-1} \\ 1 & [x_1]_1 & [x_1]_2 & \cdots & [x_1]_{h-1} \\ \vdots & & & & \\ 1 & [x_{h-1}]_1 & [x_{h-1}]_2 & \cdots & [x_{h-1}]_{h-1} \end{vmatrix} = \Delta(x_0, x_1, \ldots, x_{h-1}).$$

Proof. By elementary row and column operations, the determinant can be transformed into the Vandermonde determinant, and the result follows.

Lemma 3.3 *Let A be a nonempty, finite subset of a field F, and let $|A| = k$. For every $m \geq 0$ there exists a polynomial $g_m(x) \in F[x]$ of degree at most $k-1$ such that*

$$g_m(a) = a^m$$

for all $a \in A$.

Proof. Let $A = \{a_0, a_1, \ldots, a_{k-1}\}$. We must show that there exists a polynomial $u(x) = u_0 + u_1 x + \cdots + u_{k-1}x^{k-1} \in F[x]$ such that

$$u(a_i) = u_0 + u_1 a_i + u_2 a_i^2 + \cdots + u_{k-1}a_i^{k-1} = a_i^m$$

for $i = 0, 1, \ldots, k-1$. This is a system of k linear equations in the k unknowns $u_0, u_1, \ldots, u_{k-1}$, and it has a solution if the determinant of the coefficients of the unknowns is nonzero. The lemma follows immediately from the observation that this determinant is the Vandermonde determinant

$$\begin{vmatrix} 1 & a_0 & a_0^2 & \cdots & a_0^{k-1} \\ 1 & a_1 & a_1^2 & \cdots & a_1^{k-1} \\ \vdots & & & & \\ 1 & a_{k-1} & a_{k-1}^2 & \cdots & a_{k-1}^{k-1} \end{vmatrix} = \prod_{0 \leq i < j \leq k-1} (a_j - a_i) \neq 0.$$

3.3 Multidimensional ballot numbers

The standard basis for $\mathbf{R}^h$ is the set of vectors $\{\mathbf{e}_1, \ldots, \mathbf{e}_h\}$, where

$$\begin{aligned} \mathbf{e}_1 &= (1, 0, 0, 0, \ldots, 0) \\ \mathbf{e}_2 &= (0, 1, 0, 0, \ldots, 0) \\ &\vdots \\ \mathbf{e}_h &= (0, 0, 0, \ldots, 0, 1). \end{aligned}$$

The lattice $\mathbf{Z}^h$ is the subgroup of $\mathbf{R}^h$ generated by the set $\{\mathbf{e}_1, \ldots, \mathbf{e}_h\}$, so $\mathbf{Z}^h$ is the set of vectors in $\mathbf{R}^h$ with integral coordinates. Let

$$\mathbf{a} = (a_0, a_1, \ldots, a_{h-1}) \in \mathbf{Z}^h$$

and

$$\mathbf{b} = (b_0, b_1, \ldots, b_{h-1}) \in \mathbf{Z}^h.$$

A *path* in $\mathbf{Z}^h$ is a finite sequence of lattice points

$$\mathbf{a} = \mathbf{v}_0, \mathbf{v}_1, \ldots, \mathbf{v}_m = \mathbf{b}$$

such that

$$\mathbf{v}_j - \mathbf{v}_{j-1} \in \{\mathbf{e}_1, \ldots, \mathbf{e}_h\}$$

for $j = 1, \ldots, m$. Let $\mathbf{v}_{j-1}, \mathbf{v}_j$ be successive points on a path. We call this a *step* in the direction $\mathbf{e}_i$ if

$$\mathbf{v}_j = \mathbf{v}_{j-1} + \mathbf{e}_i.$$

The vector $\mathbf{a}$ is called *nonnegative* if $a_i \geq 0$ for $i = 0, 1, \ldots, h-1$. We write

$$\mathbf{a} \leq \mathbf{b}$$

if $\mathbf{b} - \mathbf{a}$ is a nonnegative vector.

Let $P(\mathbf{a}, \mathbf{b})$ denote the number of paths from $\mathbf{a}$ to $\mathbf{b}$. The path function $P(\mathbf{a}, \mathbf{b})$ is translation invariant in the sense that

$$P(\mathbf{a} + \mathbf{c}, \mathbf{b} + \mathbf{c}) = P(\mathbf{a}, \mathbf{b})$$

for all $\mathbf{a}, \mathbf{b}, \mathbf{c} \in \mathbf{Z}^h$. In particular,

$$P(\mathbf{a}, \mathbf{b}) = P(\mathbf{0}, \mathbf{b} - \mathbf{a}).$$

The path function satisfies the boundary conditions

$$P(\mathbf{a}, \mathbf{a}) = 1,$$

and

$$P(\mathbf{a}, \mathbf{b}) > 0 \text{ if and only if } \mathbf{a} \leq \mathbf{b}.$$

If $\mathbf{a} = \mathbf{v}_0, \mathbf{v}_1, \ldots, \mathbf{v}_m = \mathbf{b}$ is a path with $m \geq 1$, then

$$\mathbf{v}_{m-1} = \mathbf{b} - \mathbf{e}_i$$

for some $i = 1, \ldots, h$, and there is a unique path from $\mathbf{b} - \mathbf{e}_i$ to $\mathbf{b}$. It follows that the path counting function $P(\mathbf{a}, \mathbf{b})$ also satisfies the difference equation

$$P(\mathbf{a}, \mathbf{b}) = \sum_{i=1}^{h} P(\mathbf{a}, \mathbf{b} - \mathbf{e}_i).$$

Let $\mathbf{a} \leq \mathbf{b}$. For $i = 0, 1, \ldots, k-1$, every path from $\mathbf{a}$ to $\mathbf{b}$ contains exactly $b_i - a_i$ steps in the direction $\mathbf{e}_{i+1}$. Let

$$m = \sum_{i=0}^{h-1} (b_i - a_i).$$

Every path from $\mathbf{a}$ to $\mathbf{b}$ has exactly m steps, and the number of different paths is the multinomial coefficient

$$P(\mathbf{a}, \mathbf{b}) = \frac{\left(\sum_{i=0}^{h-1} (b_i - a_i)\right)!}{\prod_{i=0}^{h-1} (b_i - a_i)!} = \frac{m!}{\prod_{i=0}^{h-1} (b_i - a_i)!}. \tag{3.4}$$

Let $h \geq 2$. Suppose that there are h candidates in an election. The candidates will be labeled by the integers $0, 1, \ldots, h-1$. If m_0 votes have already been cast, and if candidate i has received a_i votes, then

$$m_0 = a_0 + a_1 + \cdots + a_{h-1}.$$

We shall call

$$\mathbf{v}_0 = \mathbf{a} = (a_0, a_1, \ldots, a_{h-1})$$

the initial *ballot vector*. Suppose that there are m remaining voters, each of whom has one vote, and these votes will be cast sequentially. Let $v_{i,k}$ denote the number of votes that candidate i has received after k additional votes have been cast. We represent the distribution of votes at step k by the ballot vector

$$\mathbf{v}_k = (v_{0,k}, v_{1,k}, \ldots, v_{h-1,k}).$$

Then

$$v_{0,k} + v_{1,k} + \cdots + v_{h-1,k} = k + m_0$$

for $k = 0, 1, \ldots, m$. Let

$$\mathbf{v}_m = \mathbf{b} = (b_0, b_1, \ldots, b_{h-1})$$

be the final ballot vector. It follows immediately from the definition of the ballot vectors that

$$\mathbf{v}_k - \mathbf{v}_{k-1} \in \{\mathbf{e}_1, \ldots, \mathbf{e}_h\}$$

for $k = 1, \ldots, m$, and so

$$\mathbf{a} = \mathbf{v}_0, \mathbf{v}_1, \ldots, \mathbf{v}_m = \mathbf{b}$$

is a path in $\mathbf{Z}^h$ from $\mathbf{a}$ to $\mathbf{b}$. Therefore, the number of distinct sequences of m votes that can lead from the initial ballot vector $\mathbf{a}$ to the final ballot vector $\mathbf{b}$ is the multinomial coefficient

$$\frac{\left(\sum_{i=0}^{h-1}(b_i - a_i)\right)!}{\prod_{i=0}^{h-1}(b_i - a_i)!} = \frac{m!}{\prod_{i=0}^{h-1}(b_i - a_i)!}.$$

Let $\mathbf{v} = (v_1, \ldots, v_h)$ and $\mathbf{w} = (w_1, \ldots, w_h)$ be vectors in $\mathbf{R}^h$. The vector $\mathbf{v}$ will be called *increasing* if

$$v_1 \leq v_2 \leq \cdots \leq v_h$$

and *strictly increasing* if

$$v_1 < v_2 < \cdots < v_h.$$

Now suppose that the initial ballot vector is

$$\mathbf{a} = (0, 0, 0, \ldots, 0)$$

and that the final ballot vector is

$$\mathbf{b} = (b_0, b_1, \ldots, b_{h-1}).$$

Let

$$m = b_0 + b_1 + \ldots + b_{h-1}.$$

Let $B(b_0, b_1, \ldots, b_{h-1})$ denote the number of ways that m votes can be cast so that all of the kth ballot vectors are nonnegative and increasing. This is the classical *h-dimensional ballot number*. Observe that

$$B(0, 0, \ldots, 0) = 1,$$

and that

$$B(b_0, b_1, \ldots, b_{h-1}) > 0$$

if and only if $(b_0, b_1, \ldots, b_{h-1})$ is a nonnegative, increasing vector. These boundary conditions and the difference equation

$$B(b_0, b_1, \ldots, b_{h-1}) = \sum_{i=0}^{h-1} B(b_0, \ldots, b_{i-1}, b_i - 1, b_{i+1}, \ldots, b_{h-1})$$

completely determine the function $B(b_0, b_1, \ldots, b_{h-1})$.

There is an equivalent combinatorial problem. Suppose that the initial ballot vector is

$$\mathbf{a}^* = (0, 1, 2, \ldots, h-1)$$

and that the final ballot vector is

$$\mathbf{b} = (b_0, b_1, \ldots, b_{h-1}).$$

Let

$$m = \sum_{i=0}^{h-1}(b_i - i) = \sum_{i=0}^{h-1} b_i - \binom{h}{2}.$$

Let $\hat{B}(b_0, b_1, \ldots, b_{h-1})$ denote the number of ways that m votes can be cast so that all of the ballot vectors $\mathbf{v}_k$ are nonnegative and strictly increasing. We shall call this the *strict h-dimensional ballot number*.

A path $\mathbf{v}_0, \mathbf{v}_1, \ldots, \mathbf{v}_m$ in $\mathbf{Z}^h$ will be called a *strictly increasing path* if every lattice point $\mathbf{v}_k$ on the path is strictly increasing. Then $\hat{B}(b_0, b_1, \ldots, b_{h-1})$ is the number of strictly increasing paths from $\mathbf{a}^*$ to $\mathbf{b} = (b_0, \ldots, b_{h-1})$.

The strict h-dimensional ballot numbers satisfy the boundary conditions

$$\hat{B}(0, 1, \ldots, h-1) = 1$$

and

$$\hat{B}(b_0, b_1, \ldots, b_{h-1}) > 0$$

if and only if $(b_0, b_1, \ldots, b_{h-1})$ is a nonnegative, strictly increasing vector. These boundary conditions and the difference equation

$$\hat{B}(b_0, b_1, \ldots, b_{h-1}) = \sum_{i=0}^{h-1} \hat{B}(b_0, \ldots, b_{i-1}, b_i - 1, b_{i+1}, \ldots, b_{h-1})$$

completely determine $\hat{B}(b_0, b_1, \ldots, b_{h-1})$.

There is a simple relationship between the numbers $B(b_0, b_1, \ldots, b_{h-1})$ and $\hat{B}(b_0, b_1, \ldots, b_{h-1})$. The lattice point

$$\mathbf{v} = (v_0, v_1, \ldots, v_{h-1})$$

is nonnegative and strictly increasing if and only if the lattice point

$$\mathbf{v}' = \mathbf{v} - (0, 1, 2, \ldots, h-1) = \mathbf{v} - \mathbf{a}^*$$

is nonnegative and increasing. It follows that

$$\mathbf{a}^* = \mathbf{v}_0, \mathbf{v}_1, \mathbf{v}_2, \ldots, \mathbf{v}_m = \mathbf{b}$$

is a path of strictly increasing vectors from $\mathbf{a}^*$ to $\mathbf{b}$ if and only if

$$\mathbf{0}, \mathbf{v}_1 - \mathbf{a}^*, \mathbf{v}_2 - \mathbf{a}^*, \ldots, \mathbf{b} - \mathbf{a}^*$$

is a path of increasing vectors from $\mathbf{0}$ to $\mathbf{b} - \mathbf{a}^*$. Thus,

$$\hat{B}(b_0, b_1, \ldots, b_{h-1}) = B(b_0, b_1 - 1, b_2 - 2, \ldots, b_{h-1} - (h-1)).$$

For $1 \leq i < j \leq h$, let $H_{i,j}$ be the hyperplane in $\mathbf{R}^h$ consisting of all vectors $(x_1, \ldots, x_h)$ such that $x_i = x_j$. There are $\binom{h}{2}$ such hyperplanes. A path

$$\mathbf{a} = \mathbf{v}_0, \mathbf{v}_1, \mathbf{v}_2, \ldots, \mathbf{v}_m = \mathbf{b}$$

will be called *intersecting* if there exists at least one vector $\mathbf{v}_k$ on the path such that $\mathbf{v}_k \in H_{i,j}$ for some hyperplane $H_{i,j}$.

The symmetric group S_h acts on $\mathbf{R}^h$ as follows. For $\sigma \in S_h$ and $\mathbf{v} = (v_0, v_1, \ldots, v_{h-1}) \in \mathbf{R}^h$, let

$$\sigma\mathbf{v} = (v_{\sigma(0)}, v_{\sigma(1)}, \ldots, v_{\sigma(h-1)}).$$

A path is intersecting if and only if there is a transposition $\tau = (i, j) \in S_h$ such that $\tau\mathbf{v}_k = \mathbf{v}_k$ for some lattice point $\mathbf{v}_k$ on the path.

Let $I(\mathbf{a}, \mathbf{b})$ denote the number of intersecting paths from $\mathbf{a}$ to $\mathbf{b}$. Let $J(\mathbf{a}, \mathbf{b})$ denote the number of paths from $\mathbf{a}$ to $\mathbf{b}$ that do not intersect any of the hyperplanes $H_{i,j}$. Then

$$P(\mathbf{a}, \mathbf{b}) = I(\mathbf{a}, \mathbf{b}) + J(\mathbf{a}, \mathbf{b}). \tag{3.5}$$

Lemma 3.4 *Let* $\mathbf{a}$ *be a lattice point in* $\mathbf{Z}^h$*, and let* $\mathbf{b} = (b_0, \ldots, b_{h-1})$ *be a strictly increasing lattice point in* $\mathbf{Z}^h$*. A path from* $\mathbf{a}$ *to* $\mathbf{b}$ *is strictly increasing if and only if it intersects none of the hyperplanes* $H_{i,j}$*, and*

$$\hat{B}(b_0, \ldots, b_{h-1}) = J(\mathbf{a}^*, \mathbf{b}).$$

Proof. Let $\mathbf{a} = \mathbf{v}_0, \mathbf{v}_1, \ldots, \mathbf{v}_m = \mathbf{b}$ be a path, and let

$$\mathbf{v}_k = (v_{0,k}, v_{1,k}, \ldots, v_{h-1,k})$$

for $k = 0, 1, \ldots, m$. If the path is strictly increasing, then every vector on the path is strictly increasing, and so the path does not intersect any of the hyperplanes $H_{i,j}$. Conversely, if the path is not strictly increasing, then there exists a greatest integer k such that the lattice point $\mathbf{v}_{k-1}$ is not strictly increasing. Then $1 \leq k \leq m$, and

$$v_{j,k-1} \leq v_{j-1,k-1}$$

for some $j = 1, \ldots, h-1$. Since the vector $\mathbf{v}_k$ is strictly increasing, we have

$$v_{j-1,k} \leq v_{j,k} - 1.$$

Since $\mathbf{v}_{k-1}$ and $\mathbf{v}_k$ are successive vectors in a path, we have

$$v_{j-1,k-1} \leq v_{j-1,k}$$

and

$$v_{j,k} - 1 \leq v_{j,k-1}.$$

Combining these inequalities, we obtain

$$v_{j,k-1} \leq v_{j-1,k-1} \leq v_{j-1,k} \leq v_{j,k} - 1 \leq v_{j,k-1}.$$

This implies that

$$v_{j,k-1} = v_{j-1,k-1}$$

and so the vector $\mathbf{v}_{k-1}$ lies on the hyperplane $H_{j-1,j}$. Therefore, if $\mathbf{b}$ is a strictly increasing vector, then a path from $\mathbf{a}$ to $\mathbf{b}$ is strictly increasing if and only if it is non-intersecting. It follows that $J(\mathbf{a}, \mathbf{b})$ is equal to the number of strictly increasing paths from $\mathbf{a}$ to $\mathbf{b}$, and so $J(\mathbf{a}^*, \mathbf{b})$ is equal to the strict ballot number $\hat{B}(b_0, \ldots, b_{h-1})$.

Lemma 3.5 *Let* $\mathbf{a}$ *and* $\mathbf{b}$ *be strictly increasing vectors. Then*

$$P(\sigma\mathbf{a}, \mathbf{b}) = I(\sigma\mathbf{a}, \mathbf{b})$$

for every $\sigma \in S_h, \sigma \neq id$.

Proof. If $\mathbf{a}$ is strictly increasing and $\sigma \in S_h, \sigma \neq \mathrm{id}$, then $\sigma\mathbf{a}$ is not strictly increasing, and so every path from $\sigma\mathbf{a}$ to $\mathbf{b}$ must intersect at least one of the hyperplanes $H_{i,j}$, and so $P(\sigma\mathbf{a}, \mathbf{b}) \leq I(\sigma\mathbf{a}, \mathbf{b})$. On the other hand, we have $I(\sigma\mathbf{a}, \mathbf{b}) \leq P(\sigma\mathbf{a}, \mathbf{b})$ by (3.5).

Lemma 3.6 *Let* $\mathbf{a}$ *and* $\mathbf{b}$ *be strictly increasing lattice points. Then*

$$\sum_{\sigma \in S_h} sign(\sigma) I(\sigma\mathbf{a}, \mathbf{b}) = 0.$$

Proof. Since $\mathbf{a}$ is strictly increasing, it follows that there are $h!$ distinct lattice points of the form $\sigma\mathbf{a}$, where $\sigma \in S_h$, and none of these lattice points lies on a hyperplane $H_{i,j}$. Let Ω be the set of all intersecting paths that start at any one of the $h!$ lattice points $\sigma\mathbf{a}$ and end at $\mathbf{b}$. We shall construct an involution from the set Ω to itself.

Let $\sigma \in S_h$, and let

$$\sigma\mathbf{a} = \mathbf{v}_0, \mathbf{v}_1, \ldots, \mathbf{v}_m = \mathbf{b}$$

be a path that intersects at least one of the hyperplanes. Let k be the least integer such that $\mathbf{v}_k \in H_{i,j}$ for some $i < j$. Then $k \geq 1$ since $\mathbf{a}$ is strictly increasing, and the hyperplane $H_{i,j}$ is uniquely determined since $\mathbf{v}_k$ lies on a path. Consider the transposition $\tau = (i, j) \in S_h$. Then

$$\tau\mathbf{v}_k = \mathbf{v}_k \in H_{i,j}$$

and

$$\tau\sigma\mathbf{a} \neq \sigma\mathbf{a}.$$

Moreover,

$$\tau\sigma\mathbf{a} = \tau\mathbf{v}_0, \tau\mathbf{v}_1, \ldots, \tau\mathbf{v}_k = \mathbf{v}_k, \mathbf{v}_{k+1}, \ldots, \mathbf{v}_m = \mathbf{b}$$

is an intersecting path in Ω from $\tau\sigma\mathbf{a}$ to $\mathbf{b}$. For $i = 0, 1, \ldots, k-1$, none of the vectors $\tau\mathbf{v}_0, \tau\mathbf{v}_1, \ldots, \tau\mathbf{v}_{k-1}$ lies on any of the hyperplanes, and $H_{i,j}$ is still the unique hyperplane containing $\mathbf{v}_k$. Since τ^2 is the identity permutation for every transposition τ, it follows that if we apply the same mapping to this path from $\tau\sigma\mathbf{a}$ to $\mathbf{b}$, we recover the original path from $\sigma\mathbf{a}$ to $\mathbf{b}$. Thus, this mapping is an involution on the set Ω of intersecting paths from the $h!$ lattice points $\sigma\mathbf{a}$ to $\mathbf{b}$. Moreover, if σ is an even (resp. odd) permutation, then an intersecting path from $\sigma\mathbf{a}$ is sent to an intersecting path from $\tau\sigma\mathbf{a}$, where τ is a transposition and so $\tau\sigma$ is an odd (resp. even) permutation. Therefore, the number of intersecting paths that start at even permutations of $\mathbf{a}$ is equal to the number of intersecting paths that start at odd permutations of $\mathbf{a}$, and so

$$\sum_{\substack{\sigma \in S_h \\ \mathrm{sign}(\sigma)=1}} I(\sigma\mathbf{a}, \mathbf{b}) = \sum_{\substack{\sigma \in S_h \\ \mathrm{sign}(\sigma)=-1}} I(\sigma\mathbf{a}, \mathbf{b}).$$

This statement is equivalent to Lemma 3.6.

Recall that $[x]_r$ denotes the polynomial $x(x-1)\cdots(x-r+1)$. If b_i and $\sigma(i)$ are nonnegative integers, then

$$\begin{aligned} [b_i]_{\sigma(i)} &= b_i(b_i-1)(b_i-2)\ldots(b_i-\sigma(i)+1) \\ &= \begin{cases} \frac{b_i!}{(b_i-\sigma(i))!} & \text{if } \sigma(i) \leq b_i \\ 0 & \text{if } \sigma(i) > b_i. \end{cases} \end{aligned}$$

Theorem 3.1 *Let $h \geq 2$, and let $b_0, b_1, \ldots, b_{h-1}$ be integers such that*

$$0 \leq b_0 < b_1 < \cdots < b_{h-1}.$$

Then

$$\hat{B}(b_0, b_1, \ldots, b_{h-1}) = \frac{(b_0 + b_1 + \cdots + b_{h-1} - \binom{h}{2})!}{b_0!b_1! \cdots b_{h-1}!} \prod_{0 \leq i < j \leq h-1} (b_j - b_i).$$

Proof. Let $\mathbf{a}^* = (0, 1, 2, \ldots, h-1)$ and $\mathbf{b} = (b_0, b_1, \ldots, b_{h-1}) \in \mathbf{Z}^h$. Applying the preceding lemmas, we obtain

$$\begin{aligned}
&\hat{B}(b_0, b_1, \ldots, b_{h-1}) \\
&= J(\mathbf{a}^*, \mathbf{b}) \\
&= P(\mathbf{a}^*, \mathbf{b}) - I(\mathbf{a}^*, \mathbf{b}) \\
&= P(\mathbf{a}^*, \mathbf{b}) + \sum_{\substack{\sigma \in S_h \\ \sigma \neq id}} \operatorname{sign}(\sigma) I(\sigma\mathbf{a}^*, \mathbf{b}) \\
&= P(\mathbf{a}^*, \mathbf{b}) + \sum_{\substack{\sigma \in S_h \\ \sigma \neq id}} \operatorname{sign}(\sigma) P(\sigma\mathbf{a}^*, \mathbf{b}) \\
&= \sum_{\sigma \in S_h} \operatorname{sign}(\sigma) P(\sigma\mathbf{a}^*, \mathbf{b}) \\
&= \sum_{\substack{\sigma \in S_h \\ \sigma a^* \leq b}} \operatorname{sign}(\sigma) \frac{(b_0 + \cdots + b_{h-1} - \binom{h}{2})!}{\prod_{i=0}^{h-1} (b_i - \sigma(i))!} \\
&= \frac{(b_0 + \cdots + b_{h-1} - \binom{h}{2})!}{b_0!b_1! \cdots b_{h-1}!} \sum_{\substack{\sigma \in S_h \\ \sigma a^* \leq b}} \operatorname{sign}(\sigma) [b_0]_{\sigma(0)} [b_1]_{\sigma(1)} \cdots [b_{h-1}]_{\sigma(h-1)} \\
&= \frac{(b_0 + \cdots + b_{h-1} - \binom{h}{2})!}{b_0!b_1! \cdots b_{h-1}!} \sum_{\sigma \in S_h} \operatorname{sign}(\sigma) [b_0]_{\sigma(0)} [b_1]_{\sigma(1)} \cdots [b_{h-1}]_{\sigma(h-1)} \\
&= \frac{(b_0 + \cdots + b_{h-1} - \binom{h}{2})!}{b_0!b_1! \cdots b_{h-1}!} \begin{vmatrix} 1 & [b_0]_1 & [b_0]_2 & \cdots & [b_0]_{h-1} \\ 1 & [b_1]_1 & [b_1]_2 & \cdots & [b_1]_{h-1} \\ \vdots & & & & \\ 1 & [b_{h-1}]_1 & [b_{h-1}]_2 & \cdots & [b_{h-1}]_{h-1} \end{vmatrix} \\
&= \frac{(b_0 + \cdots + b_{h-1} - \binom{h}{2})!}{b_0!b_1! \cdots b_{h-1}!} \prod_{0 \leq i < j \leq h-1} (b_j - b_i).
\end{aligned}$$

This completes the proof.

The following result will be used later in the proof of the Erdős–Heilbronn conjecture.

Theorem 3.2 *Let $h \geq 2$, let p be a prime number, and let $i_0, i_1, \ldots, i_{h-1}$ be integers such that*

$$0 \leq i_0 < i_1 < \cdots < i_{h-1} < p$$

and

$$i_0 + i_1 + \cdots i_{h-1} < \binom{h}{2} + p.$$

Then

$$\hat{B}(i_0, i_1, \ldots, i_{h-1}) \not\equiv 0 \pmod{p}.$$

Proof. This follows immediately from Theorem 3.1.

3.4 A review of linear algebra

Let V be a finite-dimensional vector space over a field F, and let $T : V \to V$ be a linear operator. Let $I : V \to V$ be the identity operator. For every nonnegative integer i, we define $T^i : V \to V$ by

$$\begin{aligned} T^0(\mathbf{v}) &= I(\mathbf{v}) = \mathbf{v}, \\ T^i(\mathbf{v}) &= T\left(T^{i-1}(\mathbf{v})\right) \end{aligned}$$

for all $\mathbf{v} \in V$. To every polynomial

$$p(x) = c_n x^n + c_{n-1}x^{n-1} + \cdots + c_1 x + c_0 \in F[x]$$

we associate the linear operator $p(T) : V \to V$ defined by

$$p(T) = c_n T^n + c_{n-1}T^{n-1} + \cdots + c_1 T + c_0 I.$$

The set of all polynomials $p(x)$ such that $p(T) = 0$ forms a nonzero, proper ideal J in the polynomial ring $F[x]$. Since every ideal in $F[x]$ is principal, there exists a unique monic polynomial $p_T(x) = p_{T,V}(x) \in J$ such that $p_T(x)$ divides every other polynomial in J. This polynomial is called the *minimal polynomial* of T over the vector space V.

A subspace W of V is called *invariant* with respect to T if $T(W) \subseteq W$, that is, if $T(\mathbf{w}) \in W$ for all $\mathbf{w} \in W$. Then T restricted to the subspace W is a linear operator on W with minimal polynomial $p_{T,W}(x)$. Since $p_{T,V}(T)(\mathbf{w}) = \mathbf{0}$ for all $\mathbf{w} \in W$, it follows that $p_{T,W}(x)$ divides $p_{T,V}(x)$, and so

$$\deg(p_{T,W}) \leq \deg(p_{T,V}), \tag{3.6}$$

where $\deg(p)$ denotes the degree of the polynomial p.

For $\mathbf{v} \in V$, the *cyclic subspace* with respect to T generated by $\mathbf{v}$ is the smallest subspace of V that contains $\mathbf{v}$ and is invariant under the operator T. We denote this subspace by $C_T(\mathbf{v})$. Let $\mathbf{v}_i = T^i(\mathbf{v})$ for $i = 0, 1, 2, \ldots$. Then $C_T(\mathbf{v})$ is the subspace generated by the vectors

$$\{\mathbf{v}, T(\mathbf{v}), T^2(\mathbf{v}), T^3(\mathbf{v}), \ldots\} = \{\mathbf{v}_0, \mathbf{v}_1, \mathbf{v}_2, \mathbf{v}_3, \ldots\}$$

and

$$\dim(C_T(\mathbf{v})) = l,$$

where l is the smallest integer such that the vectors $\mathbf{v}_0, \mathbf{v}_1, \ldots, \mathbf{v}_l$ are linearly dependent. This means that there exist scalars $c_0, c_1, \ldots, c_{l-1}$ in the field F such that

$$\mathbf{v}_l + c_{l-1}\mathbf{v}_{l-1} + \cdots + c_1\mathbf{v}_1 + c_0\mathbf{v}_0 = \mathbf{0}.$$

Let

$$p(x) = x^l + c_{l-1}x^{l-1} + \cdots + c_1 x + c_0.$$

Then

$$\begin{aligned} p(T)(\mathbf{v}_0) &= T^l(\mathbf{v}_0) + c_{l-1}T^{l-1}(\mathbf{v}_0) + \cdots + c_1 T(\mathbf{v}_0) + c_0 I(\mathbf{v}_0) \\ &= \mathbf{v}_l + c_{l-1}\mathbf{v}_{l-1} + \cdots + c_1\mathbf{v}_1 + c_0\mathbf{v}_0 \\ &= \mathbf{0}, \end{aligned}$$

and so

$$p(T)(\mathbf{v}_i) = p(T)(T^i(\mathbf{v}_0)) = T^i\left(p(T)(\mathbf{v}_0)\right) = T^i(\mathbf{0}) = \mathbf{0}$$

for $i = 0, 1, 2, \ldots$. Therefore, $p(T) = 0$ on the cyclic subspace $\mathcal{C}_T(\mathbf{v}) = \mathcal{C}$, and so $p(x)$ is divisible by the minimal polynomial $p_{T,\mathcal{C}}(x)$, and

$$m = \deg\left(p_{T,\mathcal{C}}\right) \leq \deg(p) = l.$$

On the other hand, since

$$p_{T,\mathcal{C}}(\mathbf{v}) = \mathbf{0},$$

it follows that the vectors $\mathbf{v}_0, \mathbf{v}_1, \ldots, \mathbf{v}_m$ are linearly dependent, and so

$$l \leq m.$$

This implies that $l = m$ and so, by inequality (3.6),

$$\dim(\mathcal{C}_T(\mathbf{v})) = \deg\left(p_{T,\mathcal{C}}\right) \leq \deg\left(p_{T,V}\right)$$

for all $\mathbf{v} \in V$.

If $T(\mathbf{f}) = a\mathbf{f}$ for some $a \in F$ and some nonzero vector $\mathbf{f} \in V$, then a is called an *eigenvalue* of T and $\mathbf{f}$ is called an *eigenvector* of T with eigenvalue a. The *spectrum* of T, denoted $\sigma(T)$, is the set of all eigenvalues of T. If V has a basis consisting entirely of eigenvectors of T, then T is called a *diagonal operator*.

The following inequality plays a central role in the proof of the Erdős–Heilbronn conjecture.

Lemma 3.7 *Let T be a diagonal linear operator on a finite-dimensional vector space V, and let $\sigma(T)$ be the spectrum of T. Then*

$$\dim\left(\mathcal{C}_T(\mathbf{v})\right) \leq |\sigma(T)| \tag{3.7}$$

for every $\mathbf{v} \in V$.

Proof. Let $a \in \sigma(T)$, and let $\mathbf{f}$ be an eigenvector with eigenvalue a. Let W be the one-dimensional subspace generated by $\mathbf{f}$. Then W is invariant with respect to T, and $p_{T,W}(x) = x - a$. It follows that $x - a$ divides $p_{T,V}(x)$, and so

$$\prod_{a \in \sigma(T)} (x - a)$$

divides $p_{T,V}(x)$. Let $\dim(V) = k$. If T is a diagonal linear operator, then V has a basis $\{\mathbf{f}_0, \mathbf{f}_1, \ldots, \mathbf{f}_{k-1}\}$ of eigenvectors, and

$$\prod_{a \in \sigma(T)} (T - aI)(\mathbf{f}_i) = \mathbf{0}$$

for $i = 0, 1, \ldots, k-1$. It follows that $\prod_{a \in \sigma(T)} (T - aI)(\mathbf{v}) = \mathbf{0}$ for all $\mathbf{v} \in V$, and so

$$p_{T,V}(x) = \prod_{a \in \sigma(T)} (x - a).$$

In particular, the degree of $p_{T,V}(x)$ is equal to the number of distinct eigenvalues of T. It follows that if T is a diagonal operator on a finite-dimensional vector space V, then

$$\dim\left(C_T(\mathbf{v})\right) \leq \deg\left(p_{T,V}\right) = |\sigma(T)|$$

for every $\mathbf{v} \in V$. This completes the proof.

Lemma 3.8 *Let $T : V \rightarrow V$ be a linear operator on the vector space V, and let $\{\mathbf{f}_0, \mathbf{f}_1, \ldots, \mathbf{f}_{k-1}\}$ be eigenvectors of T with distinct eigenvalues. Let*

$$\mathbf{v}_0 = \mathbf{f}_0 + \mathbf{f}_1 + \cdots + \mathbf{f}_{k-1},$$

and let $C_T(\mathbf{v}_0)$ be the cyclic subspace generated by $\mathbf{v}_0$. Then

$$\dim\left(C_T(\mathbf{v}_0)\right) = k$$

and

$$\{\mathbf{v}_0, T(\mathbf{v}_0), T^2(\mathbf{v}_0), \ldots, T^{k-1}(\mathbf{v}_0)\}$$

is a basis for $C_T(\mathbf{v}_0)$. If $\dim(V) = k$, then $C_T(\mathbf{v}_0) = V$.

Proof. We first show that the vectors $\mathbf{f}_0, \mathbf{f}_1, \ldots, \mathbf{f}_{k-1}$ are linearly independent. If they are linearly dependent, then there is a minimal subset of the vectors $\mathbf{f}_0, \ldots, \mathbf{f}_{k-1}$ that is linearly dependent, say, $\mathbf{f}_0, \ldots, \mathbf{f}_{l-1}$. Moreover, $l \geq 2$ since $\mathbf{f}_i \neq \mathbf{0}$ for $i = 0, 1, \ldots, k-1$. There exist nonzero scalars $c_0, c_1, \ldots, c_{l-1}$ such that $\sum_{i=0}^{l-1} c_i \mathbf{f}_i = \mathbf{0}$. Let $a_i \in \sigma(T)$ be the eigenvalue corresponding to the eigenvector $\mathbf{f}_i$. Then

$$T\left(\sum_{i=0}^{l-1} c_i \mathbf{f}_i\right) = \sum_{i=0}^{l-1} c_i T(\mathbf{f}_i) = \sum_{i=0}^{l-1} c_i a_i \mathbf{f}_i = \mathbf{0}.$$

Since

$$\sum_{i=0}^{l-1} c_i a_{l-1} \mathbf{f}_i = a_{l-1} \sum_{i=0}^{l-1} c_i \mathbf{f}_i = \mathbf{0},$$

it follows that

$$\sum_{i=0}^{l-1} c_i(a_i - a_{l-1})\mathbf{f}_i = \sum_{i=0}^{l-2} c_i(a_i - a_{l-1})\mathbf{f}_i = \mathbf{0},$$

which contradicts the minimality of l, since $c_i(a_i - a_{l-1}) \neq 0$ for $i < l - 1$. Thus, the vectors $\mathbf{f}_0, \ldots, \mathbf{f}_{k-1}$ are linearly independent and span a k-dimensional subspace W of V. Moreover, W is an invariant subspace since it has a basis of eigenvectors of T. Since

$$\mathbf{v}_0 = \mathbf{f}_0 + \cdots + \mathbf{f}_{k-1}k \in W,$$

it follows that

$$\mathcal{C}_T(\mathbf{v}_0) \subseteq W$$

and so

$$\dim(\mathcal{C}_T(\mathbf{v}_0)) \leq \dim(W) = k.$$

We have $T^i(\mathbf{v}_0) \in \mathcal{C}_T(\mathbf{v}_0)$ for every nonnegative integer i. Since

$$T^i(\mathbf{v}_0) = a_0^i\mathbf{f}_0 + a_1^i\mathbf{f}_1 + \cdots + a_{k-1}^i\mathbf{f}_{k-1},$$

the matrix of the set of vectors $\{\mathbf{v}_0, T(\mathbf{v}_0), T^2(\mathbf{v}_0), \ldots, T^{k-1}(\mathbf{v}_0)\}$ with respect to the basis $\{\mathbf{f}_0, \ldots, \mathbf{f}_{k-1}\}$ is

$$\begin{pmatrix} 1 & a_0 & a_0^2 & \cdots & a_0^{k-1} \\ 1 & a_1 & a_1^2 & \cdots & a_1^{k-1} \\ 1 & a_2 & a_2^2 & \cdots & a_2^{k-1} \\ \vdots & & & & \\ 1 & a_{k-1} & a_{k-1}^2 & \cdots & a_{k-1}^{k-1} \end{pmatrix},$$

and its determinant is the Vandermonde determinant

$$\prod_{0 \leq i < j \leq k-1} (a_j - a_i) \neq 0.$$

It follows that $\{\mathbf{v}_0, T(\mathbf{v}_0), T^2(\mathbf{v}_0), \ldots, T^{k-1}(\mathbf{v}_0)\}$ is a set of linearly independent vectors, and so

$$\dim(\mathcal{C}_T(\mathbf{v}_0)) \geq k = \dim(W).$$

Therefore, $\dim(\mathcal{C}_T(\mathbf{v}_0)) = k$. If $\dim(V) = k$, then $\mathcal{C}_T(\mathbf{v}_0) = V$. This completes the proof.

3.5 Alternating products

Let $\wedge^h V$ denote the hth *alternating product* of the vector space V. Then $\wedge^h V$ is a vector space whose elements are linear combinations of expressions of the form

$$\mathbf{v}_0 \wedge \mathbf{v}_1 \wedge \cdots \wedge \mathbf{v}_{h-1},$$

where $\mathbf{v}_0, \mathbf{v}_1, \ldots, \mathbf{v}_{h-1} \in V$. These *wedge products* have the property that

$$\mathbf{v}_0 \wedge \mathbf{v}_1 \wedge \cdots \wedge \mathbf{v}_{h-1} = \mathbf{0}$$

if $\mathbf{v}_i = \mathbf{v}_j$ for some $i \neq j$, and

$$\mathbf{v}_{\sigma(0)} \wedge \mathbf{v}_{\sigma(1)} \wedge \cdots \wedge \mathbf{v}_{\sigma(h-1)} = \text{sign}(\sigma)\mathbf{v}_0 \wedge \mathbf{v}_1 \wedge \cdots \wedge \mathbf{v}_{h-1}$$

for all $\sigma \in S_h$.

If $\dim V = k$ and $\{\mathbf{e}_0, \ldots, \mathbf{e}_{k-1}\}$ is a basis for V, then a basis for $\wedge^h V$ is the set of all vectors of the form

$$\mathbf{e}_{i_0} \wedge \mathbf{e}_{i_1} \wedge \cdots \wedge \mathbf{e}_{i_{h-1}},$$

where

$$0 \leq i_0 < i_1 < \cdots < i_{h-1} \leq k-1$$

and

$$\dim(\wedge^h V) = \binom{k}{h}.$$

Every linear operator $T : V \to V$ induces a linear operator

$$DT : \bigwedge^h V \to \bigwedge^h V$$

that acts on wedge products according to the rule

$$DT(\mathbf{v}_0 \wedge \cdots \wedge \mathbf{v}_{h-1})$$
$$\sum_{j=0}^{h-1} \mathbf{v}_0 \wedge \cdots \wedge \mathbf{v}_{j-1} \wedge T(\mathbf{v}_j) \wedge \mathbf{v}_{j+1} \cdots \wedge \mathbf{v}_{h-1}. \tag{3.8}$$

The operator DT is called the *derivative* of T.

Recall that $h^\wedge A$ denotes the set of all sums of h distinct elements of A.

Lemma 3.9 *Let T be a diagonal linear operator on V, and let $\sigma(T)$ be the spectrum of T. Let $h \geq 2$, and let $DT : \bigwedge^h V \to \bigwedge^h V$ be the derivative of T. If T has distinct eigenvalues, that is, if $|\sigma(T)| = \dim(V)$, then*

$$\sigma(DT) = h^\wedge\sigma(T)$$

and

$$|h^\wedge\sigma(T)| \geq \dim(\mathcal{C}_{DT}(\mathbf{w}))$$

for every $\mathbf{w} \in \wedge^h V$.

Proof. Let $\sigma(T) = \{a_0, a_1, \ldots, a_{k-1}\}$, and let $\{\mathbf{f}_0, \mathbf{f}_1, \cdots, \mathbf{f}_{k-1}\}$ be a basis of eigenvectors of V such that $T(\mathbf{f}_i) = a_i\mathbf{f}_i$ for $i = 0, 1, \ldots, k-1$. Then (3.8) implies that

$$DT(\mathbf{f}_{i_0} \wedge \cdots \wedge \mathbf{f}_{i_{h-1}}) = (a_{i_0} + \cdots + a_{i_{h-1}})(\mathbf{f}_{i_0} \wedge \cdots \wedge \mathbf{f}_{i_{h-1}}).$$

It follows that DT is a diagonal linear operator on $\wedge^h V$, and its spectrum $\sigma(DT)$ consists of all sums of h distinct eigenvalues of T, that is,

$$\sigma(DT) = h^\wedge \sigma(T).$$

Applying inequality (3.7) to the vector space $\wedge^h V$ and the operator DT, we obtain

$$|h^\wedge \sigma(T)| = |\sigma(DT)| \geq \dim\left(\mathcal{C}_{DT}(\mathbf{w})\right)$$

for every $\mathbf{w} \in \wedge^h V$. This completes the proof.

Theorem 3.3 *Let T be a linear operator on the finite-dimensional vector space V. Let $h \geq 2$, and let $DT : \wedge^h V \to \wedge^h V$ be the derivative of T. For $\mathbf{v}_0 \in V$, define*

$$\mathbf{v}_i = T^i(\mathbf{v}_0) \in V$$

for $i \geq 1$, and let

$$\mathbf{w} = \mathbf{v}_0 \wedge \mathbf{v}_1 \wedge \cdots \wedge \mathbf{v}_{h-1} \in \wedge^h V.$$

Then for every $r \geq 0$

$$\begin{aligned}(DT)^r(\mathbf{w}) &= (DT)^r(\mathbf{v}_0 \wedge \mathbf{v}_1 \wedge \cdots \wedge \mathbf{v}_{h-1}) \\ &= \sum \hat{B}(i_0, i_1, \ldots, i_{h-1})\mathbf{v}_{i_0} \wedge \mathbf{v}_{i_1} \wedge \cdots \wedge \mathbf{v}_{i_{h-1}},\end{aligned}$$

where the sum $\sum$ is over all integer lattice points $(i_0, i_1, \ldots, i_{h-1}) \in \mathbf{Z}^h$ such that

$$0 \leq i_0 < i_1 < \cdots < i_{h-1} \leq r + h - 1$$

and

$$i_0 + i_1 + \cdots + i_{h-1} = \binom{h}{2} + r,$$

and where $\hat{B}(i_0, i_1, \ldots, i_{h-1})$ is the strict h-dimensional ballot number corresponding to the lattice point $(i_0, i_1, \ldots, i_{h-1})$.

Proof. The proof will be by induction on r. For $r = 0$, we have

$$\begin{aligned}(DT)^0(\mathbf{w}) &= \mathbf{w} \\ &= \mathbf{v}_0 \wedge \mathbf{v}_1 \wedge \cdots \wedge \mathbf{v}_{h-1} \\ &= \hat{B}(0, 1, 2, \ldots, h-1)\mathbf{v}_0 \wedge \mathbf{v}_1 \wedge \cdots \wedge \mathbf{v}_{h-1}\end{aligned}$$

since $\hat{B}(0, 1, 2, \ldots, h-1) = 1$. Suppose the result holds for some integer $r \geq 0$. Then

$$\begin{aligned}&(DT)^{r+1}(\mathbf{w}) \\ &= DT\left((DT)^r(\mathbf{w})\right) \\ &= DT\left(\sum \hat{B}(i_0, i_1, \ldots, i_{h-1})\mathbf{v}_{i_0} \wedge \mathbf{v}_{i_1} \wedge \cdots \wedge \mathbf{v}_{i_{h-1}}\right)\end{aligned}$$

$$
\begin{aligned}
&= \sum \hat{B}(i_0, i_1, \ldots, i_{h-1}) DT \left(\mathbf{v}_{i_0} \wedge \mathbf{v}_{i_1} \wedge \cdots \wedge \mathbf{v}_{i_{h-1}}\right) \\
&= \sum \hat{B}(i_0, i_1, \ldots, i_{h-1}) \sum_{j=0}^{h-1} \left(\mathbf{v}_{i_0} \wedge \cdots \wedge \mathbf{v}_{i_{j-1}} \wedge T(\mathbf{v}_{i_j}) \wedge \mathbf{v}_{i_{j+1}} \wedge \cdots \wedge \mathbf{v}_{i_{h-1}}\right) \\
&= \sum \hat{B}(i_0, i_1, \ldots, i_{h-1}) \sum_{j=0}^{h-1} \left(\mathbf{v}_{i_0} \wedge \cdots \wedge \mathbf{v}_{i_{j-1}} \wedge \mathbf{v}_{i_j+1} \wedge \mathbf{v}_{i_{j+1}} \wedge \cdots \wedge \mathbf{v}_{i_{h-1}}\right) \\
&= \sum C(i_0, i_1, \ldots, i_{h-1}) \mathbf{v}_{i_0} \wedge \mathbf{v}_{i_1} \wedge \cdots \wedge \mathbf{v}_{i_{h-1}},
\end{aligned}
$$

where the last sum is over all integer lattice points $(i_0, i_1, \ldots, i_{h-1}) \in \mathbf{Z}^n$ such that

$$0 \leq i_0 < i_1 < \cdots < i_{h-1} \leq r + h$$

and

$$i_0 + i_1 + \cdots + i_{h-1} = \binom{h}{2} + r + 1,$$

and the integer $C(i_0, i_1, \ldots, i_{h-1})$ satisfies the difference equation

$$C(i_0, i_1, \ldots, i_{h-1}) = \sum_{j=0}^{h-1} \hat{B}(i_0, \ldots, i_{j-1}, i_j - 1, i_{j+1}, \ldots, i_{h-1}).$$

This difference equation determines the strict h-dimensional ballot numbers, and so

$$C(i_0, i_1, \ldots, i_{h-1}) = \hat{B}(i_0, i_1, \ldots, i_{h-1}).$$

Therefore, the result holds in the case $r + 1$. This completes the induction.

3.6 Erdős–Heilbronn, concluded

Theorem 3.4 (Dias da Silva–Hamidoune) *Let p be a prime number, and let $A \subseteq \mathbf{Z}/p\mathbf{Z}$, where $|A| = k$. Let $2 \leq h \leq k$. Then*

$$|h^\wedge A| \geq \min(p, hk - h^2 + 1).$$

Proof. Let $A = \{a_0, a_1, \ldots, a_{k-1}\}$. Let V be a vector space of dimension k over the field $\mathbf{Z}/p\mathbf{Z}$, and let $\{\mathbf{f}_0, \mathbf{f}_1, \ldots, \mathbf{f}_{k-1}\}$ be a basis for V. We define the diagonal linear operator $T : V \to V$ by

$$T(\mathbf{f}_i) = a_i \mathbf{f}_i$$

for $i = 0, 1, \ldots, k - 1$. The spectrum of T is

$$\sigma(T) = A.$$

Let

$$\mathbf{v}_0 = \mathbf{f}_0 + \mathbf{f}_1 + \cdots + \mathbf{f}_{k-1},$$

and define

$$\mathbf{v}_{i+1} = T(\mathbf{v}_i) = T^i(\mathbf{v}_0)$$

for $i \geq 0$. By Lemma 3.8, the cyclic subspace $\mathcal{C}_T(\mathbf{v}_0)$ generated by $\mathbf{v}_0$ is V, and the set of vectors $\{\mathbf{v}_0, \mathbf{v}_1, \ldots, \mathbf{v}_{k-1}\}$ is a basis for V. The alternating product $\bigwedge^h V$ is a vector space with a basis consisting of the $\binom{k}{h}$ wedge products of the form

$$\mathbf{v}_{i_0} \wedge \mathbf{v}_{i_1} \wedge \cdots \wedge \mathbf{v}_{i_{h-1}},$$

where

$$0 \leq i_0 < i_1 < \cdots < i_{h-1} \leq k-1.$$

Let

$$\mathbf{w} = \mathbf{v}_0 \wedge \mathbf{v}_1 \wedge \cdots \wedge \mathbf{v}_{k-1} \in \bigwedge^h V.$$

By Lemma (3.9),

$$|h^\wedge A| = |\sigma(DT)| \geq \dim \mathcal{C}_{DT}(\mathbf{w}).$$

Therefore, it suffices to prove that

$$\dim \mathcal{C}_{DT}(\mathbf{w}) \geq \min(p, hk - h^2 + 1);$$

this is equivalent to proving that the vectors

$$\mathbf{w}, (DT)(\mathbf{w}), (DT)^2(\mathbf{w}), \ldots, (DT)^n(\mathbf{w})$$

are linearly independent in the alternating product $\bigwedge^h V$, where

$$n = \min(p, hk - h^2 + 1) - 1 = \min(p - 1, hk - h^2).$$

Let $0 \leq r \leq n$. By Theorem 3.3, the vector $(DT)^r(\mathbf{w})$ is a linear combination of vectors of the form

$$\mathbf{v}_{i_0} \wedge \mathbf{v}_{i_1} \wedge \cdots \wedge \mathbf{v}_{i_{h-1}},$$

where

$$0 \leq i_0 < i_1 < \cdots < i_{h-1} \leq r + h - 1 \tag{3.9}$$

and

$$i_0 + i_1 + \cdots + i_{h-1} = \binom{h}{2} + r. \tag{3.10}$$

Let I be the interval of integers $[0, k-1]$. Since

$$h^\wedge I = \left[\binom{h}{2}, hk - \binom{h+1}{2}\right] = \binom{h}{2} + [0, hk - h^2],$$

it follows that there is at least one basis vector $\mathbf{v}_{i_0} \wedge \mathbf{v}_{i_1} \wedge \cdots \wedge \mathbf{v}_{i_{h-1}}$ in the expansion of $(DT)^r(\mathbf{w})$ such that

$$0 \leq i_0 < i_1 < \cdots < i_{h-1} \leq k - 1 < p \tag{3.11}$$

and

$$i_0 + i_1 + \cdots + i_{h-1} = \binom{h}{2} + r \leq \binom{h}{2} + n < \binom{h}{2} + p.$$

By Theorem 3.3, the coefficient of this basis vector is the strict h-dimensional ballot number $\hat{B}(i_0, i_1, \ldots, i_{h-1})$. By Theorem 3.2,

$$\hat{B}(i_0, i_1, \ldots, i_{h-1}) \not\equiv 0 \pmod{p}.$$

Since $V = C_T(\mathbf{v}_0)$ is cyclic of dimension k, every vector $\mathbf{v}_\ell \in V$ with $\ell \geq k$ is a linear combination of $\mathbf{v}_0, \mathbf{v}_1, \ldots, \mathbf{v}_{k-1}$. Let $\mathbf{v}_{i_0} \wedge \mathbf{v}_{i_1} \wedge \cdots \wedge \mathbf{v}_{i_{h-1}}$ be a vector that satisfies (3.9) and (3.10). If $i_\ell \geq k$ for some $\ell \in [0, h-1]$, then $\mathbf{v}_{i_0} \wedge \mathbf{v}_{i_1} \wedge \cdots \wedge \mathbf{v}_{i_{h-1}}$ is a linear combination of basis vectors of the form $\mathbf{v}_{j_0} \wedge \mathbf{v}_{j_1} \wedge \cdots \wedge \mathbf{v}_{j_{h-1}}$, where

$$0 \leq j_0 < j_1 < \cdots < j_{h-1} \leq k-1$$

and

$$j_0 + j_1 + \cdots + j_{h-1} < \binom{h}{2} + r.$$

It follows that $(DT)^r(\mathbf{w})$ is a linear combination of basis vectors $\mathbf{v}_{i_0} \wedge \mathbf{v}_{i_1} \wedge \cdots \wedge \mathbf{v}_{i_{h-1}}$ such that

$$0 \leq i_0 < i_1 < \cdots < i_{h-1} \leq k-1$$

and either

$$i_0 + i_1 + \cdots + i_{h-1} < \binom{h}{2} + r$$

or

$$i_0 + i_1 + \cdots + i_{h-1} = \binom{h}{2} + r.$$

Moreover, in the second case the basis vector appears with a coefficient $\hat{B}(i_0, i_1, \ldots, i_{h-1})$, and this number is nonzero modulo p. If the vectors $\mathbf{w}, (DT)(\mathbf{w}), \ldots, (DT)^n(\mathbf{w})$ are linearly dependent in the cyclic subspace $C_{DT}(\mathbf{w})$, then there exists a positive integer $m \leq n$ such that

$$(DT)^m(\mathbf{w}) = \sum_{r=0}^{m-1} c_r (DT)^r(\mathbf{w})$$

for some $c_0, \ldots, c_{m-1} \in \mathbf{Z}/p\mathbf{Z}$. The right side of this dependence relation is a linear combination of basis vectors $\mathbf{v}_{i_0} \wedge \mathbf{v}_{i_1} \wedge \cdots \wedge \mathbf{v}_{i_{h-1}}$ satisfying (3.11) and

$$i_0 + i_1 + \cdots + i_{h-1} \leq \binom{h}{2} + m - 1,$$

while the left side is a linear combination of basis vectors satisfying (3.11) and

$$i_0 + i_1 + \cdots + i_{h-1} \leq \binom{h}{2} + m,$$

and including at least one basis vector such that

$$i_0 + i_1 + \cdots + i_{h-1} = \binom{h}{2} + m.$$

This is impossible, and the proof of the Erdős–Heilbronn conjecture is complete.

Remark: This proof only requires that A be a subset of a field, and does not require that the field be $\mathbf{Z}/p\mathbf{Z}$. Let K be an arbitrary field. Let p be the characteristic of K if the characteristic is positive, and let $p = \infty$ if the characteristic is zero. Then we have, in fact, proved that if $A \subseteq K$ and $|A| = k \leq p$, then $|h^\wedge A| \geq \min(p, hk - h^2 + 1)$ for all $h \geq 1$.

3.7 The polynomial method

In the following two sections we give a second proof of the Erdős–Heilbronn conjecture that uses only elementary manipulations of polynomials. To make the idea of the method clear, we prove the conjecture first for $h = 2$. We require only the following simple property of polynomials with coefficients in a field.

Lemma 3.10 *Let $h \geq 1$, and let $A_0, A_1, \ldots, A_{h-1}$ be nonempty subsets of a field K with $|A_i| = k_i$ for $i = 0, 1, \ldots, h-1$. Let $f(x_0, x_1, \ldots, x_{h-1})$ be a polynomial with coefficients in K and of degree at most $k_i - 1$ in x_i for $i = 0, 1, \ldots, h-1$. If*

$$f(a_0, a_1, \ldots, a_{h-1}) = 0$$

for all

$$(a_0, a_1, \ldots, a_{h-1}) \in A_0 \times A_1 \times \cdots \times A_{h-1},$$

then $f(x_0, x_1, \ldots, x_{h-1})$ is the zero polynomial.

Proof. By induction on h. The case $h = 1$ follows immediately from the fact that a nonzero polynomial in $K[x]$ of degree at most $k - 1$ cannot have k distinct roots in K.

Let $h \geq 2$, and assume that the lemma is true for polynomials in at most $h - 1$ variables. We can write

$$f(x_0, x_1, \ldots, x_{h-1}) = \sum_{j=0}^{k_0-1} f_j(x_1, \ldots, x_{h-1}) x_0^j,$$

where $f_j(x_1, \ldots, x_{h-1})$ is a polynomial in the $h - 1$ variables $x_1, \ldots, x_{h-1}$ and is of degree at most $k_i - 1$ in x_i for $i = 1, \ldots, h-1$. Fix

$$(a_1, \ldots, a_{h-1}) \in A_1 \times \cdots \times A_{h-1}.$$

Then

$$g(x_0) = f(x_0, a_1, \ldots, a_{h-1}) = \sum_{j=0}^{k_0-1} f_j(a_1, \ldots, a_{h-1}) x_0^j$$

is a polynomial of degree at most $k_0 - 1$ in x_0 such that $g(a_0) = 0$ for all $a_0 \in A_0$. Since $g(x)$ has at least k_0 distinct roots, it follows that $g(x)$ is the zero polynomial, and so

$$f_j(a_1, \ldots, a_{h-1}) = 0$$

for all

$$(a_1, \ldots, a_{h-1}) \in A_1 \times \cdots \times A_{h-1}.$$

It follows from the induction hypothesis that the polynomial $f_j(x_1, \ldots, x_{h-1})$ is identically zero, and so $f(x_0, x_1, \ldots, x_{h-1})$ is the zero polynomial.

Theorem 3.5 *Let p be a prime number, and let A and B be nonempty subsets of $\mathbf{Z}/p\mathbf{Z}$ such that $|A| \neq |B|$. Let*

$$A \hat{+} B = \{a + b : a \in A, b \in B, a \neq b\}.$$

Then

$$|A \hat{+} B| \geq \min(p, |A| + |B| - 2\}.$$

Remark. This immediately implies the Erdős–Heilbronn conjecture in the case $h = 2$, as follows. Let $A \subseteq \mathbf{Z}/p\mathbf{Z}$, $|A| = k \geq 2$. Choose $a \in A$, and let $B = A \setminus \{a\}$. Then $|B| = |A| - 1$ and $2^\wedge A = A \hat{+} B$. By Theorem 3.5,

$$|2^\wedge A| = |A \hat{+} B| \geq \min(p, |A| + |B| - 2) = \min(p, 2k - 3).$$

Proof of Theorem 3.5. Let $|A| = k$ and $|B| = l$. We can assume that

$$1 \leq l < k \leq p.$$

If $k + l - 2 > p$, let $l' = p - k + 2$. Then

$$2 \leq l' < l < k$$

and

$$k + l' - 2 = p.$$

Choose $B' \subseteq B$ such that $|B'| = l'$. If the theorem holds for the sets A and B', then

$$|A \hat{+} B| \geq |A \hat{+} B'| \geq \min(p, k + l' - 2) = p = \min(p, |A| + |B| - 2).$$

Therefore, we can assume that

$$k + l - 2 \leq p.$$

Let $C = A \hat{+} B$. We must prove that $|C| \geq k + l - 2$. If

$$|C| \leq k + l - 3,$$

then we choose $r \geq 0$ so that

$$r + |C| = k + l - 3.$$

We construct three polynomials f_0, f_1, and f in $(\mathbf{Z}/p\mathbf{Z})[x, y]$ as follows. Let

$$f_0(x, y) = \prod_{c \in C}(x + y - c).$$

Then $\deg(f_0) = |C| \leq k + l - 3$ and

$$f_0(a, b) = 0 \text{ for all } a \in A, b \in B, a \neq b.$$

Let

$$f_1(x, y) = (x - y) f_0(x, y).$$

Then $\deg(f_1) = 1 + |C| \leq k + l - 2$ and

$$f_1(a, b) = 0 \text{ for all } a \in A, b \in B.$$

Multiplying f_1 by $(x + y)^r$, we obtain the polynomial

$$f(x, y) = (x - y)(x + y)^r \prod_{c \in C}(x + y - c)$$

of degree exactly $1 + r + |C| = k + l - 2$ such that

$$f(a, b) = 0 \text{ for all } a \in A, b \in B.$$

There exist coefficients $u_{m,n} \in \mathbf{Z}/p\mathbf{Z}$ such that

$$\begin{aligned} f(x, y) &= \sum_{\substack{m,n \geq 0 \\ m+n \leq k+l-2}} u_{m,n} x^m y^n \\ &= (x - y)(x + y)^r \prod_{c \in C}(x + y - c) \\ &= (x - y)(x + y)^{k+l-3} + \text{lower-order terms.} \end{aligned}$$

Since $1 \leq l < k \leq p$ and $1 \leq k+l-3 < p$, it follows that the coefficient $u_{k-1,l-1}$ of the monomial $x^{k-1}y^{l-1}$ in $f(x, y)$ is

$$\binom{k + l - 3}{k - 2} - \binom{k + l - 3}{k - 1} = \frac{(k - l)(k + l - 3)!}{(k - 1)!(l - 1)!} \not\equiv 0 \pmod{p}.$$

By Lemma 3.3, for every $m \geq k$ there exists a polynomial $g_m(x)$ of degree at most $k - 1$ such that $g_m(a) = a^m$ for all $a \in A$, and for every $n \geq l$ there exists a polynomial $h_n(y)$ of degree at most $l - 1$ such that $h_n(b) = b^n$ for all $b \in B$. We use the polynomials $g_m(x)$ and $h_n(y)$ to construct a new polynomial $f^*(x, y)$ from $f(x, y)$ as follows. If $x^m y^n$ is a monomial in $f(x, y)$ with $m \geq k$, then we replace $x^m y^n$ with $g_m(x)y^n$. Since $\deg(f(x, y)) = k + l - 2$, it follows that if $m \geq k$, then $n \leq l - 2$, and so $g_m(x)y^n$ is a sum of monomials $x^i y^j$ with $i \leq k - 1$ and $j \leq l - 2$. Similarly, if $x^m y^n$ is a monomial in $f(x, y)$ with $n \geq l$, then we replace $x^m y^n$ with $x^m h_n(y)$. If $n \geq l$, then $m \leq k - 2$, and so $x^m h_n(y)$ is a sum of monomials $x^i y^j$ with $i \leq k - 2$ and $j \leq l - 1$.

This determines a new polynomial $f^*(x, y)$ of degree exactly $k-1$ in x and $l-1$ in y. The process of constructing $f^*(x, y)$ from $f(x, y)$ does not alter the coefficient $u_{k-1,l-1}$ of the term $x^{k-1}y^{l-1}$ since the monomial $x^{k-1}y^{l-1}$ does not occur in any of the polynomials $g_m(x)y^n$ or $x^m h_n(y)$. On the other hand,

$$f^*(a, b) = f(a, b) = 0$$

for all $a \in A$ and $b \in B$. It follows immediately from Lemma 3.10 that the polynomial $f^*(x, y)$ is identically zero. This contradicts the fact that the coefficient $u_{k-1,l-1}$ of $x^{k-1}y^{l-1}$ in $f^*(x, y)$ is nonzero. This completes the proof.

3.8 Erdős–Heilbronn via polynomials

We shall again use the polynomial

$$\Delta(x_0, x_1, \ldots, x_{h-1}) = \prod_{0 \leq i < j \leq h-1} (x_j - x_i),$$

which is of degree $\binom{h}{2}$.

Theorem 3.6 *Let K be a field. Let p be equal to the characteristic of K if the characteristic is a prime number, and let p be equal to ∞ if the characteristic is zero. Let $h \geq 2$ and $t \geq 0$ be integers such that*

$$m = t + \binom{h}{2} < p.$$

Then

$$(x_0 + \ldots + x_{h-1})^t \Delta(x_0, x_1, \ldots, x_{h-1}) = \sum \hat{B}(b_0, b_1, \ldots, b_{h-1}) x_0^{b_0} \cdots x_{h-1}^{b_{h-1}},$$

where the summation runs over all h-tuples $(b_0, b_1, \ldots, b_{h-1})$ of nonnegative integers such that

$$b_0 + b_1 + \cdots + b_{h-1} = m$$

and the coefficients are the strict ballot numbers

$$\hat{B}(b_0, b_1, \ldots, b_{h-1}) = \frac{(b_0 + b_1 + \cdots + b_{h-1} - \binom{h}{2})!}{b_0! b_1! \cdots b_{h-1}!} \prod_{0 \leq i < j \leq h-1} (b_j - b_i).$$

Proof. Recall that S_h is the symmetric group of all permutations of $\{0, 1, \ldots, h-1\}$, and $\mathrm{sign}(\sigma) = \pm 1$ is the sign of the permutation $\sigma \in S_h$. Using the Vandermonde determinant (Lemmas 3.1 and 3.2), we obtain the formula for the coefficients of the polynomial by the following computation:

$$(x_0 + \cdots + x_{h-1})^t \Delta(x_0, x_1, \ldots, x_{h-1})$$

$$
\begin{aligned}
&= (x_0 + \ldots + x_{h-1})^t \prod_{0 \le i < j \le h-1} (x_j - x_i) \\
&= (x_0 + \ldots + x_{h-1})^t \begin{vmatrix} 1 & x_0 & x_0^2 & \cdots & x_0^{h-1} \\ 1 & x_1 & x_1^2 & \cdots & x_1^{h-1} \\ 1 & x_2 & x_2^2 & \cdots & x_2^{h-1} \\ \vdots & & & & \\ 1 & x_{h-1} & x_{h-1}^2 & \cdots & x_{h-1}^{h-1} \end{vmatrix} \\
&= (x_0 + \ldots + x_{h-1})^t \sum_{\sigma \in S_h} \operatorname{sign}(\sigma) \prod_{i=0}^{h-1} x_i^{\sigma(i)} \\
&= \sum_{\substack{t_i \ge 0 \\ t_0+t_1+\cdots+t_{h-1}=t}} \frac{t!}{\prod_{i=0}^{h-1} t_i!} \prod_{i=0}^{h-1} x_i^{t_i} \sum_{\sigma \in S_h} \operatorname{sign}(\sigma) \prod_{i=0}^{h-1} x_i^{\sigma(i)} \\
&= t! \sum_{\sigma \in S_h} \operatorname{sign}(\sigma) \sum_{\substack{t_i \ge 0 \\ t_0+\cdots+t_{h-1}=t}} \frac{1}{\prod_{i=0}^{h-1} t_i!} \prod_{i=0}^{h-1} x_i^{t_i+\sigma(i)} \\
&= t! \sum_{\sigma \in S_h} \operatorname{sign}(\sigma) \sum_{\substack{t_i \ge 0 \\ t_0+\cdots+t_{h-1}=t}} \prod_{i=0}^{h-1} \frac{[t_i + \sigma(i)]_{\sigma(i)}}{(t_i + \sigma(i))!} x_i^{t_i+\sigma(i)} \\
&= t! \sum_{\sigma \in S_h} \operatorname{sign}(\sigma) \sum_{\substack{b_i \ge \sigma(i) \\ b_0+b_1+\cdots+b_{h-1}=m}} \prod_{i=0}^{h-1} \frac{[b_i]_{\sigma(i)}}{b_i!} x_i^{b_i} \\
&= t! \sum_{\sigma \in S_h} \operatorname{sign}(\sigma) \sum_{\substack{b_i \ge 0 \\ b_0+b_1+\cdots+b_{h-1}=m}} \prod_{i=0}^{h-1} \frac{[b_i]_{\sigma(i)}}{b_i!} x_i^{b_i} \\
&= \sum_{\substack{b_i \ge 0 \\ b_0+\cdots+b_{h-1}=m}} t! \sum_{\sigma \in S_h} \operatorname{sign}(\sigma) \prod_{i=0}^{h-1} \frac{[b_i]_{\sigma(i)}}{b_i!} x_i^{b_i} \\
&= \sum_{\substack{b_i \ge 0 \\ b_0+\cdots+b_{h-1}=m}} \frac{t!}{\prod_{i=0}^{h-1} b_i!} \prod_{0 \le i < j \le h-1} (b_j - b_i) \prod_{i=0}^{h-1} x_i^{b_i} \\
&= \sum_{\substack{b_i \ge 0 \\ b_0+\cdots+b_{h-1}=m}} \frac{\left(b_0 + b_1 + \cdots + b_{h-1} - \binom{h}{2}\right)!}{\prod_{i=0}^{h-1} b_i!} \prod_{0 \le i < j \le h-1} (b_j - b_i) x_0^{b_0} \cdots x_{h-1}^{b_{h-1}}.
\end{aligned}
$$

By Theorem 3.1, the coefficient of the monomial $x_0^{b_0} \cdots x_{h-1}^{b_{h-1}}$ is the strict ballot number $\hat{B}(b_0, b_1, \ldots, b_{h-1})$. This completes the proof.

Theorem 3.7 (Alon–Nathanson–Ruzsa) *Let K be a field. Let p be equal to the characteristic of K if the characteristic is a prime number, and let p be equal to ∞ if the characteristic is zero. Let $h \ge 2$, and let $A_0, A_1, \ldots, A_{h-1}$ be nonempty,* *finite subsets of* K. *Let* $|A_i| = k_i$ *for* $i = 0, 1, \ldots, k-1$, *and suppose that* $k_i \neq k_j$ *for* $i \neq j$. *Let*

$$\begin{aligned} C &= \{a_0 + a_1 + \cdots + a_{h-1} : a_i \in A_i \text{ for } i = 0, 1, \ldots, h-1 \\ &\quad \text{and } a_i \neq a_j \text{ for } i \neq j\}. \end{aligned}$$

Then

$$|C| \geq \min\{p, \sum_{i=0}^{h-1} k_i - \binom{h+1}{2} + 1\}.$$

Proof. Without loss of generality, we can assume that

$$1 \leq k_0 < k_1 < \cdots < k_{h-1}.$$

Then $k_i \geq i+1$ for $i = 0, 1, \ldots, h-1$. Let $\ell_i = k_i - i - 1$ for $i = 0, 1, \ldots, h-1$. Then

$$0 \leq \ell_0 \leq \ell_1 \leq \cdots \leq \ell_{h-1}$$

and

$$t = \sum_{i=0}^{h-1} k_i - \binom{h+1}{2} = \sum_{i=0}^{h-1}(k_i - i - 1) = \sum_{i=0}^{h-1} \ell_i.$$

We shall show that is the theorem holds for $t < p$, then it also holds for $t \geq p$. If

$$t = \sum_{i=0}^{h-1} \ell_i \geq p,$$

then we choose integers $\ell_i' \leq \ell_i$ such that

$$0 \leq \ell_0' \leq \ell_1' \cdots \leq \ell_{h-1}'$$

and

$$\sum_{i=0}^{h-1} \ell_i' = p - 1.$$

Let

$$k_i' = \ell_i' + i + 1$$

for $i = 0, 1, \ldots, h-1$. Then $i + 1 \leq k_i' \leq k_i$ and

$$\sum_{i=0}^{h-1} k_i' - \binom{h+1}{2} = \sum_{i=0}^{h-1}(k_i' - i - 1) = \sum_{i=0}^{h-1} \ell_i' = p - 1.$$

Choose $A_i' \subseteq A_i$ such that $|A_i'| = k_i'$ for $i = 0, 1, \ldots, h-1$, and let

$$\begin{aligned} C' &= \{a_0 + \cdots + a_{h-1} : a_i \in A_i' \text{ for } i = 0, 1, \ldots, h-1 \\ &\quad \text{and } a_i \neq a_j \text{ for } i \neq j\}. \end{aligned}$$

Then $C' \subseteq C$ and

$$\begin{aligned}
|C| &\geq |C'| \\
&\geq \min\left\{p, \sum_{i=0}^{h-1} k_i' - \binom{h+1}{2} + 1\right\} \\
&= p \\
&= \min\{p, t+1\} \\
&= \min\left\{p, \sum_{i=0}^{h-1} k_i - \binom{h+1}{2} + 1\right\}.
\end{aligned}$$

Therefore, we can assume that

$$t = \sum_{i=0}^{h-1} k_i - \binom{h+1}{2} < p.$$

We must prove that $|C| \geq t+1$.

Suppose that

$$|C| \leq t = \sum_{i=0}^{h-1} k_i - \binom{h+1}{2}.$$

Choose the nonnegative integer r so that

$$r + |C| = \sum_{i=0}^{h-1} k_i - \binom{h+1}{2} = t.$$

We define the polynomial $f \in K[x_0, x_1, \ldots, x_{h-1}]$ defined as follows:

$$\begin{aligned}
& f(x_0, x_1, \ldots, x_{h-1}) \\
&= \Delta(x_0, x_1, \ldots, x_{h-1})(x_0 + \ldots + x_{h-1})^r \prod_{c \in C}(x_0 + \ldots + x_{h-1} - c).
\end{aligned}$$

The degree of this polynomial is

$$\begin{aligned}
m &= \binom{h}{2} + r + |C| \\
&= t + \binom{h}{2} \\
&= \sum_{i=0}^{h-1} k_i - \binom{h+1}{2} + \binom{h}{2} \\
&= \sum_{i=0}^{h-1} k_i - h \\
&= \sum_{i=0}^{h-1} (k_i - 1).
\end{aligned}$$

Moreover,

$$f(a_0, a_1, \ldots, a_{h-1}) = 0$$

for all

$$(a_0, a_1, \ldots, a_{h-1}) \in A_0 \times A_1 \times \cdots \times A_{h-1}.$$

Since

$$\begin{aligned} & f(x_0, x_1, \ldots, x_{h-1}) \\ = \quad & (x_0 + \ldots + x_{h-1})^t \Delta(x_0, x_1, \ldots, x_{h-1}) + \text{ lower order terms}, \end{aligned}$$

it follows from Theorem 3.6 that the coefficient of the monomial

$$x_0^{k_0-1} x_1^{k_1-1} \cdots x_{h-1}^{k_{h-1}-1}$$

is

$$\begin{aligned} & \hat{B}(k_0 - 1, k_1 - 1, \ldots, k_{h-1} - 1) \\ = \quad & \frac{(k_0 + k_1 + \cdots + k_{h-1} - h - \binom{h}{2})!}{(k_0 - 1)!(k_1 - 1)! \cdots (k_{h-1} - 1)!} \prod_{0 \leq i < j \leq h-1} (k_j - k_i) \\ = \quad & \frac{t!}{(k_0 - 1)!(k_1 - 1)! \cdots (k_{h-1} - 1)!} \prod_{0 \leq i < j \leq h-1} (k_j - k_i), \end{aligned}$$

and this number is nonzero in the field K.

By Lemma 3.3, for $i = 0, 1, \ldots, h - 1$ and for every $m \geq k_i$ there exists a polynomial $g_{i,m}(x_i)$ of degree at most $k_i - 1$ such that $g_{i,m}(a_i) = a_i^m$ for all $a_i \in A_i$. We use the polynomials $g_{i,m}(x_i)$ to construct a new polynomial $f^*(x_0, \ldots, x_{h-1})$ from $f(x_0, \ldots, x_{h-1})$, exactly as in Theorem 3.5. If $x_0^{b_0} \cdots x_{h-1}^{b_{h-1}}$ is a monomial in $f(x_0, \ldots, x_{h-1})$, then we replace $x_i^{b_i}$ with $g_{i,b_i}(x_i)$ for all i such that $b_i \geq k_i$. Since $\deg(f) = \sum_{i=0}^{h-1}(k_i - 1)$, it follows that if $b_i \geq k_i$ for some i in a given monomial, then $b_j < k_j - 1$ for some $j \neq i$ in the same monomial. It follows that the coefficient of the monomial $x_0^{k_0-1} \cdots x_{h-1}^{k_{h-1}-1}$ in f^* is exactly the same as the coefficient of this monomial in f, and this coefficient is

$$\hat{B}(k_0 - 1, k_1 - 1, \ldots, k_{h-1} - 1) \neq 0.$$

On the other hand, the polynomial f^* has degree exactly $k_i - 1$ in the variable x_i, and

$$f^*(a_0, \ldots, a_{h-1}) = f(a_0, \ldots, a_{h-1}) = 0$$

for all $(a_0, \ldots, a_{h-1}) \in A_0 \times \cdots \times A_{h-1}$. By Lemma 3.10, the polynomial f^* must be identically zero. This is a contradiction, and so $|C| \geq t + 1$. This completes the proof.

Theorem 3.8 *Let K be a field, let p denote the characteristic of K if the characteristic is a prime, and let $p = \infty$ if the characteristic is zero. Let $h \geq 2$, and let A be a subset of K such that $|A| = k \geq h$. Then*

$$|h^\wedge A| \geq \min\{p, hk - h^2 + 1\}.$$

Proof. Let $A_0, A_1, \ldots, A_{h-1}$ be subsets of A such that

$$|A_i| = k_i = k - i.$$

Then

$$\sum_{i=0}^{h-1} k_i = hk - \binom{h}{2}.$$

Let

$$\begin{aligned} C &= \{a_0 + \cdots + a_{h-1} : a_i \in A_i \text{ for } i = 0, 1, \ldots, h-1 \\ &\quad \text{and } a_i \neq a_j \text{ for } i \neq j\}. \end{aligned}$$

Then

$$C \subseteq h^\wedge A.$$

It follows from Theorem 3.7 than

$$\begin{aligned} |h^\wedge A| &\geq |C| \\ &\geq \min\{p, \sum_{i=0}^{h-1} k_i - \binom{h+1}{2} + 1\} \\ &= \min\{p, hk - \binom{h}{2} - \binom{h+1}{2} + 1\} \\ &= \min\{p, hk - h^2 + 1\}. \end{aligned}$$

This completes the polynomial proof of the Erdős–Heilbronn conjecture.

3.9 Notes

The Erdős–Heilbronn conjecture originated in the 1960s. Erdős and Heilbronn did not include it in their paper on sums of sets of congruence classes [43], but Erdős [36, pages 16–17] stated the conjecture at a number theory conference at the University of Colorado in 1963 and subsequently has often mentioned the problem in his lectures and papers (see, for example, Erdős [37] and the book of Erdős and Graham [42, page 95]. Partial results on the Erdős–Heilbronn conjecture were obtained by Rickert [110], Mansfield [85], Rödseth [111], Pyber [109], and Freiman, Low, and Pitman [59].

Dias da Silva and Hamidoune [29] proved the complete conjecture by using results from representation theory and linear algebra. This algebraic technique had previously been applied to additive number theory by Dias da Silva and Hamidoune [28]. Spigler [120] had earlier applied additive number theory to problems in linear algebra.

Nathanson [93] simplified the Dias da Silva–Hamidoune method by replacing the representation theory with simple properties of the ballot numbers. The proof of the formula in Theorem 3.1 for the strict ballot number $\hat{B}(b_0, \ldots, b_{n-1})$ follows a paper of Zeilberger [129]. There is a vast literature on ballot numbers and other lattice path counting problems (see, for example, Mohanty [87] and Narayana [89]).

The polynomial proof of the Erdős–Heilbronn conjecture is due to Alon, Nathanson, and Ruzsa [2, 3].

3.10 Exercises

1. Let $\sigma, \tau \in S_h$, let $f, g \in F[x_0, x_1, \ldots, x_{h-1}]$, and let $c \in F$. Prove that

$$(\sigma\tau)f = \sigma(\tau f),$$

$$\sigma(f+g) = \sigma(f) + \sigma(g),$$

$$\sigma(cf) = c\sigma(f),$$

and

$$\sigma(fg) = \sigma(f)\sigma(g).$$

2. Let $\tau \in S_h$ be a transposition. Prove that $\mathrm{sign}(\tau) = -1$.

3. Use the polynomial method to prove the Cauchy–Davenport theorem.

4. Let A and B be nonempty subsets of $\mathbf{Z}/p\mathbf{Z}$, and let

$$C = \{a + b : a \in A, b \in B, ab \not\equiv 1\}.$$

Prove that

$$|C| \geq \min\{p, |A| + |B| - 3\}.$$

5. The division algorithm for polynomials with coefficients in a field F states that if $u(x), v(x) \in F[x]$ and $v(x) \neq 0$, then there exist $g(x), h(x) \in F[x]$ such that $\deg(g) < \deg(v)$ and

$$u(x) = h(x)v(x) + g(x).$$

Use the division algorithm to prove Lemma 3.3.
Hint: Let $v(x) = \prod_{a\in A}(x - a)$.

6. Let $A = \{a_0, a_1, \ldots, a_{k-1}\}$ be a set of k distinct elements of a field F, and let $b_0, b_1, \ldots, b_{k-1}$ be a sequence of k not necessarily distinct elements of F. Consider the polynomial $g(x) \in F[x]$ defined by

$$g(x) = \sum_{i=0}^{k-1} b_i \prod_{\substack{j=0 \\ j\neq i}}^{k-1} \left(\frac{x - a_j}{a_i - a_j}\right).$$

Check that $\deg(g) = k - 1$ and $g(a_i) = b_i$ for $i = 0, 1, \ldots, k-1$. This is the Lagrange interpolation formula. Use this to give another proof of Lemma 3.3.

4

Kneser's theorem for groups

4.1 Periodic subsets

The principal goal of this chapter is to prove a beautiful theorem of Kneser about sums of finite subsets of an abelian group G. We need the following definitions.

Let S be a nonempty subset of the abelian group G. The *stabilizer* of S is the set

$$H(S) = \{g \in G \mid g + S = S\}.$$

Then $0 \in H(S)$, and $H(S)$ is the largest subgroup of G such that

$$H(S) + S = S.$$

In particular, $H(S) = G$ if and only if $S = G$. An element $g \in H(S)$ is called a *period* of S, and S is called a *periodic set*, if $H(S) \neq \{0\}$. For example, if S is an infinite arithmetic progression in $\mathbf{Z}$ with difference d, then $H(S) = d\mathbf{Z}$.

Kneser proved that if A and B are nonempty, finite subsets of an abelian group G, then either $|A + B| \geq |A| + |B|$ or

$$|A + B| = |A + H| + |B + H| - |H|,$$

where $H = H(A + B)$ is the stabilizer of $A + B$. In the special case when G is a finite cyclic group, Kneser's theorem implies the theorems of Cauchy–Davenport and I. Chowla (see Exercises 5 and 6).

Kneser's theorem has many applications in additive number theory. We shall use it to generalize the inverse theorem for sumsets of the form $2A$ (Theorem 1.16) to sumsets of the form $A + B$, where A and B are nonempty, finite sets of integers. We shall also use Kneser's theorem to obtain a density criterion for the order of an additive basis for a σ-finite abelian group.

4.2 The addition theorem

We begin by proving a special case of the main theorem. This simple result is sufficient in many applications.

Theorem 4.1 (Kneser) *Let G be an abelian group, $G \neq \{0\}$, and let A and B be nonempty, finite subsets of G. If $|A| + |B| \leq |G|$, then there exists a proper subgroup H of G such that*

$$|A + B| \geq |A| + |B| - |H|.$$

Proof. By induction on $|B|$. If $|B| = 1$, then

$$|A + B| = |A| = |A| + |B| - 1 \geq |A| + |B| - |H|$$

for every subgroup H.

Let $|B| > 1$, and suppose that the theorem holds for all pairs A', B' of finite, nonempty subsets of G such that $|B'| < |B|$. There are two cases.

Case 1. Suppose that

$$a_1 + b_2 - b_1 \in A$$

for all $a_1 \in A$ and $b_1, b_2 \in B$. Then

$$A + b_2 - b_1 = A$$

for all $b_1, b_2 \in B$. Let H be the subgroup of G generated by all elements of the form $b_2 - b_1$, where $b_1, b_2 \in B$. Then

$$|B| \leq |H|$$

and

$$A + H = A \neq G.$$

Therefore, H is a proper subgroup of G, and

$$|A + B| \geq |A| \geq |A| + |B| - |H|.$$

Case 2. Suppose that there exist $a_1 \in A$ and $b_1, b_2 \in B$ such that

$$a_1 + b_2 - b_1 \notin A.$$

Let

$$e = a_1 - b_1.$$

Then

$$b_2 \notin A - (a_1 - b_1) = A - e, \tag{4.1}$$

but

$$b_1 \in A - (a_1 - b_1) = A - e \tag{4.2}$$

since $0 \in A - a_1$. Applying the e-transform (Section 2.2) to the pair (A, B), we obtain a new pair $(A(e), B(e))$ of subsets of G defined by

$$\begin{aligned} A(e) &= A \cup (B + e) \\ B(e) &= B \cap (A - e). \end{aligned}$$

It follows from (4.1) that $b_2 \notin B(e)$, and so $B(e)$ is a proper subset of B. Also, $b_1 \in B(e)$ by (4.2), so $B(e)$ is a nonempty subset of B. Therefore, the induction hypothesis applies to the pair $(A(e), B(e))$. Using properties (2.1) and (2.3) of the e-transform, we conclude that there exists a proper subgroup H of G such that

$$\begin{aligned} |A + B| &\geq |A(e) + B(e)| \\ &\geq |A(e)| + |B(e)| - |H| \\ &= |A| + |B| - |H|. \end{aligned}$$

This completes the proof.

The proof of Theorem 4.2 uses the following three lemmas.

Lemma 4.1 *Let G be an abelian group, and let $C = C_1 \cup C_2$ be a finite subset of G, where C_1 and C_2 are nonempty, proper subsets of C. Then*

$$|C_i| + |H(C_i)| \leq |C| + |H(C)|$$

for $i = 1$ or $i = 2$.

Proof. If $|C_i| + |H(C_i)| \leq |C|$ for $i = 1$ or $i = 2$, then we are done. Therefore, we can assume that

$$|C| < |C_i| + |H(C_i)| \tag{4.3}$$

for $i = 1$ and $i = 2$. Let $H(C_i) = H_i$, and let m_i denote the index of the subgroup H_i in $H_1 + H_2$. Let $H = H_1 \cap H_2$ and $|H| = h$. By a standard isomorphism theorem of group theory,

$$(H_1 + H_2)/H_1 \cong H_2/H$$

and

$$(H_1 + H_2)/H_2 \cong H_1/H.$$

Therefore, $|H_1| = m_2 h$, $|H_2| = m_1 h$, and $|H_1 + H_2| = m_1 m_2 h$. Since $H \subseteq H_i$, it follows that $H + C_i = C_i$ and C_i is a union of cosets of H in G. Therefore, $C_1 \setminus C_2$ and $C_2 \setminus C_1$ are unions of H-cosets, and

$$|C_1| \equiv |C_2| \equiv |C_1 \setminus C_2| \equiv |C_2 \setminus C_1| \equiv 0 \pmod{h}.$$

Since C is the union of the proper subsets C_1 and C_2, it follows that $C_1 \setminus C_2$ and $C_2 \setminus C_1$ are nonempty. By (4.3),

$$0 < |C_1 \setminus C_2| = |C \setminus C_2| = |C| - |C_2| < |H_2| = m_1 h,$$

and so

$$h \leq |C_1 \setminus C_2| \leq (m_1 - 1)h. \tag{4.4}$$

Similarly,

$$h \leq |C_2 \setminus C_1| \leq (m_2 - 1)h. \tag{4.5}$$

Choose $c^* \in C_1 \setminus C_2$, and let

$$D = c^* + H_1 + H_2.$$

Then D is a union of H_1-cosets of the form

$$D_1 = c^* + h_2 + H_1, \tag{4.6}$$

where $h_2 \in H_2$, and D is also a union of H_2-cosets of the form

$$D_2 = c^* + h_1 + H_2, \tag{4.7}$$

where $h_1 \in H_1$. Let D_1 be an H_1-coset of the form (4.6), and let D_2 be an H_2-coset of the form (4.7). Since $h_2 + H \subseteq H_2$ and $h_1 + H \subseteq H_1$, it follows that

$$c^* + h_1 + h_2 + H \subseteq D_1 \cap D_2.$$

Conversely, if $g \in D_1 \cap D_2$, then there exist $h_1' \in H_1$ and $h_2' \in H_2$ such that

$$g = c^* + h_1 + h_2' = c^* + h_1' + h_2.$$

This implies that

$$g - (c^* + h_1 + h_2) = h_1' - h_1 \in H_1$$

and

$$g - (c^* + h_1 + h_2) = h_2' - h_2 \in H_2,$$

and so

$$g - (c^* + h_1 + h_2) \in H_1 \cap H_2 = H.$$

It follows that

$$g \in c^* + h_1 + h_2 + H$$

and so

$$D_1 \cap D_2 \subseteq c^* + h_1 + h_2 + H.$$

Thus,

$$D_1 \cap D_2 = c^* + h_1 + h_2 + H,$$

and the intersection of an H_1-coset in D with an H_2-coset in D is an H-coset.

Since the index of H_i in $H_1 + H_2$ is m_i, the subgroup $H_1 + H_2$ is the union of m_i pairwise disjoint H_i-cosets, and so $D = c^* + H_1 + H_2$ is also the union of m_i pairwise disjoint H_i-cosets. Since $H_i + C_i = C_i$ is a union of H_i-cosets, it follows that $C_i \cap D$ is the union of, say, u_i pairwise disjoint H_i-cosets, and so $\overline{C_i} \cap D$ is the union of $m_i - u_i$ pairwise disjoint H_i-cosets. Since the intersection of an H_1-coset in D and an H_2-coset in D is an H-coset, it follows that

$$(C_2 \setminus C_1) \cap D = (C_2 \cap D) \cap (\overline{C_1} \cap D)$$

is the union of $u_2(m_1 - u_1)$ pairwise disjoint H-cosets, and so

$$|(C_2 \setminus C_1) \cap D| = u_2(m_1 - u_1)h. \tag{4.8}$$

Similarly,

$$(C_1 \setminus C_2) \cap D = (C_1 \cap D) \cap (\overline{C_2} \cap D)$$

is the union of $u_1(m_2 - u_2)$ pairwise disjoint H-cosets, and

$$|(C_1 \setminus C_2) \cap D| = u_1(m_2 - u_2)h. \tag{4.9}$$

Since

$$c^* \in (C_1 \setminus C_2) \cap D \subseteq C_1 \setminus C_2,$$

it follows that

$$0 < |(C_1 \setminus C_2) \cap D| = u_1(m_2 - u_2)h \leq |C_1 \setminus C_2| \leq (m_1 - 1)h.$$

Therefore, $1 \leq u_1(m_2 - u_2) \leq m_1 - 1$, and so

$$1 \leq u_1 \leq m_1 - 1$$

and

$$1 \leq u_2 \leq m_2 - 1.$$

It follows from (4.4), (4.5), (4.8), and (4.9) that

$$\begin{aligned} 0 &\leq (m_1 - u_1 - 1)(u_2 - 1)h + (m_2 - u_2 - 1)(u_1 - 1)h \\ &= u_2(m_1 - u_1)h - (m_2 - 1)h + u_1(m_2 - u_2)h - (m_1 - 1)h \\ &= |(C_2 \setminus C_1) \cap D| - (m_2 - 1)h + |(C_1 \setminus C_2) \cap D| - (m_1 - 1)h \\ &= |C_2 \setminus C_1| - (m_2 - 1)h - |(C_2 \setminus C_1) \cap \overline{D}| \\ &\quad + |C_1 \setminus C_2| - (m_1 - 1)h - |(C_1 \setminus C_2) \cap \overline{D}| \\ &\leq 0, \end{aligned}$$

and so $|C_2 \setminus C_1| = (m_2 - 1)h$ and $|C_1 \setminus C_2| = (m_1 - 1)h$. Since $H = H_1 \cap H_2$, it follows that

$$H + C = H + (C_1 \cup C_2) = (H + C_1) \cup (H + C_2) = C_1 \cup C_2 = C,$$

and so $H \subseteq H(C)$. Therefore,

$$\begin{aligned} |C| - |C_2| &= |C \setminus C_2| \\ &= |C_1 \setminus C_2| \\ &= m_1 h - h \\ &= |H_2| - |H| \\ &\geq |H_2| - |H(C)|. \end{aligned}$$

Similarly,

$$|C| - |C_1| \geq |H_1| - |H(C)|,$$

and so

$$|C_i| + |H(C_i)| \leq |C| + |H(C)|$$

for both $i = 1$ and $i = 2$. This completes the proof of the lemma.

Lemma 4.2 *Let $n \geq 2$, and let G be an abelian group. Let C be a finite subset of G such that*

$$C = C_1 \cup C_2 \cup \cdots \cup C_n,$$

where $C_1, \ldots, C_n$ are nonempty, proper subsets of C. Then

$$|C_i| + |H(C_i)| \leq |C| + |H(C)|$$

for some $i = 1, \ldots, n$.

Proof. By induction on n. Lemma 4.1 is the case $n = 2$. Let $n \geq 3$, and suppose that the result holds for $n - 1$. If $|C_i| + |H(C_i)| \leq |C|$ for some i, then $|C_i| + |H(C_i)| \leq |C| < |C| + |H(C)|$, and we are done. If not, then

$$|C| < |C_i| + |H(C_i)|$$

for all $i = 1, \ldots, n$. If C is the union of $n - 1$ of the subsets $C_1, \ldots, C_n$, we are done, since the result follows from the case $n - 1$. Thus, we can assume that C is not the union of $n - 1$ of the sets $C_1, \ldots, C_n$. Let

$$C' = C_1 \cup \cdots \cup C_{n-1}.$$

Then the sets $C_1, \ldots, C_{n-1}$ are proper subsets of C', and C' is a proper subset of C. It follows from the case $n - 1$ of the lemma that

$$|C_i| + |H(C_i)| \leq |C'| + |H(C')| \tag{4.10}$$

for some $i = 1, \ldots, n - 1$. Since

$$C = C' \cup C_n,$$

Lemma 4.1 implies that either

$$|C_n| + |H(C_n)| \leq |C| + |H(C)|$$

or

$$|C'| + |H(C')| \leq |C| + |H(C)|,$$

and the result follows from (4.10).

Lemma 4.3 *Let $C_1, \ldots, C_n$ be finite, nonempty subsets of the abelian group G, and let*

$$C = C_1 \cup \cdots \cup C_n.$$

Then

$$\min (|C_i| + |H(C_i)| \ : \ i = 1, \ldots, n) \leq |C| + |H(C)|.$$

Proof. If $C_i = C$ for some i, we are done. If not, then each set C_i is a proper, nonempty subset of C, and Lemma 4.2 implies that

$$|C_i| + |H(C_i)| \leq |C| + |H(C)|$$

for some $i = 1, \ldots, n$.

Theorem 4.2 (Kneser) *Let G be an abelian group, and let A and B be finite, nonempty subsets of G. Let*

$$H = H(A + B) = \{g \in G \mid g + A + B = A + B\}$$

be the stabilizer of $A + B$. If

$$|A + B| < |A| + |B|, \tag{4.11}$$

then

$$|A + B| = |A + H| + |B + H| - |H|. \tag{4.12}$$

Proof. Let $C = A + B$ satisfy inequality (4.11). Let $B = \{b_1, \ldots, b_n\}$. For each $b_i \in B$, we consider the collection of all pairs of finite subsets (A_i, B_i) of G such that

$$\begin{aligned} A &\subseteq A_i, \\ b_i &\in B_i, \\ A_i + B_i &\subseteq A + B, \\ |A_i| + |B_i| &= |A + H| + |B + H|. \end{aligned}$$

This collection is nonempty since the sets $A_i = A + H$ and $B_i = B + H$ satisfy these conditions. Fix a pair (A_i, B_i) for which $|A_i|$ is maximal, and let $C_i = A_i + B_i$. Then $|A_i| \leq |C_i|$, and

$$A + b_i \subseteq A_i + B_i = C_i \subseteq C. \tag{4.13}$$

Let $a \in A_i$ and $e = a - b_i$. Applying the e-transform to the sets A_i, B_i, we obtain the sets

$$A_i(e) = A_i \cup (B_i + e) = A_i \cup (a + B_i - b_i)$$

and

$$B_i(e) = B_i \cap (A_i - e) = B_i \cap (-a + A_i + b_i).$$

Then $A_i \subseteq A_i(e)$ and $b_i \in B_i(e)$. By Lemma 2.3,

$$A_i(e) + B_i(e) \subseteq A_i + B_i \subseteq C = A + B$$

and

$$|A_i(e)| + |B_i(e)| = |A_i| + |B_i| = |A + H| + |B + H|.$$

It follows from the maximality of $|A_i|$ that $A_i(e) = A_i$, and so

$$a \in a + B_i - b_i \subseteq A_i$$

for each $a \in A_i$. Therefore,

$$A_i \subseteq A_i + B_i - b_i = C_i - b_i \subseteq A_i$$

and so $A_i = C_i - b_i$. Then $|A_i| = |C_i|$, $H(A_i) = H(C_i)$, and

$$B_i - b_i \subseteq H(A_i) = H(C_i);$$

hence $|B_i| \leq |H(C_i)|$. We obtain

$$|A + H| + |B + H| = |A_i| + |B_i| \leq |C_i| + |H(C_i)|$$

for all $i = 1, \ldots, n$. Since

$$\bigcup_{i=1}^{n} C_i = C = A + B$$

by (4.13), it follows from Lemma 4.3 that

$$\begin{aligned} |A + H| + |B + H| &\leq \min(|C_i| + |H(C_i)|) \\ &\leq |C| + |H(C)| \\ &= |A + B| + |H|. \end{aligned}$$

Since each of the integers $|A + H|$, $|B + H|$, and $|A + B|$ is a multiple of $|H|$, it follows that if

$$|A + H| + |B + H| < |A + B| + |H|,$$

then

$$|A| + |B| \leq |A + H| + |B + H| \leq |A + B|,$$

which contradicts (4.11). Therefore, $|A + H| + |B + H| = |A + B| + |H|$. This completes the proof of the theorem.

Theorem 4.3 *Let G be an abelian group, and let A and B be finite, nonempty subsets of G. Let $H = H(A + B)$. Then*

$$|A + B| \geq |A + H| + |B + H| - |H|. \tag{4.14}$$

Proof. We apply Kneser's theorem to the sets $A + H$ and $B + H$. Then either

$$\begin{aligned} |A + B| &= |(A + H) + (B + H)| \\ &\geq |A + H| + |B + H| \\ &\geq |A + H| + |B + H| - |H| \end{aligned}$$

or

$$|(A + H) + (B + H)| < |A + H| + |B + H|,$$

and so

$$|A + B| = |(A + H) + (B + H)| = |A + H| + |B + H| - |H|.$$

This proves (4.14).

Theorem 4.4 *Let $h \geq 2$, let $A_1, A_2, \ldots, A_h$ be finite, nonempty subsets of an abelian group G, and let $H = H(A_1 + \cdots + A_h)$. Then*

$$|A_1 + \cdots + A_h| \geq |A_1| + \cdots + |A_h| - (h-1)|H|.$$

Proof. By induction on h. The case $h = 2$ comes from inequality (4.14).

Let $h \geq 3$, and suppose that the theorem holds for some $h-1$. Let $H' = H(A_1 + \cdots + A_{h-1})$. It follows from Exercise 1 that $H' \subseteq H$, and so

$$\begin{aligned} |A_1 + \cdots + A_h| &\geq |A_1 + \cdots + A_{h-1}| + |A_h| - |H| \\ &\geq |A_1| + \cdots + |A_{h-1}| - (h-2)|H'| + |A_h| - |H| \\ &\geq |A_1| + \cdots + |A_{h-1}| + |A_h| - (h-1)|H|. \end{aligned}$$

Theorem 4.5 *Let G be an abelian group, and let A be a finite, nonempty subset of G. Let hA be the h-fold sumset of A, and let*

$$H_h = H(hA) = \{g \in G \mid g + hA = hA\}$$

be the stabilizer of hA. Then

$$|hA| \geq h|A + H_h| - (h-1)|H_h|$$

for all $h \geq 1$.

Proof. By Theorem 4.4, for any finite, nonempty subset B of the group G we have

$$|hB| \geq h|B| - (h-1)|H(hB)|$$

for all $h \geq 2$. Let

$$B = A + H_h.$$

Then

$$hB = h(A + H_h) = hA$$

and so $H(hB) = H(hA) = H_h$. Therefore,

$$|hA| = |hB| \geq h|A + H_h| - (h-1)|H_h|.$$

4.3 Application: The sum of two sets of integers

Let A and B be nonempty, finite sets of integers. Then $|A + B| \geq |A| + |B| - 1$. By Theorem 1.3, $|A + B| = |A| + |B| - 1$ if and only A and B are arithmetic progressions with the same common difference. Our goal in this section is to show that if $|A+B|$ is "small," then A and B are "large" subsets of arithmetic progressions with the same common difference. This inverse theorem for the sumset $A + B$ is a generalization of Theorem 1.16.

Theorem 4.6 *Let $k, \ell \geq 2$, and let $A = \{a_0, a_1, \ldots, a_{k-1}\}$ and $B = \{b_0, b_1, \ldots, b_{\ell-1}\}$ be nonempty, finite sets of integers such that*

$$0 = a_0 < a_1 < \cdots < a_{k-1},$$

$$0 = b_0 < b_1 < \cdots < b_{\ell-1},$$

$$b_{\ell-1} \leq a_{k-1},$$

and

$$(a_1, a_2, \ldots, a_{k-1}) = 1.$$

Let

$$\delta = \begin{cases} 0 & \text{if } b_{\ell-1} < a_{k-1} \\ 1 & \text{if } b_{\ell-1} = a_{k-1}. \end{cases}$$

Then

$$|A + B| \geq \min\{a_{k-1} + \ell, k + 2\ell - \delta - 2\}.$$

Proof. If $|A+B| \geq k+2\ell-\delta-2$, we are done. Therefore, we can assume that

$$|A + B| \leq k + 2\ell - \delta - 3. \tag{4.15}$$

We shall prove that

$$|A + B| \geq a_{k-1} + \ell.$$

Let $G = \mathbf{Z}/a_{k-1}\mathbf{Z}$, and let

$$\pi : \mathbf{Z} \to G$$

be the canonical homomorphism onto the cyclic group G. Then

$$\pi(A + B) = \pi(A) + \pi(B),$$

$$|\pi(A)| = k - 1$$

since $\pi(a_0) = \pi(a_{k-1}) = 0$, and

$$|\pi(B)| = \ell - \delta$$

since $\pi(b_{\ell-1}) = 0$ if and only if $b_{\ell-1} = a_{k-1}$. We can rewrite (4.15) in the form

$$|A + B| \leq |\pi(A)| + |\pi(B)| + \ell - 2. \tag{4.16}$$

We shall show that there are at least ℓ integers in $A + B$ that lie in the same congruence classes modulo a_{k-1} as other integers in $A + B$. If $b_{\ell-1} < a_{k-1}$, then

$$\pi(a_0 + b_i) = \pi(a_{k-1} + b_i)$$

for $i = 0, 1, \ldots, \ell - 1$, and

$$\begin{aligned} a_0 + b_0 < a_0 + b_1 < \cdots < a_0 + b_{\ell-1} \\ < \quad a_{k-1} + b_0 < a_{k-1} + b_1 < \cdots < a_{k-1} + b_{\ell-1}. \end{aligned}$$

If $b_{\ell-1} = a_{k-1}$, then

$$\begin{aligned} a_0 + b_0 < a_0 + b_1 < \cdots < a_0 + b_{\ell-2} < a_0 + b_{\ell-1} = a_{k-1} + b_0 \\ < a_{k-1} + b_1 < \cdots < a_{k-1} + b_{\ell-2} < a_{k-1} + b_{\ell-1}, \end{aligned}$$

while

$$\pi(a_0 + b_0) = \pi(a_0 + b_{\ell-1}) = \pi(a_{k-1} + b_{\ell-1})$$

and

$$\pi(a_0 + b_i) = \pi(a_{k-1} + b_i)$$

for $i = 1, \ldots, \ell - 2$. Therefore, by inequality (4.16), we have

$$|\pi(A) + \pi(B)| \leq |A + B| - \ell \leq |\pi(A)| + |\pi(B)| - 2. \tag{4.17}$$

We can apply Kneser's theorem to the sumset $\pi(A) + \pi(B)$ in the group G. Let $H = H(\pi(A) + \pi(B)$ be the stabilizer of $\pi(A) + \pi(B))$. By Theorem 4.2, we have

$$|\pi(A) + \pi(B)| = |\pi(A) + H| + |\pi(B) + H| - |H|. \tag{4.18}$$

Since every subgroup of a cyclic group is cyclic, there is a divisor d of a_{k-1} such that $H = dG = d\mathbf{Z}/a_{k-1}\mathbf{Z}$. We shall prove that $d = 1$.

Let $\sigma : G \to G/H$ be the canonical homomorphism from G onto the quotient group G/H. We partition $A + B$ as follows:

$$A + B = C_1 \cup C_2,$$

where

$$\begin{aligned} C_1 &= \{c \in A + B : \sigma\pi(c) \in \sigma\pi(B)\} \\ &= \{c \in A + B : \pi(c) \in \pi(B) + H\} \end{aligned}$$

and

$$\begin{aligned} C_2 &= \{c \in A + B : \sigma\pi(c) \in \sigma\pi(A + B) \setminus \sigma\pi(B)\} \\ &= \{c \in A + B : \pi(c) \notin \pi(B) + H\}. \end{aligned}$$

Then $C_1 \cap C_2 = \emptyset$, and

$$|A + B| = |C_1| + |C_2|.$$

We shall estimate the cardinalities of the sets C_1 and C_2. Since

$$\pi(B) + H \subseteq \pi(A) + \pi(B) + H = \pi(A + B)$$

and

$$\pi(a_0 + b_i) = \pi(a_{k-1} + b_i) = \pi(b_i) \in \pi(B) \subseteq \pi(B) + H$$

for $i = 0, 1, \ldots, \ell - 1$, it follows (by the same argument used to derive (4.17)) that

$$\begin{aligned} |C_1| &= |\{c \in A + B : \pi(c) \in \pi(B) + H\}| \\ &\geq \ell + |\{\pi(c) \in \pi(A + B) : \pi(c) \in \pi(B) + H\}| \\ &= \ell + |\pi(B) + H| \\ &= \ell + |\sigma\pi(B)||H|. \end{aligned}$$

Next we estimate $|C_2|$. Let

$$r = |\sigma\pi(A + B) \setminus \sigma\pi(B)|.$$

It follows from (4.18) that

$$|\sigma\pi(A + B)| = |\sigma\pi(A)| + |\sigma\pi(B)| - 1$$

and so

$$r = |\sigma\pi(A)| - 1.$$

Choose $c_1, \ldots, c_r \in C_2$ such that

$$\sigma\pi(A + B) \setminus \sigma\pi(B) = \{\sigma\pi(c_1), \ldots, \sigma\pi(c_r)\},$$

and choose $a_i \in A$ and $b_i \in B$ such that

$$a_i + b_i = c_i$$

for $i = 1, \ldots, r$. For each $i = 1, \ldots, r$, we have

$$\begin{aligned} &|\{c \in A + B : \sigma\pi(c) = \sigma\pi(c_i)\}| \\ &\geq \quad |\{a \in A : \sigma\pi(a) = \sigma\pi(a_i)\} + \{b \in B : \sigma\pi(b) = \sigma\pi(b_i)\}| \\ &\geq \quad |\{a \in A : \sigma\pi(a) = \sigma\pi(a_i)\}| + |\{b \in B : \sigma\pi(b) = \sigma\pi(b_i)\}| - 1. \end{aligned}$$

Since

$$\begin{aligned} &|\{a \in A : \sigma\pi(a) = \sigma\pi(a_i)\}| \\ &\geq \quad |(\pi(a_i) + H) \cap \pi(A)| \\ &= \quad |\pi(a_i) + H| + |\pi(A)| - |(\pi(a_i) + H) \cup \pi(A)| \\ &\geq \quad |H| + |\pi(A)| - |\pi(A) + H| \end{aligned}$$

and

$$|\{b \in B : \sigma\pi(b) = \sigma\pi(b_i)\}| \geq |H| + |\pi(B)| - |\pi(B) + H|,$$

it follows from (4.18) that

$$\begin{aligned} &|\{c \in A + B : \sigma\pi(c) = \sigma\pi(c_i)\}| \\ &\geq \quad 2|H| + |\pi(A)| + |\pi(B)| - |\pi(A) + H| - |\pi(B) + H| - 1 \\ &= \quad |H| + |\pi(A)| + |\pi(B)| - |\pi(A + B)| - 1, \end{aligned}$$

and so

$$\begin{aligned}|C_2| &= \sum_{i=1}^{r} |\{c \in A + B : \sigma\pi(c) = \sigma\pi(c_i)\}| \\ &\geq r\left(|H| + |\pi(A)| + |\pi(B)| - |\pi(A+B)| - 1\right).\end{aligned}$$

Using our estimates for $|C_1|$ and $|C_2|$, we obtain

$$\begin{aligned}|A+B| &= |C_1| + |C_2| \\ &\geq \ell + |\sigma\pi(B)||H| + r\left(|H| + |\pi(A)| + |\pi(B)| - |\pi(A+B)| - 1\right) \\ &= \ell + |\sigma\pi(B)||H| + (|\sigma\pi(A)| - 1)|H| \\ &\quad + r\left(|\pi(A)| + |\pi(B)| - |\pi(A+B)| - 1\right) \\ &= \ell + |\sigma\pi(A+B)||H| + r\left(|\pi(A)| + |\pi(B)| - |\pi(A+B)| - 1\right) \\ &= \ell + |\pi(A+B)| + r\left(|\pi(A)| + |\pi(B)| - |\pi(A+B)| - 1\right).\end{aligned}$$

On the other hand, from (4.16) we have

$$|A+B| \leq |\pi(A)| + |\pi(B)| + \ell - 2.$$

Combining these upper and lower bounds for $|A+B|$, we obtain

$$\begin{aligned}&\ell + |\pi(A+B)| + r\left(|\pi(A)| + |\pi(B)| - |\pi(A+B)| - 1\right) \\ &\quad \leq |A+B| \\ &\quad \leq |\pi(A)| + |\pi(B)| + \ell - 2,\end{aligned}$$

and so

$$(r-1)(|\pi(A)| + |\pi(B)| - |\pi(A+B)|) \leq r - 2.$$

By (4.17), we have

$$|\pi(A+B)| \leq |\pi(A)| + |\pi(B)| - 2,$$

and so

$$2(r-1) \leq r - 2.$$

Therefore,

$$r = |\sigma\pi(A)| - 1 = 0.$$

Then $\sigma\pi(A) = H$ in G/H since $0 \in A$, and so $\pi(A) \subseteq H$, that is,

$$a_i \equiv 0 \pmod{d}$$

for every $a_i \in A$. Since $(a_1, \ldots, a_{k-1}) = 1$, we must have $d = 1$, hence $H = \mathbf{Z}/a_{k-1}\mathbf{Z} = G$ and

$$\pi(A+B) = \pi(A+B) + H = \mathbf{Z}/a_{k-1}\mathbf{Z}.$$

Suppose that $\delta = 0$. Then the congruence classes $\pi(a_0 + b_i)$ are pairwise distinct for $i = 0, 1, \ldots, \ell - 1$. Since $\pi(a_0 + b_i) = \pi(a_{k-1} + b_i)$ for $i = 0, 1, \ldots, \ell - 1$, it follows that there are at least two distinct integers in $A + B$ that belong to each of the ℓ congruence classes $\pi(a_0 + b_i)$, and there is at least one integer in $A + B$ in each of the remaining $a_{k-1} - \ell$ congruence classes in $\mathbf{Z}/a_{k-1}\mathbf{Z}$. Therefore,

$$|A + B| \geq 2\ell + (a_{k-1} - \ell) = a_{k-1} + \ell.$$

Similarly, if $\delta = 1$, then the $\ell - 1$ congruence classes $\pi(a_0 + b_i)$ are pairwise distinct for $i = 0, 1, \ldots, \ell - 2$. Since $\pi(a_0 + b_i) = \pi(a_{k-1} + b_i)$ for $i = 1, \ldots, \ell - 2$, and

$$\pi(a_0 + b_0) = \pi(a_{k-1} + b_0) = \pi(a_{k-1} + b_{\ell-1}) = \pi(0),$$

it follows that there are at least two distinct integers in $A + B$ that belong to each of the $\ell - 2$ congruence classes $\pi(a_0 + b_i)$ for $i = 1, \ldots, \ell - 2$, that there are at least three distinct integers in $A + B$ in the congruence class $\pi(0)$, and there is at least one integer in $A + B$ in each of the remaining $a_{k-1} - \ell + 1$ congruence classes. Therefore,

$$|A + B| \geq 2(\ell - 2) + 3 + (a_{k-1} - \ell + 1) = a_{k-1} + \ell.$$

This completes the proof.

Theorem 4.7 *Let $k, \ell \geq 2$, and let $A = \{a_0, a_1, \ldots, a_{k-1}\}$ and $B = \{b_0, b_1, \ldots, b_{\ell-1}\}$ be nonempty, finite sets of integers such that*

$$0 = a_0 < a_1 < \cdots < a_{k-1},$$

$$0 = b_0 < b_1 < \cdots < b_{\ell-1},$$

$$b_{\ell-1} \leq a_{k-1},$$

and

$$(a_1, \ldots, a_{k-1}, b_1, \ldots, b_{\ell-1}) = 1. \tag{4.19}$$

Let

$$\delta = \begin{cases} 0 & \text{if } b_{\ell-1} < a_{k-1} \\ 1 & \text{if } b_{\ell-1} = a_{k-1} \end{cases}$$

and let

$$m = \min(k, \ell - \delta).$$

Then

$$|A + B| \geq \min\left(a_{k-1} + \ell, k + \ell + m - 2\right). \tag{4.20}$$

Proof. If $(a_1, \ldots, a_{k-1}) = 1$, then inequality (4.20) follows immediately form Theorem 4.6.

Let

$$d = (a_1, \ldots, a_{k-1}) \geq 2.$$

For $i = 0, 1, \ldots, d-1$, let

$$B_i = \{b \in B : b \equiv i \pmod{d}\}$$

and let

$$\ell_i = |B_i| = |[0, a_{k-1} - 1] \cap B_i| + \delta_i,$$

where $\delta_i = 0$ for $i \neq 0$ and $\delta_0 = \delta$. Then $B_0 \neq \emptyset$ since $0 \in B$. Let s denote the number of nonempty sets B_i, or, equivalently, the number of congruence classes modulo d that contain at least one element of B. Then (4.19) implies that $B_i \neq \emptyset$ for some $i \neq 0$, and so $2 \leq s \leq d$. If $c \in A + B_i$, then $c \equiv i \pmod{d}$, and so the sumsets $A + B_i$ are pairwise disjoint. Moreover,

$$A + B = \bigcup_{\substack{i=0 \\ B_i \neq \emptyset}}^{d-1} (A + B_i).$$

It follows that

$$\begin{aligned} |A + B| &= \sum_{\substack{i=0 \\ B_i \neq \emptyset}}^{d-1} |A + B_i| \\ &\geq \sum_{\substack{i=0 \\ B_i \neq \emptyset}}^{d-1} (k + \ell_i - 1) \\ &= s(k-1) + \ell \\ &\geq 2k + \ell - 2 \\ &\geq k + \ell + m - 2 \\ &\geq \min(a_{k-1} + \ell, k + \ell + m - 2). \end{aligned}$$

This completes the proof of inequality (4.20).

Recall that the *diameter* of a set A is

$$\mathrm{diam}(A) = \sup(|a - a'| : a, a' \in A).$$

If A is finite and $A = \{a_0, a_1, \ldots, a_{k-1}\}$, where $a_0 < a_1 < \cdots, a_{k-1}$, then $\mathrm{diam}(A) = a_{k-1} - a_0$.

Theorem 4.8 *Let A and B be nonempty, finite sets of integers such that*

$$diam(B) \leq diam(A).$$

Let

$$\delta = \begin{cases} 0 & \text{if } diam(B) < diam(A) \\ 1 & \text{if } diam(B) = diam(A) \end{cases}$$

Let $|A| = k$, $|B| = \ell$, and $m = \min(k, \ell - \delta)$. If

$$|A + B| = k + \ell - 1 + b \leq k + \ell + m - 3,$$

then A and B are subsets of arithmetic progressions of length at most $k + b$ with the same common difference.

Proof. Let $A = \{a_0, a_1, \ldots, a_{k-1}\}$ and $B = \{b_0, b_1, \ldots, b_{\ell-1}\}$, where

$$a_0 < a_1 < \cdots < a_{k-1},$$
$$b_0 < b_1 < \cdots < b_{\ell-1},$$

and let

$$d = (a_1 - a_0, a_2 - a_0, \ldots, a_{k-1} - a_0, b_1 - b_0, \ldots, b_{k-1} - b_0).$$

Let

$$a_i^{(N)} = \frac{a_i - a_0}{d} \qquad \text{for } i = 0, 1, \ldots, k-1$$

and

$$b_j^{(N)} = \frac{b_j - b_0}{d} \qquad \text{for } j = 0, 1, \ldots, \ell - 1.$$

Let

$$A^{(N)} = \left\{a_i^{(N)} : i = 0, 1, \ldots, k-1\right\}$$

and

$$B^{(N)} = \left\{b_j^{(N)} : j = 0, 1, \ldots, \ell-1\right\}.$$

Then

$$\min(A^{(N)}) = \min(B^{(N)}) = 0$$

and

$$(a_1^{(N)}, \ldots, a_{k-1}^{(N)}, b_1^{(N)}, \ldots, b_{\ell-1}^{(N)}) = 1.$$

Since $\operatorname{diam}(B) \leq \operatorname{diam}(A)$, it follows that

$$b_{\ell-1}^{(N)} \leq a_{k-1}^{(N)}.$$

The sets $A^{(N)}$ and $B^{(N)}$ are constructed from A and B, respectively, by affine transformations, and

$$|A^{(N)} + B^{(N)}| = |A + B| \leq k + \ell + m - 3.$$

It follows from Theorem 4.7 that

$$|A^{(N)} + B^{(N)}| \geq a_{k-1}^{(N)} + \ell,$$

or, equivalently,

$$b_{\ell-1}^{(N)} \leq a_{k-1}^{(N)} \leq |A^{(N)} + B^{(N)}| - \ell = k - 1 + b.$$

Since $a_i = a_0 + a_i^{(N)} d$ for $i = 0, 1, \ldots, k-1$, it follows that

$$A \subseteq \left\{a_0 + xd : x = 0, \ldots, a_{k-1}^{(N)}\right\} \subseteq \{a_0 + xd : x = 0, \ldots, k - 1 + b\}.$$

Similarly,

$$B \subseteq \left\{b_0 + yd : y = 0, \ldots, b_{\ell-1}^{(N)}\right\} \subseteq \{b_0 + yd : y = 0, \ldots, k - 1 + b\}.$$

This completes the proof.

Theorem 4.9 *Let $k, \ell \geq 2$, and let $A = \{a_0, a_1, \ldots, a_{k-1}\}$ and $B = \{b_0, b_1, \ldots, b_{\ell-1}\}$ be nonempty, finite sets of integers such that*

$$0 = a_0 < a_1 < \cdots < a_{k-1},$$

$$0 = b_0 < b_1 < \cdots < b_{\ell-1},$$

$$b_{\ell-1} \leq a_{k-1},$$

$$d = (a_1, \ldots, a_{k-1}) > 1,$$

and

$$(a_1, \ldots, a_{k-1}, b_1, \ldots, b_{\ell-1}) = 1.$$

Let

$$\delta = \begin{cases} 0 & \text{if } b_{\ell-1} < a_{k-1} \\ 1 & \text{if } b_{\ell-1} = a_{k-1}. \end{cases}$$

If

$$a_{k-1} \leq k + \ell - \delta - 2, \tag{4.21}$$

then

$$|A + B| \geq a_{k-1} + \ell.$$

Proof. Since d divides a_i for all $i = 1, \ldots, k-1$, we have

$$d(k-1) \leq a_{k-1}. \tag{4.22}$$

The interval $[0, a_{k-1} - 1]$ contains exactly a_{k-1}/d integers in each congruence class modulo d. Let s denote the number of congruence classes modulo d that contain at least one element of B. Since

$$B \subseteq [0, a_{k-1} - 1 + \delta],$$

it follows that

$$\ell = |B| \leq \frac{s a_{k-1}}{d} + \delta.$$

Inequalities (4.21) and (4.22) imply that

$$\begin{aligned} a_{k-1} &\leq k + \ell - \delta - 2 \\ &\leq k + \frac{s a_{k-1}}{d} - 2, \end{aligned}$$

and so

$$d(k-1)(d-s) \leq a_{k-1}(d-s) \leq d(k-2).$$

It follows that

$$s = d,$$

that is, B intersects every congruence class modulo d. Let

$$B_i = \{b \in B : b \equiv i \pmod{d}\}$$

and let

$$\ell_i = |B_i| = |[0, a_{k-1} - 1] \cap B_i| + \delta_i,$$

where $\delta_i = 0$ for $i \neq 0$ and $\delta_0 = \delta$. By (4.21),

$$|[0, a_{k-1} - 1] \setminus B| = a_{k-1} - \ell + \delta \leq k - 2,$$

and so

$$\begin{aligned} \ell_i &= |[0, a_{k-1} - 1] \cap B_i| + \delta_i \\ &\geq \frac{a_{k-1}}{d} - |[0, a_{k-1} - 1] \setminus B| + \delta_i \\ &\geq \frac{a_{k-1}}{d} - k + 2 + \delta_i. \end{aligned}$$

Therefore,

$$\min\left(\frac{a_{k-1}}{d}, k + \ell_i - \delta_i - 2\right) = \frac{a_{k-1}}{d} \tag{4.23}$$

for $i = 0, 1, \ldots, d - 1$.

Let $b_{i,0} = \min(B_i)$ for $i = 0, \ldots, d - 1$. Let

$$A^{(N)} = \left\{\frac{a}{d} : a \in A\right\}$$

and let

$$B_i^{(N)} = \left\{\frac{b - b_{i,0}}{d} : b \in B_i\right\}.$$

Since the elements of $A^{(N)}$ are relatively prime, and since

$$\min\left(A^{(N)} \cup B_i^{(N)}\right) = 0$$

and

$$\max\left(A^{(N)} \cup B_i^{(N)}\right) = \frac{a_{k-1}}{d},$$

it follows from Theorem 4.6 and (4.23) that

$$\begin{aligned} |A + B_i| &= |A^{(N)} + B_i^{(N)}| \\ &\geq \min\left(\frac{a_{k-1}}{d}, k + \ell_i - \delta_i - 2\right) + \ell_i \\ &= \frac{a_{k-1}}{d} + \ell_i. \end{aligned}$$

Since the sets $A + B_i$ are pairwise disjoint for $i = 0, 1, \ldots, d - 1$, and $A \cup B = \cup_{i=0}^{d-1}(A + B_i)$, we have

$$\begin{aligned} |A + B| &= \sum_{i=0}^{d-1} |A + B_i| \\ &\geq \sum_{i=0}^{d-1} \left(\frac{a_{k-1}}{d} + \ell_i\right) \\ &= a_{k-1} + \ell. \end{aligned}$$

This completes the proof.

4.4 Application: Bases for finite and σ-finite groups

Let G be an abelian group, written additively, and let $A \subseteq G$. The set A is a *basis of order* h for G if $hA = G$.

Theorem 4.10 *Let G be a finite abelian group, and let A be a nonempty subset of G. Let G' be the subgroup of G generated by A. Then A is a basis for G' of order at most*

$$\max\left(2, \frac{2|G'|}{|A|} - 1\right).$$

Proof. Without loss of generality we can assume that $G' = G$. Since $h(A - \{g_0\}) = hA - \{hg_0\}$ for any $g_0 \in G$, it follows that $hA = G$ if and only if $h(A - \{g_0\}) = G$, and so we can assume that $0 \in A$.

Since A generates G and G is finite and abelian, it follows that A is a basis for G of some finite order. Let h be the smallest positive integer such that $hA = G$. If $h = 1$ or 2, we are done. Suppose that

$$h \geq 3.$$

Then

$$(h-1)A = A + (h-2)A \neq G,$$

and so, by Lemma 2.2,

$$|G| \geq |A| + |(h-2)A|.$$

Let $H_{h-2} = H((h-2)A)$ be the stabilizer of $(h-2)A$. Then H_{h-2} is the largest subgroup of G such that

$$(h-2)A + H_{h-2} = (h-2)A.$$

For some $r \geq 1$, the set $A + H_{h-2}$ is a union of r pairwise disjoint cosets of H_{h-2}, Since $0 \in A \cap H_{h-2}$, we have

$$H_{h-2} \subseteq A + H_{h-2}$$

and

$$A \subseteq A + H_{h-2}.$$

Therefore,

$$|A| \leq |A + H_{h-2}| = r|H_{h-2}|. \tag{4.24}$$

If $r = 1$, then

$$A + H_{h-2} = H_{h-2}.$$

Since H_{h-2} is a subgroup, it follows that

$$G = hA \subseteq H_{h-2} \subseteq G,$$

and so $H_{h-2} = G$. This implies that $(h-2)A = G$, which contradicts the minimality of h. Therefore,

$$r \geq 2.$$

By Theorem 4.5 and (4.24), we have

$$\begin{aligned}
|(h-2)A| &\geq (h-2)|A + H_{h-2}| - (h-3)|H_{h-2}| \\
&= (h-2)r|H_{h-2}| - (h-3)|H_{h-2}| \\
&= \left((h-2) - \frac{h-3}{r}\right) r|H_{h-2}| \\
&\geq \left((h-2) - \frac{h-3}{r}\right) |A| \\
&\geq \left((h-2) - \frac{h-3}{2}\right) |A| \\
&= \left(\frac{h-1}{2}\right) |A|,
\end{aligned}$$

and so

$$|G| \geq |A| + |(h-2)A| \geq \left(\frac{h+1}{2}\right) |A|.$$

Solving for h, we obtain

$$h \leq \frac{2|G|}{|A|} - 1.$$

This completes the proof.

Exercise 10 shows that the upper bound in Theorem 4.10 is sharp.

Let G be a countable abelian torsion group, and let $G_1 \subseteq G_2 \subseteq \cdots$ be an increasing sequence of finite subgroups of G. Then G is *σ-finite with respect to the sequence $\{G_n\}$* if

$$G = \bigcup_{n=1}^{\infty} G_n.$$

Let A be a subset of G, and define $A_n = A \cap G_n$. Then $A = \bigcup_{n=1}^{\infty} A_n$. The set A is called a *σ-basis of order h* for G with respect to the sequence $\{G_n\}$ if $hA_n = G_n$ for $n = 1, 2, \ldots$. Clearly, every σ-basis of order h for G is a basis of order h for G. The converse is not true (see Exercise 11.)

Let $G = \bigcup_{n=1}^{\infty} G_n$ be a σ-finite abelian group, and let A be a subset of G. Let $A_n = A \cap G_n$. The *upper asymptotic density* of the set A is defined by

$$d_U^{(G)}(A) = d_U(A) = \limsup_{n \to \infty} \frac{|A_n|}{|G_n|}.$$

Clearly, $0 \leq d_U(A) \leq 1$ for every subset A of G.

Theorem 4.11 *Let $G = \bigcup_{n=1}^{\infty} G_n$ be a σ-finite abelian group. Let A be a subset of G such that $0 \in A$, and let G' be the subgroup of G generated by A. If $d_U(A) > 0$, then A is a basis for G' of order at most*

$$\max\left(2, \frac{2}{d_U(A)} - 1\right).$$

Proof. Let $A_n = A \cap G_n$, and let G'_n be the subgroup generated by A_n for $n = 1, 2, \ldots$. Then

$$A_n \subseteq G'_n \subseteq G' \cap G_n$$

and

$$A \cap G'_n = A \cap (G'_n \cap G_n) = (A \cap G_n) \cap G'_n = A_n \cap G'_n = A_n.$$

If $g \in G'$, then g is generated by some finite subset of A. Since this finite subset belongs to A_n for some n, it follows that $g \in G'_n$ and so

$$G' = \bigcup_{n=1}^{\infty} G'_n.$$

Thus, the group G' is a σ-finite with respect to the sequence $\{G'_n\}$. Let

$$d_U^{(G)}(A) = \limsup_{n\to\infty} \frac{|A_n|}{|G_n|}$$

and

$$d_U^{(G')}(A) = \limsup_{n\to\infty} \frac{|A_n|}{|G'_n|}.$$

Since $G'_n \subseteq G_n$, it follows that

$$0 < d_U^{(G)}(A) \leq d_U^{(G')}(A).$$

Choose ε such that

$$0 < \varepsilon < d_U^{(G')}(A).$$

It follows from the definition of upper asymptotic density that there exists an infinite sequence $n_1 < n_2 < \cdots$ of positive integers such that

$$\frac{|A_{n_i}|}{|G'_{n_i}|} > d_U^{(G')}(A) - \varepsilon > 0$$

for all $i = 1, 2, \ldots$. Moreover,

$$G' = \cup_{i=1}^{\infty} G'_{n_i}.$$

If $g \in G'$, then $g \in G'_{n_i}$ for some i. By Theorem 4.10,

$$g \in hA_{n_i} \subseteq hA$$

for some h such that

$$h \leq \max\left(2, \frac{2|G'_{n_i}|}{|A_{n_i}|} - 1\right) \leq \max\left(2, \frac{2}{d_U^{(G)}(A) - \varepsilon} - 1\right).$$

Therefore,

$$hA = G'$$

for some

$$h \leq \max\left(2, \frac{2}{d_U^{(G')}(A) - \varepsilon} - 1\right).$$

Since this is true for all sufficiently small positive ε, it follows that A is a basis of order h for G, where

$$h \leq \max\left(2, \frac{2}{d_U^{(G')}(A)} - 1\right) \leq \max\left(2, \frac{2}{d_U^{(G)}(A)} - 1\right).$$

This completes the proof.

Exercise 12 shows that the upper bound in Theorem 4.11 is sharp.

4.5 Notes

The proof of Kneser's theorem [76] for abelian groups in Section 4.2 follows Kemperman [74]. Mann [84] gives a condensed proof of this result.

If A and B be nonempty, finite subsets of an abelian group such that $|A + B| < |A| + |B|$, then (A, B) is called a *critical pair*. Vosper (Theorem 2.7) classified the critical pairs in the finite cyclic groups $\mathbf{Z}/p\mathbf{Z}$, where p is a prime number. It is an open problem to classify the critical pairs of subsets of an arbitrary abelian group. Important partial results are due to Kemperman [74], who used Kneser's addition theorem for abelian groups to study this problem, and to Hamidoune [65, 66], who used graph theory.

There are a few results about critical pairs in nonabelian groups. Diderrich [30] extended Kneser's theorem to certain special pairs of subsets of nonabelian groups. Hamidoune [62] showed that Diderrich's result followed from Kneser's theorem. Brailovsky and Freiman [12] completely classified the critical pairs in arbitrary torsion-free groups. Hamidoune [63] found a short proof of a theorem that includes the Brailovsky–Freiman result as a special case.

The results in Section 4.3 on inverse theorem for sumsets of the form $A + B$ were originally obtained by Freiman [52]. Another version of Freiman's proof is due to Steinig [121]. The proofs in this chapter use Kneser's theorem and are due to Lev and Smeliansky [81]. Similar proofs were obtained by Hamidoune [64]. Theorem 4.9 has been applied recently to different problems in number theory, for example, the structure theory of sum-free sets (see Deshouillers, Freiman, Sós, and Temkin [26] and Freiman [57]).

Lev [80] also used Kneser's theorem to prove that if $A = \{a_0, a_1, \ldots, a_{k-1}\}$ is a finite set of integers in normal form, then

$$|hA| \geq |(h-1)A| + \min(a_{k-1}, h(k-2)+1).$$

for all $h \geq 2$. For $h = 2$, this is Theorem 1.15. For large h this is weaker than Theorem 1.1.

Theorem 4.11 is due independently to Deshouillers and Wirsing [27] and Hamidoune and Rödseth [68]. It generalizes a result of Jia and Nathanson [73] for arbitrary σ-finite abelian groups, and of Cherly and Deshouillers [17] in the special case of the σ-finite group $F_q[x]$ of polynomials over a finite field. Hamidoune and Rödseth [68] prove their theorem for σ-finite groups that are not necessarily abelian.

4.6 Exercises

1. Let A and B be subsets of an abelian group G. Prove that $H(A) \subseteq H(A+B)$.

2. Let A be a nonempty subset of an abelian group G. Prove that A is a subgroup if and only if $H(A) = A$.

3. Let G be an abelian group, and let $A_1, A_2, \ldots, A_h$ be finite, nonempty subsets of G. Prove that if $A_1 + \cdots + A_h$ is not periodic, then

$$|A_1 + \cdots + A_h| \geq |A_1| + \cdots + |A_h| - (h-1).$$

4. Let G be an abelian group. For any subgroup H of G and for any subset S of G, we define

$$S/H = \{s + H \mid s \in S\} \subseteq G/H.$$

Let A and B be subsets of G, and let $H = H(A+B)$. Prove that either $|A+B| \geq |A| + |B|$ or

$$|(A+B)/H| = |A/H| + |B/H| - 1.$$

5. Prove that Theorem 4.1 implies the Cauchy–Davenport theorem.

6. Prove that Kneser's theorem (Theorem 4.2) implies Chowla's theorem (Theorem 2.1).

7. Let G be an abelian group. For $A, B \subseteq G$, let $H(A+B)$ be the stabilizer of the sumset $A + B$ in G. Suppose that

$$|A+B| \geq |A| + |B| - |H(A+B)|,$$

holds for all finite, nonempty subsets A, B of G. Show that this implies Kneser's theorem (Theorem 4.2).

8. Prove that Theorem 4.6 implies Theorem 1.16.

9. Let A and B be finite, nonempty sets of integers, and let $|A| = k$ and $|B| = \ell$. Prove that if
$$|A + B| \leq k + \ell + \min(k, \ell) - 4,$$
then A and B are subsets of arithmetic progressions of length at most $|A + B| - \min(k, \ell) + 1$ and with the same common difference.

10. Let $h \geq 2$ and $m = h + 1$. Let $G = \mathbf{Z}/m\mathbf{Z}$ and $A = \{0, 1\} \subseteq G$. Show that A is a basis for G of exact order
$$h = \frac{2|G|}{|A|} - 1.$$
This example shows that the upper bound in Theorem 4.10 is best possible.

11. Let q be a power of the prime p, and let $G = \mathbf{F}_q[x]$ denote the ring of polynomials with coefficients in the finite field $\mathbf{F}_q$. Let G_n be the subgroup of $\mathbf{F}_q[x]$ consisting of all polynomials f of degree at most n. Then $\mathbf{F}_q[x] = \bigcup_{n=1}^{\infty} G_n$. Choose $N \geq 2$, and let
$$A = \{0\} \cup \{f \in \mathbf{F}[x] \mid \deg f \geq N\}.$$
Prove that A is a basis of order 2 for G, but not a σ-basis of order 2 for G.

12. Let $m \geq 3$, and let $G = \mathbf{Z}_m[x]$ be the additive abelian group of polynomials with coefficients in the ring $\mathbf{Z}_m$ of integers modulo m. Then $G = \bigcup_{n=1}^{\infty} G_n$, where G_n is the subgroup of G consisting of all polynomials of degree less than n. Let A be the subset of G consisting of all polynomials with constant term 0 or 1. Prove that $d_U(A) = 2/n$ and that A is a σ-basis for G of exact order
$$h = m - 1 = \frac{2}{d_U(A)} - 1.$$
This example shows that the upper bound in Theorem 4.11 is best possible.

13. Prove that there exists a basis A of order 2 for $\mathbf{Z}$ such that every integer has a *unique* representation as the sum of two elements of A.
Hint: Construct the set A inductively. Let $a_1 = 0$. Suppose that integers $a_1, \ldots, a_k$ have been chosen so that the $k(k+1)/2$ sums $a_i + a_j$ are distinct for $1 \leq i \leq j \leq k$. Choose n so that $n \neq a_i + a_j$ and $|n|$ is minimal. Let $a_{k+1} = n + b$ and $a_{k+2} = -b$. Then $n = (n + b) - b = a_{k+1} + a_{k+2}$. Show that it is possible to choose b so large that the $(k+2)(k+3)/2$ sums $a_i + a_j$ are distinct for $1 \leq i \leq j \leq k + 2$.

5

Sums of vectors in Euclidean space

5.1 Small sumsets and hyperplanes

The "philosophy" of inverse problems is that if a finite set A has a small sumset $2A$, then A must have "structure." We have already obtained simple results of this kind, for example, Theorems 1.16 and 2.7. In Chapter 8, we shall prove Freiman's theorem, which states that if a finite set A satisfies $|2A| \leq c|A|$, then A must have an arithmetical structure in the sense that A is a large subset of a multi-dimensional arithmetic progression. In this chapter, we shall prove that if A is a finite set of vectors in Euclidean space $\mathbf{R}^n$ and if the cardinality of the sumset $2A$ is very small, then A will have a geometrical structure. More precisely, if $|2A| \leq c|A|$, where $1 < c < 2^n$, then a positive proportion of the elements of A must lie on a hyperplane, or, equivalently, A is a subset of a bounded number of parallel hyperplanes. This result is independent of Freiman's theorem and, indeed, played an essential role in the original proof of that theorem.

Let $n \geq 2$, and let V be an n-dimensional Euclidean space with inner product $(\ ,\ )$. Let h be a nonzero vector in V, and let $\gamma \in \mathbf{R}$. The *hyperplane* H defined by h and γ is the set

$$H = \{v \in V \mid (h, v) = \gamma\}.$$

The vector h is called a *normal vector* to the hyperplane H. Let A be a finite subset of V. Denote the cardinality of A by $|A|$. The *h-fold sumset* hA is the set

$$hA = \{a_1 + a_2 + \ldots + a_h \mid a_i \in A \text{ for } i = 1, 2, \ldots, h\}.$$

Theorem 5.1 *Let $n \geq 2$, and let*

$$1 < c < 2^n.$$

There exist constants $k_0^* = k_0^*(n, c)$ and $\varepsilon_0^* = \varepsilon_0^*(n, c) > 0$ such that, if A is a finite subset of an n-dimensional Euclidean space V and if A satisfies

$$|A| \geq k_0^*$$

and

$$|2A| \leq c|A|,$$

then there exists a hyperplane H in V such that

$$|A \cap H| > \varepsilon_0^*|A|.$$

The following example, in the case $h = 2$, shows that the upper bound for c in Theorem 5.1 is best possible.

Theorem 5.2 *Let $n \geq 2$, and let V be an n-dimensional Euclidean space. Let $h \geq 2$. For any numbers k_0 and $\varepsilon_0 > 0$, there exists a finite subset A of V such that*

$$|A| \geq k_0$$

and

$$|hA| < h^n|A|,$$

but

$$|A \cap H| \leq \varepsilon_0|A|$$

for every hyperplane H in V.

Proof. Choose an orthonormal basis $\{\mathbf{e}_1, \ldots, \mathbf{e}_n\}$ for V, and let $u, v \in V$, where $u = \sum_{i=1}^n u_i\mathbf{e}_i$ and $v = \sum_{i=1}^n v_i\mathbf{e}_i$. Then $(u, v) = \sum_{i=1}^n u_iv_i$ is the inner product on V. Choose k_0 and $\varepsilon_0 > 0$, and let $t \in \mathbf{Z}$ satisfy

$$t \geq \max\left\{k_0^{1/n}, \varepsilon_0^{-1}\right\}.$$

Let

$$A = \left\{\sum_{i=1}^n v_i\mathbf{e}_i \in V \;\middle|\; v_i \in \{0, 1, \ldots, t-1\} \text{ for } i = 1, 2, \ldots, n\right\}.$$

Then

$$|A| = t^n \geq k_0$$

and

$$hA = \left\{\sum_{i=1}^n v_i\mathbf{e}_i \in V \;\middle|\; v_i \in \{0, 1, \ldots, ht-h\} \text{ for } i = 1, 2, \ldots, n\right\}.$$

Therefore,

$$|hA| = (ht - h + 1)^n < h^nt^n = h^n|A|.$$

Let $h = \sum_{i=1}^{n} h_i \mathbf{e}_i \neq 0$ and let $\gamma \in \mathbf{R}$. Let H be the hyperplane defined by h and γ. Then $h_j \neq 0$ for some j. If $v = \sum_{i=1}^{n} v_i \mathbf{e}_i \in A \cap H$, then $\sum_{i=1}^{n} h_i v_i = \gamma$, and the integer v_j is uniquely determined by the $n-1$ integers $v_1, \ldots, v_{j-1}, v_{j+1}, \ldots, v_n$. Since $v_i \in \{0, 1, \ldots, t-1\}$ for $i = 1, \ldots, n$, it follows that

$$|A \cap H| \leq t^{n-1} \leq \varepsilon_0 t^n = \varepsilon_0 |A|.$$

This completes the proof.

5.2 Linearly independent hyperplanes

Let $n \geq 2$, and let V be an n-dimensional Euclidean space with inner product $(\ , \ )$. Let h be a nonzero vector in V, and let $\gamma \in \mathbf{R}$. Define H, $H^{(+1)}$, and $H^{(-1)}$ as follows:

$$\begin{aligned} H &= \{v \in V \mid (h, v) = \gamma\} \\ H^{(+1)} &= \{v \in V \mid (h, v) > \gamma\} \\ H^{(-1)} &= \{v \in V \mid (h, v) < \gamma\}. \end{aligned}$$

The sets $H^{(+1)}$ and $H^{(-1)}$ are, respectively, the *upper and lower open half-spaces* determined by H. The vector h is called a *normal vector* to the hyperplane H. If $0 \in H$, then H is an $(n-1)$-dimensional subspace of V. Note that $0 \in H$ if and only if $\gamma = 0$.

The set K in $\mathbf{R}^n$ is *convex* if $\mathbf{a}, \mathbf{b} \in K$ implies that $t\mathbf{a} + (1-t)\mathbf{b} \in K$ for all $t \in [0, 1]$. The sets H, $H^{(+1)}$, and $H^{(-1)}$ are convex. For any subset S of a Euclidean space V, the *convex hull* of S, denoted $\mathrm{conv}(S)$, is the smallest convex subset of V that contains S. Since the intersection of convex sets is convex, it follows that the convex hull of S is the intersection of all convex sets containing S. This intersection is nonempty since the Euclidean space V is convex and contains S.

Let $H_1, \ldots, H_m$ be hyperplanes, and let $H^* = \bigcup_{i=1}^{m} H_i$. Let $\{1, -1\}^m$ denote the set of all m-tuples $(\mu_1, \ldots, \mu_m)$ such that $\mu_i \in \{1, -1\}$ for $i = 1, \ldots, m$. For $(\mu_1, \ldots \mu_m) \in \{1, -1\}^m$, let

$$H(\mu_1, \ldots, \mu_m) = \bigcap_{i=1}^{m} H_i^{(\mu_i)}.$$

The 2^m sets $H(\mu_1, \ldots, \mu_m)$ are pairwise disjoint, and

$$V \setminus H^* = \bigcup_{(\mu_1, \ldots, \mu_m) \in \{1, -1\}^m} H(\mu_1, \ldots, \mu_m).$$

Let V be an n-dimensional Euclidean space, and let $H_1, \ldots, H_m$ be hyperplanes with normal vectors $h_1, \ldots, h_m$, respectively. The hyperplanes $H_1, \ldots, H_m$ are *linearly independent* if the vectors $h_1, \ldots, h_m$ are linearly independent. The hyperplanes $H_1, \ldots, H_m$ are *linearly dependent* if the vectors $h_1, \ldots, h_m$ are linearly dependent.

Lemma 5.1 *Let $H_1, \ldots, H_m$ be hyperplanes in an n-dimensional Euclidean space V. Suppose that $0 \in H_i$ for all $i = 1, \ldots, m$. The hyperplanes $H_1, \ldots, H_m$ are linearly independent if and only if*

$$H(\mu_1, \ldots, \mu_m) \neq \emptyset$$

for all $(\mu_1, \ldots, \mu_m) \in \{1, -1\}^m$.

Proof. There exist nonzero vectors $h_1, \ldots, h_m$ in V such that

$$H_i = \left\{ v \in V \mid (h_i, v) = 0 \right\}$$

for $i = 1, \ldots, m$. Suppose that $H_1, \ldots, H_m$ are linearly independent hyperplanes. Then the vectors $h_1, \ldots, h_m$ are independent, and there exists a dual set of vectors $h_1^*, \ldots, h_m^*$ such that

$$\left(h_i, h_j^*\right) = \delta_{i,j} = \begin{cases} 1 \text{ if } i = j \\ 0 \text{ if } i \neq j \end{cases}$$

for $i, j = 1, \ldots, m$. Let $(\mu_1, \ldots, \mu_m) \in \{1, -1\}^m$. Let

$$v = \sum_{j=1}^{m} \mu_j h_j^* \in V.$$

Then

$$(h_i, v) = \sum_{j=1}^{m} \mu_j (h_i, h_j^*) = \mu_i$$

and so

$$v \in H_i^{(\mu_i)}$$

for all $i = 1, \ldots, m$. It follows that $H(\mu_1, \ldots \mu_m) \neq \emptyset$.

Suppose that $H_1, \ldots H_m$ are linearly dependent hyperplanes. Then the vectors $h_1, \ldots, h_m$ are dependent, and there exist scalars $\alpha_1, \ldots, \alpha_m$ not all zero such that $\sum_{i=1}^{m} \alpha_i h_i = 0$. Define $(\mu_1, \ldots, \mu_m) \in \{1, -1\}^m$ by

$$\mu_i = \begin{cases} +1 & \text{if } \alpha_i > 0 \\ -1 & \text{if } \alpha_i \leq 0. \end{cases}$$

Then $\mu_i \alpha_i \geq 0$ for all $i = 1, \ldots, m$, and $\mu_j \alpha_j > 0$ for some j.

We shall show that $H(\mu_1, \ldots, \mu_m) = \emptyset$. If not, choose $v \in H(\mu_1, \ldots, \mu_m)$. Then $v \in H_i^{(\mu_i)}$ implies that

$$\alpha_i (h_i, v) \geq 0$$

for $i = 1, \ldots, m$, and

$$\alpha_j (h_j, v) > 0.$$

It follows that

$$0 = (0, v) = \left(\sum_{i=1}^{m} \alpha_i h_i, v \right) = \sum_{i=1}^{m} \alpha_i (h_i, v) > 0,$$

which is impossible. This proves that $H(\mu_1, \ldots, \mu_m) = \emptyset$.

Lemma 5.2 *Let V be an n-dimensional Euclidean space, and let $H_1, \ldots, H_r$ be linearly independent hyperplanes in V such that $0 \in H_i$ for $i = 1, \ldots, r$. Then*

$$\dim\left(\bigcap_{i=1}^{r} H_i\right) = n - r.$$

In particular, if $r = n$, then

$$\bigcap_{i=1}^{n} H_i = \{0\}.$$

Proof. Let $h_1, \ldots, h_r$ be normal vectors for the hyperplanes $H_1, \ldots H_r$, respectively. Then the set of vectors $\{h_1, \ldots, h_r\}$ is linearly independent, and $(h_i, v) = 0$ for all $v \in H_i$. Let $W = \bigcap_{i=1}^{r} H_i$, and let

$$W^{\perp} = \left\{v \in V \mid (v, w) = 0 \text{ for all } w \in W\right\}.$$

Since $h_i \in W^{\perp}$ for $i = 1, \ldots, r$, it follows that $\dim(W^{\perp}) \geq r$, and so

$$\dim(W) = n - \dim(W^{\perp}) \leq n - r.$$

We shall prove that $\dim(W) \geq n - r$ by induction on r. If $r = 1$, then $W = H_1$ and $\dim(W) = \dim(H_1) = n - 1$. Let $2 \leq r \leq n$, and assume that the assertion is true for $r - 1$. Let $W' = \bigcap_{i=1}^{r-1} H_i$. Then $\dim(W') \geq n - r + 1$. Since $W = W' \cap H_r$, it follows that

$$\begin{aligned}
\dim(W) &= \dim(W' \cap H_r) \\
&= \dim(W') + \dim(H_r) - \dim\left(W' + H_r\right) \\
&\geq \dim(W') + \dim(H_r) - \dim V \\
&\geq (n - r + 1) + (n - 1) - n \\
&= n - r.
\end{aligned}$$

Therefore, $\dim(W) = n - r$. If $r = n$, then $\dim\left(\bigcap_{i=1}^{n} H_i\right) = 0$, and so $\bigcap_{i=1}^{n} H_i = \{0\}$.

Lemma 5.3 *Let V be an n-dimensional Euclidean space and let $H_1, \ldots, H_m$ be hyperplanes in V such that $0 \in H_i$ for $i = 1, \ldots, m$. Let S be a subset of V such that*

$$S \cap H(\mu_1, \ldots, \mu_m) \neq \emptyset \tag{5.1}$$

for all $(\mu_1, \ldots, \mu_m) \in \{1, -1\}^m$. Then

$$conv(S) \bigcap \left(\bigcap_{i=1}^{m} H_i\right) \neq \emptyset.$$

Proof. By induction on m. Let $m = 1$, and let h_1 be a normal vector for the hyperplane H_1. By (5.1), there exist vectors $s_1 \in S \cap H_1^{(+1)}$ and $s_2 \in S \cap H_1^{(-1)}$ such that

$$(h_1, s_1) = \alpha_1 > 0$$

and

$$(h_1, s_2) = -\alpha_2 < 0.$$

Then

$$s = \left(\frac{\alpha_2}{\alpha_1 + \alpha_2}\right) s_1 + \left(\frac{\alpha_1}{\alpha_1 + \alpha_2}\right) s_2 \in \mathrm{conv}(S).$$

Since

$$(h_1, s) = \frac{\alpha_2(h_1, s_1)}{\alpha_1 + \alpha_2} + \frac{\alpha_1(h_1, s_2)}{\alpha_1 + \alpha_2} = \frac{\alpha_1\alpha_2 - \alpha_1\alpha_2}{\alpha_1 + \alpha_2} = 0,$$

it follows that $s \in H_1$, and so

$$\mathrm{conv}(S) \cap H_1 \neq \emptyset.$$

Let $m \geq 2$, and suppose that the lemma holds for $m - 1$. Define $S^{(+1)}$ and $S^{(-1)}$ by

$$S^{(+1)} = S \cap H_m^{(+1)}$$

and

$$S^{(-1)} = S \cap H_m^{(-1)}.$$

Then (5.1) implies that

$$S^{(+1)} \cap \left(\bigcap_{i=1}^{m-1} H_i^{(\mu_i)}\right) = S \cap H(\mu_1, \ldots, \mu_{m-1}, +1) \neq \emptyset$$

and

$$S^{(-1)} \cap \left(\bigcap_{i=1}^{m-1} H_i^{(\mu_i)}\right) = S \cap H(\mu_1, \ldots, \mu_{m-1}, -1) \neq \emptyset$$

for all $(\mu_1, \ldots, \mu_{m-1}) \in \{1, -1\}^{m-1}$. By the induction hypothesis, the lemma holds for the $m - 1$ hyperplanes $H_1, \ldots, H_{m-1}$, and so

$$\mathrm{conv}\left(S^{(+1)}\right) \cap \left(\bigcap_{i=1}^{m-1} H_i\right) \neq \emptyset.$$

This means that there exist vectors $s_1, \ldots, s_k \in S^{(+1)}$ and scalars $\alpha_1, \ldots \alpha_k \in \mathbf{R}$ with $\alpha_i \geq 0$ for $i = 1, \ldots, k$ and $\alpha_1 + \ldots + \alpha_k = 1$ such that

$$s^{(+1)} = \sum_{i=1}^{k} \alpha_i s_i \in \mathrm{conv}(S^{(+1)}) \cap \left(\bigcap_{i=1}^{m-1} H_i\right) \subseteq \mathrm{conv}(S) \cap \left(\bigcap_{i=1}^{m-1} H_i\right).$$

Moreover, $s^{(+1)} \in H_m^{(+1)}$ since $s_i \in H_m^{(+1)}$ for $i = 1, \ldots, k$ and $H_m^{(+1)}$ is convex. Similarly, there exists

$$s^{(-1)} \in \mathrm{conv}\left(S^{(-1)}\right) \cap \left(\bigcap_{i=1}^{m-1} H_i\right) \subseteq \mathrm{conv}(S) \cap \left(\bigcap_{i=1}^{m-1} H_i\right),$$

and $s^{(-1)} \in H_m^{(-1)}$.

Let $T = \{s^{(+1)}, s^{(-1)}\}$. Then

$$T \subseteq \operatorname{conv}(S) \cap \left(\bigcap_{i=1}^{m-1} H_i \right),$$

and so

$$\operatorname{conv}(T) \subseteq \operatorname{conv}(S) \cap \left(\bigcap_{i=1}^{m-1} H_i \right).$$

Since $T \cap H_m^{(\mu_m)} \neq \emptyset$ for $\mu_m \in \{1, -1\}$, it follows that there exists

$$s \in \operatorname{conv}(T) \cap H_m \subseteq \operatorname{conv}(S) \cap \left(\bigcap_{i=1}^{m} H_i \right).$$

This completes the proof.

Lemma 5.4 *Let V be an n-dimensional Euclidean space and let $H_1, \ldots, H_n$ be hyperplanes in V such that $0 \in H_i$ for all $i = 1, \ldots, n$. Let S be a subset of V such that*

$$S \cap H(\mu_1, \ldots, \mu_n) = S \cap \left(\bigcap_{i=1}^{n} H_i^{(\mu_i)} \right) \neq \emptyset$$

for all $(\mu_1, \ldots, \mu_n) \in \{1, -1\}^n$. Then

$$0 \in \operatorname{conv}(S).$$

Proof. It follows from Lemma 5.1 that the hyperplanes $H_1, \ldots, H_n$ are linearly independent and so, by Lemmas 5.2 and 5.3,

$$\operatorname{conv}(S) \cap \{0\} = \operatorname{conv}(S) \cap \left(\bigcap_{i=1}^{n} H_i \right) \neq \emptyset.$$

Lemma 5.5 *Let V be an n-dimensional vector space, and let $H_1, \ldots, H_n$ be linearly independent hyperplanes in V with normal vectors $h_1, \ldots, h_n$, respectively, and with $0 \in H_i$ for all $i = 1, \ldots, n$. Let*

$$L_j = \bigcap_{\substack{i=1 \\ i \neq j}}^{n} H_i$$

for $j = 1, \ldots, n$. Then

$$V = H_j \oplus L_j$$

for $j = 1, \ldots, n$. Moreover, there exists a dual basis $\{h_1^*, \ldots, h_n^*\}$ for V such that

$$\left(h_i, h_j^*\right) = \delta_{i,j} = \begin{cases} 1 & \text{if } i = j \\ 0 & \text{if } i \neq j \end{cases}$$

and H_j is the $(n-1)$-dimensional subspace spanned by $h_1^*, \ldots, h_{j-1}^*, h_{j+1}^*, \ldots, h_n^*$ and L_j is the one-dimensional subspace spanned by h_j^*.

Proof. It follows from Lemma 5.2 that $\dim(L_j) = 1$. Let f_j^* be a basis vector for L_j. Then $f_j^* \in L_j$ implies that $f_j^* \in H_i$ for all $i \neq j$, and so $\left(h_i, f_j^*\right) = 0$ for $i \neq j$. Moreover,

$$\left(h_j, f_j^*\right) = 0$$

if and only if

$$f_j^* \in H_j \cap L_j = \bigcap_{i=1}^{n} H_i = \{0\},$$

which is impossible since $f_j^* \neq 0$. Therefore, $\left(h_j, f_j^*\right) \neq 0$ and

$$h_j^* = \frac{f_j^*}{\left(h_j, f_j^*\right)} \in L_j \setminus H_j.$$

Then $(h_i, h_j^*) = \delta_{i,j}$ for $i, j = 1, \ldots, n$, and L_j is spanned by h_j^*. The vectors $h_1^*, \ldots, h_n^*$ are linearly independent, since $\sum_{j=1}^{n} x_j h_j^* = 0$ implies that

$$0 = (h_i, 0) = (h_i, \sum_{j=1}^{n} x_j h_j^*) = \sum_{j=1}^{n} x_j (h_i, h_j^*) = x_i$$

for $i = 1, \ldots, n$. Moreover, $(h_j, h_i^*) = 0$ for all $i \neq j$ implies that $h_i^* \in H_j$ for $i \neq j$. Since H_j is a vector subspace of dimension $n - 1$, it follows that the set $\{h_1^*, \ldots, h_{j-1}^*, h_{j+1}^*, \ldots, h_n^*\}$ is a basis for H_j, and $V = H_j \oplus L_j$. This completes the proof.

Lemma 5.6 *Let V be an n-dimensional vector space, and let $H_1, \ldots, H_n$ be linearly independent hyperplanes in V with $0 \in H_i$ for $i = 1, \ldots, n$. Let $Q_i = H_i \cap H_n$ for $i = 1, \ldots, n-1$. Then* dim $Q_i = n - 2$, *and $Q_1, \ldots, Q_{n-1}$ are linearly independent hyperplanes in H_n. Let*

$$L_n = \bigcap_{i=1}^{n-1} H_i,$$

and let $\pi : V \to H_n$ be the projection corresponding to the direct sum decomposition

$$V = H_n \oplus L_n.$$

Let $(\mu_1, \ldots, \mu_{n-1}) \in \{1, -1\}^{n-1}$, and let $v \in V$. If

$$v \in \bigcap_{i=1}^{n-1} H_i^{(\mu_i)},$$

then

$$\pi(v) \in \bigcap_{i=1}^{n-1} Q_i^{(\mu_i)}.$$

Let S be a subset of V, and let $\pi(S) = \{\pi(s) \mid s \in S\} \subseteq H_n$. If

$$S \cap \left(\bigcap_{i=1}^{n-1} H_i^{(\mu_i)} \right) \neq \emptyset$$

for all $(\mu_1, \ldots, \mu_{n-1}) \in \{1, -1\}^{n-1}$, then

$$0 \in conv\big(\pi(S)\big).$$

There exists a basis vector h_n^* for L_n such that if $S \subseteq H_n^{(+1)}$, then

$$\alpha h_n^* \in conv(S)$$

for some $\alpha > 0$.

Proof. It follows from Lemma 5.2 that $\dim(Q_i) = n - 2$, and so Q_i is a hyperplane in H_n for $i = 1, \ldots, n-1$.

Let $h_1, \ldots, h_n$ be normal vectors for $H_1, \ldots, H_n$, respectively, and let

$$q_i = h_i - \frac{(h_n, h_i)}{(h_n, h_n)} h_n.$$

Since $h_1, \ldots, h_n$ are linearly independent vectors, it follows that the vectors $q_1, \ldots, q_{n-1}$ are linearly independent. Moreover, $(h_n, q_i) = 0$, and so $q_i \in H_n$ for $i = 1, \ldots, n-1$.

Let $w \in H_n$. Then $(q_i, w) = (h_i, w)$, and so $w \in Q_i$ if and only if $(q_i, w) = 0$. Thus, q_i is a normal vector for Q_i in the vector subspace H_n, and the hyperplanes $Q_1, \ldots, Q_{n-1}$ are independent in H_n.

By Lemma 5.5, there is a basis vector h_n^* for L_n such that $(h_i, h_n^*) = \delta_{i,n}$. If $v \in V$, then

$$v = \pi(v) + \varphi(v) h_n^*,$$

where $\pi(v) \in H_n$ and $\varphi(v) \in \mathbf{R}$. Moreover,

$$\begin{aligned} \big(q_i, \pi(v)\big) &= \big(h_i, \pi(v)\big) - \frac{(h_n, h_i)}{(h_n, h_n)} \big(h_n, \pi(v)\big) \\ &= \big(h_i, \pi(v)\big) \\ &= (h_i, v) - \varphi(v)(h_i, h_n^*) \\ &= (h_i, v) \end{aligned}$$

for $i = 1, \ldots, n-1$. It follows that

$$v \in \bigcap_{i=1}^{n-1} H_i^{(\mu_i)}$$

if and only if

$$\pi(v) \in \bigcap_{i=1}^{n-1} Q_i^{(\mu_i)}.$$

If

$$S \cap \left(\bigcap_{i=1}^{n-1} H_i^{(\mu_i)} \right) \neq \emptyset$$

for all $(\mu_1, \ldots, \mu_{n-1}) \in \{1, -1\}^{n-1}$, then

$$\pi(S) \cap \left(\bigcap_{i=1}^{n-1} Q_i^{(\mu_i)} \right) \neq \emptyset$$

for all $(\mu_1, \ldots, \mu_{n-1}) \in \{1, -1\}^{n-1}$. Lemma 5.4 implies that

$$0 \in \operatorname{conv}(\pi(S)).$$

This means that there exist vectors $s_1, \ldots, s_k \in S$ with the property that

$$\sum_{i=1}^{k} \alpha_i \pi(s_i) = 0$$

for some nonnegative scalars $\alpha_1, \ldots, \alpha_k$ such that $\alpha_1 + \ldots + \alpha_k = 1$. Let

$$s_i = \pi(s_i) + \varphi(s_i) h_n^*$$

and

$$\alpha = \sum_{i=1}^{n} \alpha_i \varphi(s_i).$$

Then

$$\sum_{i=1}^{k} \alpha_i s_i = \sum_{i=1}^{k} \alpha_i \pi(s_i) + \sum_{i=1}^{k} \alpha_i \varphi(s_i) h_n^* = \alpha h_n^* \in \operatorname{conv}(S).$$

If $S \subseteq H_n^{(+1)}$, then

$$(h_n, s_i) = (h_n, \pi(s_i)) + \varphi(s_i)(h_n, h_n^*) = \varphi(s_i) > 0$$

for all $i = 1, \ldots, k$, and so $\alpha > 0$ and $\alpha h_n^* \in \operatorname{conv}(S)$. This completes the proof.

5.3 Blocks

Let V be an n-dimensional Euclidean space. Let $\mathbf{e}_0 \in V$, and let $\{\mathbf{e}_1, \ldots, \mathbf{e}_n\}$ be a basis for V. The *block* with center $\mathbf{e}_0$ and basis $\{\mathbf{e}_1, \ldots, \mathbf{e}_n\}$ is the set

$$B(\mathbf{e}_0; \mathbf{e}_1, \ldots, \mathbf{e}_n) = \left\{ \mathbf{e}_0 + \sum_{i=1}^{n} x_i \mathbf{e}_i \;\middle|\; -1 \leq x_i \leq 1 \text{ for } i = 1, \ldots, n \right\}.$$

A subset B of V is a *block* if $B = B(\mathbf{e}_0; \mathbf{e}_1, \ldots, \mathbf{e}_n)$ for some vector $\mathbf{e}_0 \in V$ and some basis $\{\mathbf{e}_1, \ldots \mathbf{e}_n\}$ of V.

Let $B = B(\mathbf{e}_0; \mathbf{e}_1, \ldots, \mathbf{e}_n)$. The *vertices* of B are the 2^n vectors in the set

$$\operatorname{vert}(B) = \left\{ \mathbf{e}_0 + \sum_{i=1}^{n} \mu_i \mathbf{e}_i \;\middle|\; (\mu_1, \ldots, \mu_n) \in \{1, -1\}^n \right\}.$$

The block B is the convex hull of its vertices. The *interior* of B is the set

$$\operatorname{int}(B) = \left\{ \mathbf{e}_0 + \sum_{i=1}^{n} x_i \mathbf{e}_i \;\middle|\; -1 < x_i < 1 \text{ for } i = 1, \ldots, n \right\}.$$

Corresponding to each block $B = B(\mathbf{e}_0; \mathbf{e}_1, \ldots, \mathbf{e}_n)$ are the $2n$ *facial hyperplanes*

$$F_{j,\mu_j} = \left\{ \mathbf{e}_0 + \mu_j \mathbf{e}_j + \sum_{\substack{i=1 \\ i \neq j}}^{n} x_i \mathbf{e}_i \;\middle|\; x_i \in \mathbf{R} \text{ for } i \neq j \right\}$$

where $j = 1, \ldots, n$ and $\mu_j \in \{1, -1\}$.

Let $\{0, 1, -1\}^n$ denote the set of n-tuples $(\lambda_1, \ldots, \lambda_n)$, where $\lambda_i \in \{0, 1, -1\}$ for $i = 1, \ldots, n$. To each $(\lambda_1, \ldots, \lambda_n) \in \{0, 1, -1\}^n$ there is associated the set $D(\lambda_1, \ldots, \lambda_n)$ consisting of all vectors $\mathbf{e}_0 + \sum_{i=1}^{n} x_i \mathbf{e}_i \in V$ such that

$$\begin{array}{rl} x_i > 1 & \text{if } \lambda_i = +1 \\ -1 < x_i < 1 & \text{if } \lambda_i = 0 \\ x_i < -1 & \text{if } \lambda_i = -1. \end{array}$$

Let $F^* = \bigcup_{j=1}^{n} \bigcup_{\mu_j = \pm 1} F_{j,\mu_j}$. Then $V \setminus F^*$ is the disjoint union of the 3^n open convex sets $D(\lambda_1, \ldots, \lambda_n)$. In particular,

$$D(0, 0, \ldots, 0) = \operatorname{int}(B).$$

For example, in the vector space $V = \mathbf{R}^2$, let $\mathbf{e}_1 = (4, 1)$, $\mathbf{e}_2 = (2, -2)$, and $B = B(0; \mathbf{e}_1, \mathbf{e}_2)$. Then

$$\operatorname{vert(B)} = \{\pm(2, 3), \pm(6, -1)\}.$$

The block B, its four facial hyperplanes, and the nine convex sets $D(\lambda_1, \lambda_2)$ are indicated in the following diagram:

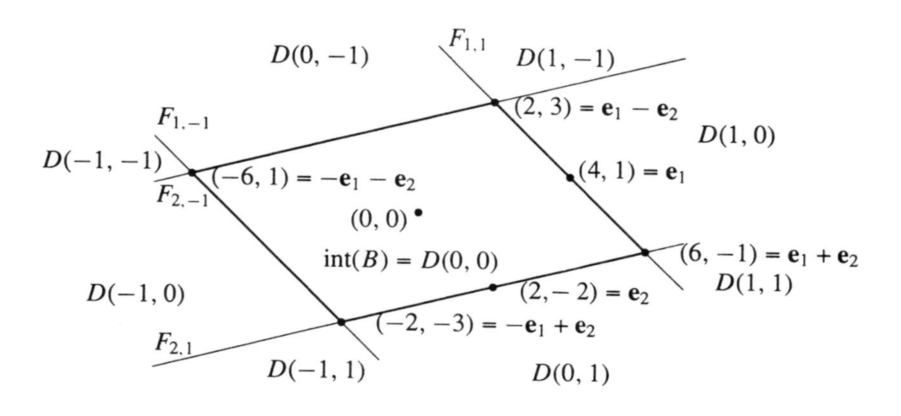

Let f and a be vectors in V. Then

$$a = f + (a - f).$$

The *reflection* of a through f is the vector

$$2f - a = f - (a - f).$$

For example, if $f = (3, 2)$ and $a = (4, 1)$, then the reflection of a through f is the vector $2f - a = (2, 3)$, as shown in the following diagram:

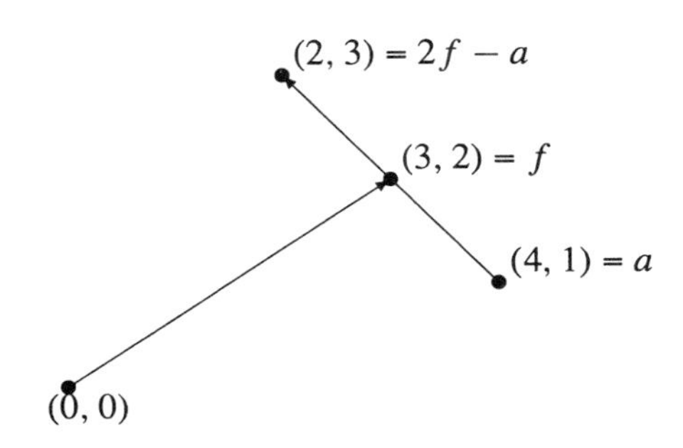

If $S \subseteq V$, the reflection of S through f is the set

$$\{2f\} - S = \{2f - s \mid s \in S\}.$$

Let $f_0, f_1, \ldots, f_{n-1}$ be n vectors in V. For any vector $f_n \in V$, we shall consider the sequence of finite sets $S_n, S_{n-1}, \ldots, S_0$ obtained by successive reflections as follows: $S_n = \{f_n\}$ and

$$S_{k-1} = S_k \cup (\{2f_{k-1}\} - S_k)$$

for $k = 1, \ldots, n$. Thus,

$$\begin{aligned} S_n &= \{f_n\}, \\ S_{n-1} &= \{f_n, 2f_{n-1} - f_n\}, \\ S_{n-2} &= \{f_n, 2f_{n-1} - f_n, 2f_{n-2} - f_n, 2f_{n-2} - 2f_{n-1} + f_n\}, \end{aligned}$$

and so on. For example, in the vector space $V = \mathbf{R}^2$, let $f_0 = (0, 0)$, $f_1 = (4, 1)$, and $f_2 = (6, -1)$. Then

$$\begin{aligned} S_2 &= \{(6, -1)\}, \\ S_1 &= \{(2, 3), (6, -1)\}, \\ S_0 &= \{\pm(2, 3), \pm(6, -1)\}, \end{aligned}$$

and $\mathrm{conv}(S_0)$ is the block B constructed earlier.

Lemma 5.7 *Let V be an n-dimensional Euclidean space, and let $\{f_0, f_1, \ldots, f_n\} \subseteq V$, where $f_0 = 0$. Let*

$$S_n \subseteq S_{n-1} \subseteq \ldots \subseteq S_0$$

be the sequence of sets constructed inductively by setting $S_n = \{f_n\}$ and

$$S_{k-1} = S_k \cup (\{2f_{k-1}\} - S_k)$$

for $k = 1, 2, \ldots, n$. Let $\mathbf{e}_i = f_i - f_{i-1}$ for $i = 1, \ldots, n$. Then

$$S_0 = \left\{ \sum_{i=1}^{n} \mu_i \mathbf{e}_i \;\middle|\; (\mu_1, \ldots, \mu_n) \in \{1, -1\}^n \right\}.$$

Let $B = conv(S_0)$. If the vectors $f_1, \ldots, f_n$ are linearly independent, then

(i) *the vectors $\mathbf{e}_1, \ldots \mathbf{e}_n$ are linearly independent,*

(ii) *$B = conv(S_0) = B(0; \mathbf{e}_1, \ldots, \mathbf{e}_n)$,*

(iii) *$vert(B) = S_0$, and*

(iv) *$|S_k| = 2^{n-k}$ for $k = 0, 1, \ldots, n$.*

Proof. It will be shown inductively that

$$\begin{aligned} S_k &= \{f_k\} + \left\{ \sum_{i=k+1}^{n} \mu_i \mathbf{e}_i \;\middle|\; \mu_i \in \{1, -1\} \text{ for } i = k+1, k+2, \ldots, n \right\} \\ &= \{f_{k-1}\} + \{e_k\} + \left\{ \sum_{i=k+1}^{n} \mu_i \mathbf{e}_i \;\middle|\; \mu_i \in \{1, -1\} \text{ for } i = k+1, k+2, \ldots, n \right\} \end{aligned}$$

for $k = 0, 1, \ldots, n$. This clearly holds for $k = n$ since $S_n = \{f_n\}$. Assume that the relationship holds for k. Then

$$\begin{aligned} &\{2f_{k-1}\} - S_k \\ &= \{2f_{k-1} - f_k\} - \left\{ \sum_{i=k+1}^{n} \mu_i \mathbf{e}_i \;\middle|\; \mu_i \in \{1, -1\} \text{ for } i = k+1, \ldots, n \right\} \\ &= \{f_{k-1} - \mathbf{e}_k\} - \left\{ \sum_{i=k+1}^{n} \mu_i \mathbf{e}_i \;\middle|\; \mu_i \in \{1, -1\} \text{ for } i = k+1, \ldots, n \right\} \\ &= \{f_{k-1}\} + \{-\mathbf{e}_k\} + \left\{ \sum_{i=k+1}^{n} \mu_i \mathbf{e}_i \;\middle|\; \mu_i \in \{1, -1\} \text{ for } i = k+1, \ldots, n \right\}. \end{aligned}$$

It follows that

$$\begin{aligned} S_{k-1} &= S_k \cup (\{2f_{k-1}\} - S_k) \\ &= \{f_{k-1}\} + \left\{ \sum_{i=k}^{n} \mu_i \mathbf{e}_i \;\middle|\; \mu_i \in \{1, -1\}^n \text{ for } i = k, \ldots, n \right\}. \end{aligned}$$

This completes the induction. For $k = 0$, we obtain

$$S_0 = \left\{ \sum_{i=1}^{n} \mu_i \mathbf{e}_i \;\middle|\; (\mu_1, \ldots, \mu_n) \in \{1, -1\}^n \right\}.$$

If $\{f_1, \ldots, f_n\}$ is a basis for V, then $\{\mathbf{e}_1, \ldots, \mathbf{e}_n\}$ is also a basis for V, and

$$|S_k| = 2^{n-k}$$

for $k = 0, 1, \ldots, n$. In particular, $|S_0| = 2^n$. Clearly,

$$\text{conv}(S_0) = B(0; \mathbf{e}_1, \ldots, \mathbf{e}_n) = B$$

and

$$\text{vert}(B) = S_0.$$

This completes the proof.

Lemma 5.8 *Let V be an n-dimensional Euclidean space, let $H_1, \ldots, H_n$ be hyperplanes in V, and let $\{f_0, f_1, \ldots, f_n\} \subseteq V$, with $f_0 = \mathbf{0}$. If*

$$f_i \in H_j \qquad \text{for } 0 \le i < j \le n$$
$$f_i \in H_j^{(+1)} \qquad \text{for } 1 \le j \le i \le n,$$

then the vectors $f_1, \ldots f_n$ are linearly independent. Let $S_n \subseteq S_{n-1} \subseteq \ldots \subseteq S_0$ be the sequence of finite subsets of V constructed inductively from $\{f_0, f_1, \ldots, f_n\}$ by setting $S_n = \{f_n\}$ and

$$S_{k-1} = S_k \cup (\{2f_{k-1}\} - S_k)$$

for $k = 1, \ldots, n$. Suppose that

$$S_k \subseteq \bigcap_{j=1}^{k} H_j^{(+1)}$$

for $k = 1, \ldots, n$. Then

$$S_0 \cap H(\mu_1, \ldots, \mu_n) \neq \emptyset$$

for all $(\mu_1, \ldots, \mu_n) \in \{1, -1\}^n$, and the hyperplanes $H_1, \ldots, H_n$ are linearly independent. Let $s \in S_0 \cap H(\mu_1, \ldots, \mu_n)$ and $S_{n+1} = \emptyset$. Then $s \in S_k \backslash S_{k+1}$ if and only if $\mu_i = 1$ for $i = 1, \ldots, k$ and $\mu_{k+1} = -1$.

Proof. If the vectors $f_1, \ldots, f_n$ are dependent, then

$$f_k = \sum_{i=1}^{k-1} x_i f_i$$

for some $k \ge 1$ and $x_1, \ldots, x_{k-1} \in \mathbf{R}$. Since $\{f_1, \ldots, f_{k-1}\} \subseteq H_k$ and the hyperplane H_k is a subspace (since $0 = f_0 \in H_k$), it follows that $f_k \in H_k$, which is impossible because $f_k \in H_k^{(+1)}$ and $H_k \cap H_k^{(+1)} = \emptyset$. Therefore, the vectors $f_1, \ldots, f_n$ are independent. By Lemma 5.7,

$$|S_k| = 2^{n-k}$$

for $k = 0, 1, \ldots, n$.

We shall show by induction on k that

$$S_k \cap H(1, \ldots, 1, \mu_{k+1}, \ldots, \mu_n) \neq \emptyset \tag{5.2}$$

for all $(\mu_{k+1}, \ldots, \mu_n) \in \{1, -1\}^{n-k}$. Let $k = n$. Since $f_n \in H_j^{(+1)}$ for all $j = 1, \ldots, n$, and $S_n = \{f_n\}$, it follows that

$$S_n \cap H(1, \ldots, 1) \neq \emptyset.$$

Suppose that (5.2) holds for some $k \in \{1, \ldots, n\}$. For every $(\mu_{k+1}, \ldots, \mu_n) \in \{1, -1\}^{n-k}$, there exists

$$s \in S_k \cap H_k^{(+1)} \cap \left(\bigcap_{j=k+1}^{n} H_j^{(\mu_j)} \right).$$

Since $2f_{k-1} - s \in S_{k-1}$ and $f_{k-1} \in H_j$ for $j = k, k+1, \ldots, n$, it follows that $2f_{k-1} - s \in H_k^{(-1)}$ and $2f_{k-1} - s \in H_j^{(-\mu_j)}$ for $j = k+1, \ldots, n$. Thus,

$$\left(\{2f_{k-1}\} - S_k\right) \cap H_k^{(-1)} \cap \left(\bigcap_{j=k+1}^{n} H_j^{(\mu_j)} \right) \neq \emptyset$$

for all $(\mu_{k+1}, \ldots, \mu_n) \in \{1, -1\}^{n-k}$. Since $S_{k-1} = S_k \cup \left(\{2f_{k-1}\} - S_k\right)$ and $S_{k-1} \subseteq \bigcap_{j=1}^{k-1} H_j^{(+1)}$, it follows that

$$\begin{aligned} & S_{k-1} \cap H(1, \ldots, 1, \mu_k, \mu_{k+1}, \ldots, \mu_n) \\ = \quad & S_{k-1} \cap \left(\bigcap_{j=1}^{k-1} H_j^{(+1)} \right) \cap \left(\bigcap_{j=k}^{n} H_j^{(\mu_j)} \right) \\ \neq \quad & \emptyset \end{aligned}$$

for all $(\mu_k, \ldots, \mu_n) \in \{1, -1\}^{n-k+1}$. This completes the induction.

In particular, it follows that

$$S_0 \cap H(\mu_1, \ldots, \mu_n) \neq \emptyset$$

for all $(\mu_1, \ldots, \mu_n) \in \{1, -1\}^n$, and so, by Lemma 5.1, the hyperplanes $H_1, \ldots, H_n$ are linearly independent in V.

Since $|S_k| = 2^{n-k}$ and $S_k \cap H(1, \ldots, 1, \mu_{k+1}, \ldots, \mu_n) \neq \emptyset$ for every $k = 0, 1, \ldots, n$ and every $(\mu_{k+1}, \ldots, \mu_n) \in \{1, -1\}^n$, it follows that

$$|S_k \cap H(1, \ldots, 1, \mu_{k+1}, \ldots, \mu_n)| = 1$$

for every $k = 0, 1, \ldots, n$ and $(\mu_{k+1}, \ldots, \mu_n) \in \{1, -1\}^{n-k}$. Since each of the 2^n sets $H(\mu_1, \ldots, \mu_n)$ contains exactly one element of S_0, it follows that if

$$s \in S_0 \cap H(\mu_1, \ldots, \mu_n)$$

and $\mu_i = 1$ for $i = 1, \ldots, k$, then $s \in S_k$. If $\mu_{k+1} = -1$, then $s \notin S_{k+1}$. Thus

$$S_0 \cap H(1, \ldots, 1, -1, \mu_{k+2}, \ldots, \mu_n) \subseteq S_k \setminus S_{k+1}.$$

Since

$$\left|S_0 \cap H(1, \ldots, 1, -1, \mu_{k+2}, \ldots, \mu_n)\right| = \left|S_k \setminus S_{k+1}\right| = 2^{n-k-1},$$

it follows that

$$S_0 \cap H(1, \ldots, 1, -1, \mu_{k+2}, \ldots, \mu_n) = S_k \setminus S_{k+1}.$$

This completes the proof.

Lemma 5.9 *Let the n-dimensional vector space V, the hyperplanes $H_1, \ldots, H_n$ and the set $F = \{f_0, f_1, \ldots, f_{n-1}\}$ satisfy the conditions of Lemma 5.8. For $a \in H(1, \ldots, 1)$, let $\{S_k(a)\}_{k=0}^n$ be the sequence of subsets of V constructed inductively from the set $F \cup \{a\}$ by the method of Lemma 5.7. If $a, a' \in H(1, \ldots, 1)$ and $a \neq a'$, then*

$$S_0(a) \cap S_0(a') = \emptyset.$$

Proof. Let $a \neq a'$. Then $S_n(a) = \{a\}$, $S_n(a') = \{a'\}$, and so $S_n(a) \cap S_n(a') = \emptyset$. Suppose that $S_0(a) \cap S_0(a') \neq \emptyset$. Let k be the greatest integer such that $S_k(a) \cap S_k(a') \neq \emptyset$. Then $0 \leq k \leq n - 1$. Choose

$$s \in S_k(a) \cap S_k(a').$$

Then

$$s \notin S_{k+1}(a) \cap S_{k+1}(a').$$

If $s \notin S_{k+1}(a)$, then

$$s \in S_k(a) \setminus S_{k+1}(a) = \{2f_k\} - S_{k+1}(a).$$

It follows from Lemma 5.8 that

$$s \in H(1, \ldots, 1, -1, \mu_{k+2}, \ldots, \mu_n)$$

and so, again by Lemma 5.8,

$$s \in S_k(a') \setminus S_{k+1}(a') = \{2f_k\} - S_{k+1}(a').$$

Therefore, there exist vectors $v \in S_{k+1}(a)$ and $v' \in S_{k+1}(a')$ such that

$$s = 2f_k - v = 2f_k - v',$$

and so

$$v = v' \in S_{k+1}(a) \cap S_{k+1}(a'),$$

which contradicts the maximality of k. This completes the proof.

Lemma 5.10 *Let $\{\mathbf{e}_1, \ldots \mathbf{e}_n\}$ be a basis for the n-dimensional vector space V. Consider the block $B = B(0; \mathbf{e}_1, \ldots, \mathbf{e}_n)$. Let $u, v \in V$. If*

$$0 \in B(u; \mathbf{e}_1, \ldots, \mathbf{e}_n) = B + \{u\},$$

then

$$0 \in B + \{tu\}$$

for all $t \in [0, 1]$. If

$$\left(B + \{u\}\right) \cap \left(B + \{v\}\right) \neq \emptyset,$$

then

$$\left(vert(B) + \{u\}\right) \cap \left(B + \{v\}\right) \neq \emptyset.$$

Proof. If $0 \in B + \{u\}$, then $0 = b + u$ for some $b \in B$. Let $t \in [0, 1]$. Then $tb \in B$ since B is convex, and so

$$0 = b + u = tb + tu \in B + \{tu\}.$$

Let $u = \sum_{i=1}^n u_i \mathbf{e}_i$ and $v = \sum_{i=1}^n v_i \mathbf{e}_i$. If $\left(B+\{u\}\right)\cap\left(B+\{v\}\right) \neq \emptyset$, then there exist scalars $x_i, y_i \in [-1, 1]$ for $i = 1, \ldots, n$ such that $\sum_{i=1}^n x_i \mathbf{e}_i \in B$, $\sum_{i=1}^n y_i \mathbf{e}_i \in B$, and

$$\sum_{i=1}^n (u_i + x_i)\mathbf{e}_i = \sum_{i=1}^n (v_i + y_i)\mathbf{e}_i,$$

and so

$$u_i + x_i = v_i + y_i$$

for $i = 1, \ldots, n$. Then

$$v_i - 1 \leq v_i + y_i = u_i + x_i \leq u_i + 1.$$

If $u_i \leq v_i$, then

$$v_i - 1 \leq u_i + 1 \leq v_i + 1,$$

and we let $\mu_i = 1$. If $v_i < u_i$, then

$$v_i - 1 < u_i - 1 \leq u_i + x_i = v_i + y_i \leq v_i + 1$$

and we let $\mu_i = -1$. In both cases,

$$v_i - 1 \leq u_i + \mu_i \leq v_i + 1$$

for $i = 1, \ldots, n$, and so

$$\sum_{i=1}^n (u_i + \mu_i)\mathbf{e}_i = u + \sum_{i=1}^n \mu_i \mathbf{e}_i \in \left(\text{vert}(B) + \{u\}\right) \cap \left(B + \{v\}\right) \neq \emptyset.$$

This completes the proof.

Theorem 5.3 *Let V be an n-dimensional vector space, let $H_1, \ldots, H_n$ be hyperplanes in V, and let $\{f_0, f_1, \ldots, f_{n-1}\} \subseteq V$. Suppose that $f_0 = 0$ and*

$$\begin{aligned} f_i &\in H_j \qquad \text{for } 0 \leq i < j \leq n \\ f_i &\in H_j^{(+1)} \qquad \text{for } 1 \leq j \leq i \leq n-1. \end{aligned}$$

Let A_n be a finite subset of $H(1, \ldots, 1)$. To each $a \in A_n$, let $S_0(a)$ be the set constructed inductively from $\{f_0, f_1, \ldots, f_{n-1}, a\}$ by letting $S_n(a) = \{a\}$ and

$$S_{k-1}(a) = S_k(a) \cup \left(\{2f_{k-1}\} - S_k(a) \right)$$

for $k = 1, \ldots, n$. Let $B(a)$ denote the block

$$B(a) = conv(S_0(a)).$$

Suppose that

$$S_k(a) \subseteq \bigcap_{j=1}^{k} H_j^{(+1)}$$

for $k = 1, \ldots, n$ and all $a \in A_n$. Then there exists a vector $a^* \in A_n$ such that

$$S_0(a) \cap B(a^*) \neq \emptyset$$

for all $a \in A_n$, and

$$\left| \left(\bigcup_{a \in A_n} S_0(a) \right) \cap B(a^*) \right| \geq |A_n|.$$

Proof. Let $\mathbf{e}_i = f_i - f_{i-1}$ for $i = 1, \ldots, n-1$. For $a \in A_n$, let

$$\mathbf{e}_n(a) = a - f_{n-1}.$$

By Lemma 5.8, the vectors $f_1, \ldots, f_{n-1}, a$ are linearly independent, and so the vectors $\mathbf{e}_1, \ldots, \mathbf{e}_{n-1}, \mathbf{e}_n(a)$ are also linearly independent. Since

$$\{\mathbf{e}_1, \ldots, \mathbf{e}_{n-1}\} \subseteq H_n,$$

it follows that the set $\{\mathbf{e}_1, \ldots, \mathbf{e}_{n-1}\}$ is a basis for H_n. Moreover,

$$B = B(0; \mathbf{e}_1, \ldots, \mathbf{e}_{n-1})$$

and

$$B(a) = B(0; \mathbf{e}_1, \ldots, \mathbf{e}_{n-1}, \mathbf{e}_n(a)) = \bigcup_{t \in [-1,1]} (B + t\mathbf{e}_n(a))$$

are blocks in H_n and V, respectively. By Lemma 5.7,

$$\text{vert}\big(B(a)\big) = \left\{ \sum_{i=1}^{n-1} \mu_i \mathbf{e}_i + \mu_n \mathbf{e}_n(a) \;\middle|\; (\mu_1, \ldots, \mu_n) \in \{1, -1\}^n \right\} = S_0(a)$$

and

$$B(a) = \operatorname{conv}\Big(\operatorname{vert}\big(B(a)\big)\Big) = \operatorname{conv}\big(S_0(a)\big).$$

For $\mu_n \in \{1, -1\}$, let

$$S_0^{(\mu_n)}(a) = \left\{ \sum_{i=1}^{n-1} \mu_i \mathbf{e}_i + \mu_n \mathbf{e}_n(a) \;\middle|\; (\mu_1, \ldots, \mu_{n-1}) \in \{1, -1\}^{n-1} \right\}.$$

Then

$$S_0^{(+1)}(a) = \operatorname{vert}(B) + \{e_n(a)\},$$

and so

$$\operatorname{conv}\left(S_0^{(+1)}(a)\right) = B + \{\mathbf{e}_n(a)\} \subseteq B(a).$$

Similarly,

$$\operatorname{conv}\left(S_0^{(-1)}(a)\right) = B - \{\mathbf{e}_n(a)\} \subseteq B(a).$$

Let $h_1, \ldots, h_n$ be normal vectors to the hyperplanes $H_1, \ldots, H_n$, respectively. Let $L_n = \bigcap_{i=1}^{n-1} H_i$, and let h_n^* be the dual basis vector for L_n such that $(h_i, h_n^*) = 0$ if $i \neq n$ and $(h_n, h_n^*) = 1$. Let $\pi : V \rightarrow H_n$ be the projection corresponding to the direct sum decomposition

$$V = H_n \oplus L_n.$$

Let $a \in A_n \subseteq H_n^{(+1)}$. Since $f_{n-1} \in H_n$, it follows that $\mathbf{e}_n(a) = a - f_{n-1} \in H_n^{(+1)}$ and so $\mathbf{e}_n(a)$ can be written uniquely in the form

$$\mathbf{e}_n(a) = \pi(\mathbf{e}_n(a)) + \varphi(a) h_n^*,$$

where

$$(h_n, \mathbf{e}_n(a)) = (h_n, \pi(\mathbf{e}_n(a))) + \varphi(a)\left(h_n, h_n^*\right) = \varphi(a) > 0.$$

By Lemma 5.8,

$$S_0(a) \cap H(\mu_1, \ldots, \mu_{n-1}, +1) = S_0^{(+1)}(a) \cap \left(\bigcap_{j=1}^{n-1} H_j^{(\mu_j)} \right) \neq \emptyset$$

for all $(\mu_1, \ldots, \mu_{n-1}) \in \{1, -1\}^{n-1}$. By Lemma 5.6,

$$\begin{aligned} 0 \in \operatorname{conv}\Big(\pi\left(S_0^{(+1)}(a)\right)\Big) &= \pi\Big(\operatorname{conv}\big(S_0^{(+1)}(a)\big)\Big) \\ &= \pi\big(B + \mathbf{e}_n(a)\big) \\ &= B + \pi(\mathbf{e}_n(a)) \end{aligned}$$

for all $a \in A_n$. Since the set A_n is finite, we can choose $a^* \in A_n$ such that

$$\varphi(a^*) = \max\Big\{\varphi(a) \;\Big|\; a \in A_n\Big\}.$$

Let $a \in A_n$, and let $t = \varphi(a)/\varphi(a^*)$. Then $0 < t \leq 1$ and

$$t\mathbf{e}_n(a^*) = \pi(t\mathbf{e}_n(a^*)) + t\varphi(a^*)h_n^* = \pi(t\mathbf{e}_n(a^*)) + \varphi(a)h_n^*.$$

Since $0 \in B + \pi\left(\mathbf{e}_n(a^*)\right)$, it follows from Lemma 5.10 that

$$0 \in B + t\pi\left(\mathbf{e}_n(a^*)\right) = B + \pi\left(t\mathbf{e}_n(a^*)\right).$$

Then

$$0 \in \left(B + \pi(\mathbf{e}_n(a))\right) \cap \left(B + t\pi(\mathbf{e}_n(a^*))\right) \neq \emptyset$$

and so, by Lemma 5.10,

$$\left(\text{vert}(B) + \pi(\mathbf{e}_n(a))\right) \cap \left(B + \pi(t\mathbf{e}_n(a^*))\right) \neq \emptyset.$$

It follows that there exists $(\mu_1, \ldots, \mu_{n-1}) \in \{1, -1\}^{n-1}$ such that

$$\sum_{i=1}^{n-1} \mu_i \mathbf{e}_i + \pi(\mathbf{e}_n(a)) \in B + \pi(t\mathbf{e}_n(a^*)).$$

Then

$$\sum_{i=1}^{n-1} \mu_i \mathbf{e}_i + \pi(\mathbf{e}_n(a)) + \varphi(a)h_n^* = \sum_{i=1}^{n-1} \mu_i \mathbf{e}_i + \mathbf{e}_n(a) \in S_0^+(a)$$

and

$$\sum_{i=1}^{n-1} \mu_i \mathbf{e}_i + \mathbf{e}_n(a) \in B + t\pi(\mathbf{e}_n(a^*)) + \varphi(a)h_n^* = B + t\mathbf{e}_n(a^*) \subseteq B(a^*).$$

This proves that for all $a \in A_n$

$$S_0^{(+)}(a) \cap B(a^*) \neq \emptyset$$

and so

$$S_0(a) \cap B(a^*) \neq \emptyset.$$

By Lemma 5.9, if $a, a' \in A_n$ and $a \neq a'$, then $S_0(a) \cap S_0(a') = \emptyset$. This implies that

$$\left| \left(\bigcup_{a \in A_n} S_0(a) \right) \cap B(a^*) \right| \geq |A_n|.$$

This completes the proof of Theorem 5.3.

5.4 Proof of the theorem

Let A_1, A_2, and A be subsets of V. The *set of midpoints* of A_1 and A_2 is the set

$$\text{mid}(A_1, A_2) = \left\{ \frac{a_1 + a_2}{2} \mid a_1 \in A_1, a_2 \in A_2 \right\}.$$

If K is convex and $A_1, A_2 \subseteq K$, then $\mathrm{mid}(A_1, A_2) \subseteq K$. Let

$$\mathrm{mid}(A) = \left\{ \frac{a + a'}{2} \mid a, a' \in A \right\}$$

denote the set of midpoints of A. Then $A \subseteq \mathrm{mid}(A)$ and $|2A| = |\mathrm{mid}(A)|$.

Lemma 5.11 *Let V be an n-dimensional Euclidean space. Let B be a block in V, and let $W \subseteq int(B)$. Then*

$$mid\big(W, vert(B)\big) \subseteq int(B)$$

and

$$\big|mid\big(W, vert(B)\big)\big| = 2^n |W|.$$

Proof. Let $B = B(\mathbf{e}_0; \mathbf{e}_1, \ldots, \mathbf{e}_n)$. For $j = 1, 2$, let

$$w_j = \mathbf{e}_0 + \sum_{i=1}^{n} x_{ij}\mathbf{e}_i \in W \subseteq \mathrm{int}(B)$$

and

$$b_j = \mathbf{e}_0 + \sum_{i=1}^{n} \mu_{ij}\mathbf{e}_i \in \mathrm{vert}(B).$$

Then $-1 < x_{ij} < 1$ and $\mu_{ij} \in \{1, -1\}$ for $i = 1, \ldots, n$ and $j = 1, 2$. If

$$\frac{w_1 + b_1}{2} = \frac{w_2 + b_2}{2},$$

then

$$\sum_{i=1}^{n}(x_{i1} + \mu_{i1})\mathbf{e}_i = \sum_{i=1}^{n}(x_{i2} + \mu_{i2})\mathbf{e}_i,$$

and so

$$x_{i1} - x_{i2} = \mu_{i2} - \mu_{i1}$$

for $i = 1, \ldots, n$. Since

$$-2 < x_{i1} - x_{i2} < 2$$

and

$$\mu_{i2} - \mu_{i1} \in \{0, 2, -2\},$$

it follows that

$$x_{i1} - x_{i2} = \mu_{i2} - \mu_{i1} = 0$$

for $i = 1, \ldots, n$, and so $w_1 = w_2$ and $b_1 = b_2$. Therefore,

$$\big|\mathrm{mid}\big(W, \mathrm{vert}(B)\big)\big| = |W|\,|\mathrm{vert}(B)| = 2^n |W|.$$

Also, $(x_{ij} + \mu_{ij})/2 \in (-1, 1)$ for $i = 1, \ldots, n$, and so $(w_j + b_j)/2 \in \mathrm{int}(W)$; hence $\mathrm{mid}\big(W, \mathrm{vert}(B)\big) \subseteq \mathrm{int}(B)$.

Lemma 5.12 *Let $n \geq 2$ and $1 < c < 2^n$. Define $\varepsilon_0 = \varepsilon_0(n,c) > 0$ and $k^* = k^*(n,c)$ by*

$$\varepsilon_0 = \frac{2^n - c}{4n(3^n c + 2nc + 1)(4c)^{2^n - 1}}$$

and

$$k^* = k^*(n,c) = (4c)^{2^n - 1}.$$

If V is an n-dimensional vector space, and if $A \subseteq V$ satisfies

$$|A| \geq k^*,$$

$$|2A| \leq c|A|,$$

and

$$|A \cap H| \leq \varepsilon_0 |A|$$

for every hyperplane H in V, then there exists $W \subseteq A$ such that

$$|W| \geq \varepsilon_0 |A|$$

and

$$|2W| \leq (c - \varepsilon_0)|W|.$$

Proof. We defined

$$\text{mid}(W) = \left\{ \frac{w_1 + w_2}{2} \;\middle|\; w_1, w_2 \in W \right\} = \left\{ \frac{v}{2} \;\middle|\; v \in 2W \right\}$$

for any set $W \subseteq V$, and so $|\text{mid}(W)| = |2W|$. Let $r_W(v)$ denote the number of sets $\{w_1, w_2\} \subseteq W$ such that $(w_1 + w_2)/2 = v$. Then

$$\begin{aligned} \frac{|W|^2}{2} &< \frac{|W|(|W|+1)}{2} = \sum_{v \in \text{mid}(W)} r_W(v) \\ &\leq |\text{mid}(W)| \max\{r_W(v) \mid v \in \text{mid}(W)\}. \end{aligned}$$

Let $A \subseteq V$ satisfy the three conditions of the lemma. We shall prove by induction that there exist hyperplanes $H_1, \ldots, H_n$ in V, vectors $f_0, f_1, \ldots, f_{n-1} \in V$ with $f_0 = 0$, and sets $A_1, \ldots, A_n$ such that

(i) $f_i \in H_j$ for $0 \leq i < j \leq n$,

(ii) $f_i \in H_j^{(+1)}$ for $1 \leq j \leq i \leq n-1$,

(iii) $A_n \subseteq A_{n-1} \subseteq \ldots \subseteq A_1 \subseteq A_0 = A$,

(iv) $k_j = |A_j| > \frac{k_0}{(4c)^{2^j - 1}}$ for $j = 1, \ldots, n$,

(v) $A_j \subseteq \bigcap_{i=1}^{j} H_i^{(+1)}$ for $j = 1, \ldots, n$, and

(vi) $A_{j+1} \cup \left(\{2f_j\} - A_{j+1}\right) \subseteq A_j$ for $j = 0, 1, \ldots, n-1$.

The proof is by induction. Let $A_0 = A$ and $k_0 = |A_0|$. Choose $f_0 \in A_0$ such that

$$r_{A_0}(f_0) = \max\left\{r_{A_0}(v) \mid v \in \text{mid}(A_0)\right\}.$$

Then

$$\begin{aligned}\frac{k_0^2}{2} &< \sum_{v \in \text{mid}(A_0)} r_{A_0}(v) \\ &\leq r_{A_0}(f_0)\left|\text{mid}(A_0)\right| \\ &= r_{A_0}(f_0)|2A_0| \\ &\leq r_{A_0}(f_0)ck_0,\end{aligned}$$

and so

$$r_{A_0}(f_0) > \frac{k_0}{2c}.$$

Replacing A by $A - \{f_0\}$, we can assume without loss of generality that $f_0 = 0$.

Let H_1 be any hyperplane in V such that $\mathbf{0} = f_0 \in H_1$. If $(w_1 + w_2)/2 = f_0$, then either $\{w_1, w_2\} \subseteq H_1$ or

$$\left|\{w_1, w_2\} \cap H_1^{(+1)}\right| = \left|\{w_1, w_2\} \cap H_1^{(-1)}\right| = 1.$$

Let

$$A_1 = \left\{w \in A_0 \mid w \in H_1^{(+1)} \text{ and } 2f_0 - w \in A_0\right\}.$$

Then

$$A_1 \subseteq A_0 \cap H_1^{(+1)} \subseteq H_1^{(+1)}$$

and

$$A_1 \cup \left(\{2f_0\} - A_1\right) \subseteq A_0.$$

Since $|A_0 \cap H_1| \leq \varepsilon_0 k_0$ and

$$\varepsilon_0 < \frac{1}{(4c)^{4^n - 1}} \leq \frac{1}{4c},$$

it follows that

$$\begin{aligned}k_1 = |A_1| &\geq r_{A_0}(f_0) - |A_0 \cap H_1| \\ &> \frac{k_0}{2c} - \varepsilon_0 k_0 \\ &> \frac{k_0}{4c}.\end{aligned}$$

Thus, H_1, f_0, and A_1 satisfy conditions (i)–(vi).

Suppose that $1 \leq m \leq n-1$, and that we have constructed hyperplanes $H_1, \ldots, H_m$, vectors $f_0, f_1, \ldots, f_{m-1}$, and sets $A_1, \ldots, A_m$ that satisfy conditions (i)–(vi). Since $A_m \subseteq \bigcap_{i=1}^m H_i^{(+1)}$ and $\bigcap_{i=1}^m H_i^{(+1)}$ is convex, it follows that

$$\mathrm{mid}(A_m) \subseteq \bigcap_{i=1}^{m} H_i^{(+1)}.$$

Choose $f_m \in \mathrm{mid}(A_m)$ such that

$$r_{A_m}(f_m) = \max\left\{ r_{A_m}(v) \mid v \in \mathrm{mid}(A_m) \right\}.$$

Then

$$\begin{aligned} \frac{1}{2}\left(\frac{k_0}{(4c)^{2^m-1}}\right)^2 &< \frac{k_m^2}{2} \\ &< r_{A_m}(f_m)\left|\mathrm{mid}(A_m)\right| \\ &\leq r_{A_m}(f_m)\left|\mathrm{mid}(A_0)\right| \\ &\leq r_{A_m}(f_m) c k_0, \end{aligned}$$

and so

$$r_{A_m}(f_m) > \frac{2k_0}{(4c)^{2^{m+1}-1}}.$$

Let H_{m+1} be any hyperplane in V such that

$$\{f_0, f_1, \ldots, f_m\} \subseteq H_{m+1}.$$

Let

$$A_{m+1} = \left\{ w \in A_m \mid w \in H_{m+1}^{(+1)} \text{ and } 2f_m - w \in A_m \right\}.$$

Then

$$A_{m+1} \subseteq A_m \cap H_{m+1}^{(+1)} \subseteq \bigcap_{i=1}^{m+1} H_i^{(+1)}$$

and

$$A_{m+1} \cup \left(\{2f_m\} - A_{m+1}\right) \subseteq A_m.$$

Since

$$\left|A_m \cap H_{m+1}\right| \leq \left|A_0 \cap H_{m+1}\right| \leq \varepsilon_0 k_0$$

and

$$\varepsilon_0 < \frac{1}{(4c)^{2^{m+1}-1}},$$

it follows that

$$\begin{aligned} k_{m+1} &= |A_{m+1}| \\ &\geq r_{A_m}(f_m) - |A_m \cap H_{m+1}| \\ &> \frac{2k_0}{(4c)^{2^{m+1}-1}} - \varepsilon_0 k_0 \\ &> \frac{k_0}{(4c)^{2^{m+1}-1}}. \end{aligned}$$

Thus, the hyperplanes $H_1, \ldots, H_{m+1}$, the vectors $f_0, f_1, \ldots, f_m$, and the sets $A_1, \ldots, A_{m+1}$ satisfy conditions (i)–(vi) . This completes the induction.

Let $a \in A_n$. Construct sets

$$S_n(a) \subseteq S_{n-1}(a) \subseteq \ldots \subseteq S_0(a) \subseteq A$$

by setting $S_n(a) = \{a\}$ and

$$S_{j-1}(a) = S_j(a) \cup \left(\{2f_{j-1}\} - S_j(a)\right)$$

for $j = 0, 1, \ldots, n-1$. We shall prove by induction that

$$S_m(a) \subseteq A_m \subseteq \bigcap_{i=1}^{m} H_i^{(+1)}$$

for $m = 0, 1, \ldots, n$. Clearly,

$$S_n(a) = \{a\} \subseteq A_n \subseteq \bigcap_{i=1}^{n} H_i^{(+1)}.$$

Suppose that

$$S_{m+1}(a) \subseteq A_{m+1} \subseteq \bigcap_{i=1}^{m+1} H_i^{(+1)}$$

where $0 \leq m \leq n-1$. Then

$$\{2f_m\} - S_{m+1}(a) \subseteq \{2f_m\} - A_{m+1} \subseteq A_m \subseteq \bigcap_{i=1}^{m} H_i^{(+1)},$$

and so

$$S_m(a) = S_{m+1}(a) \cup \left(\{2f_m\} - S_{m+1}(a)\right) \subseteq A_m \subseteq \bigcap_{i=1}^{m} H_i^{(+1)}.$$

This completes the induction.

We have shown that the hyperplanes $H_1, \ldots, H_n$, the vectors $f_0, f_1, \ldots, f_{n-1}$, the set $A_n \subseteq H(1, \ldots, 1)$, and the sets

$$\left\{S_k(a) \mid a \in A_n \text{ and } k = 0, 1, \ldots, n\right\}$$

satisfy the hypotheses of Theorem 5.3. Therefore, there exists a vector $a^* \in A_n$ such that the block

$$B(a^*) = \operatorname{conv}\left(S_0(a^*)\right)$$

has the property that

$$\begin{aligned} \left|A \cap B(a^*)\right| &\geq \left|\left(\bigcup_{a \in A_n} S_0(a)\right) \cap B(a^*)\right| \\ &\geq |A_n| \\ &= k_n \\ &> \frac{k_0}{(4c)^{2^n-1}}. \end{aligned}$$

The block $B(a^*)$ determines the $2n$ facial hyperplanes F_{j,μ_j}, where $j = 1, \ldots, n$ and $\mu_j \in \{1, -1\}$. Let

$$F^* = \bigcup_{j=1}^{n} \bigcup_{\mu_j=1,-1} F_{j,\mu_j}.$$

Let $W_0 = A \cap \operatorname{int}\big(B(a^*)\big)$. Because $|A \cap H| \leq \varepsilon_0 k_0$ for every hyperplane H and

$$\varepsilon_0 < \frac{1}{4n(4c)^{2^n-1}},$$

it follows that

$$|A \cap F^*| \leq 2n\varepsilon_0 k_0,$$

and so

$$\begin{aligned} |W_0| &= \big|A \cap \operatorname{int}\big(B(a^*)\big)\big| \\ &\geq \big|A \cap B(a^*)\big| - \big|A \cap F^*\big| \\ &> \frac{k_0}{(4c)^{2^n-1}} - 2n\varepsilon_0 k_0 \\ &> \frac{k_0}{2(4c)^{2^n-1}}. \end{aligned}$$

Since

$$\operatorname{vert}\big(B(a^*)\big) = S_0(a^*) \subseteq A,$$

it follows from Lemma 5.11 that

$$\begin{aligned} \operatorname{mid}\big(W_0, S_0(a^*)\big) &= \left\{\frac{w+s}{2} \mid w \in W_0, s \in S_0(a^*)\right\} \\ &\subseteq \operatorname{int}\big(B(a^*)\big) \cap \operatorname{mid}(A) \end{aligned}$$

and

$$\big|\operatorname{mid}\big(W_0, S_0(a^*)\big)\big| = 2^n |W_0|.$$

The $2n$ facial hyperplanes F_{j,μ_j} partition $V \setminus F^*$ into 3^n pairwise disjoint open convex sets $D(\lambda_1, \ldots, \lambda_n)$, where $(\lambda_1, \ldots, \lambda_n) \in \{0, 1, -1\}^n$ and

$$D(0, \ldots, 0) = \operatorname{int}\big(B(a^*)\big).$$

Then

$$W_0 = A \cap \operatorname{int}\big(B(a^*)\big) = A \cap D(0, \ldots, 0)$$

and

$$\operatorname{mid}\big(W_0, S_0(a^*)\big) \subseteq \operatorname{int}\big(B(a^*)\big) \cap \operatorname{mid}(A) = D(0, \ldots, 0) \cap \operatorname{mid}(A).$$

Let $W_1, \ldots, W_{3^n-1}$ be the pairwise disjoint sets

$$A \cap D(\lambda_1, \ldots, \lambda_n),$$

where $(\lambda_1, \ldots, \lambda_n) \in \{0, 1, -1\}^n$ and $(\lambda_1, \ldots, \lambda_n) \neq (0, \ldots, 0)$. Since the sets $D(\lambda_1, \ldots, \lambda_n)$ are convex, it follows that

$$\mathrm{mid}(W_i) \cap \mathrm{mid}(W_j) = \emptyset$$

for $1 \leq i < j \leq 3^n - 1$, and

$$M\big(W_0, S_0(a^*)\big) \cap \mathrm{mid}(W_i) = \emptyset$$

for $i = 1, \ldots, 3^n - 1$.

We shall prove that there exists $i \in \{1, 2, \ldots, 3^n - 1\}$ such that the set W_i satisfies the conditions

$$|W_i| \geq \varepsilon_0 |A| = \varepsilon k_0$$

and

$$|2W_i| = \big|\mathrm{mid}(W_i)\big| \leq (c - \varepsilon_0)|W_i|.$$

Suppose not. Then $\big|\mathrm{mid}(W_i)\big| > (c - \varepsilon_0)|W_i|$ for every set W_i satisfying $|W_i| \geq \varepsilon_0 k_0$. Let $\sum'$ denote the sum over all $i \in [1, 3^n - 1]$ such that $|W_i| \geq \varepsilon_0 k_0$. Then

$$\begin{aligned}
ck_0 &\geq |2A| \\
&= \big|\mathrm{mid}(A)\big| \\
&\geq \big|\mathrm{mid}\big(W_0, S(a^*)\big)\big| + \sum_{i=1}^{3^n-1} \big|\mathrm{mid}(W_i)\big| \\
&\geq 2^n |W_0| + \sum{}' \big|\mathrm{mid}(W_i)\big| \\
&\geq 2^n |W_0| + (c - \varepsilon_0) \sum{}' |W_i| \\
&> 2^n |W_0| + c \sum{}' |W_i| - \varepsilon_0 k_0 \\
&> 2^n |W_0| + c \sum_{i=1}^{3^n-1} |W_i| - 3^n c \varepsilon_0 k_0 - \varepsilon_0 k_0 \\
&= (2^n - c)|W_0| + c \sum_{i=0}^{3^n-1} |W_i| - 3^n c \varepsilon_0 k_0 - \varepsilon_0 k_0 \\
&= (2^n - c)|W_0| + c\big(k_0 - |A \cap F^*|\big) - 3^n c \varepsilon_0 k_0 - \varepsilon_0 k_0 \\
&\geq (2^n - c)|W_0| + ck_0 - (3^n c + 2nc + 1)\varepsilon_0 k_0.
\end{aligned}$$

This implies that

$$(3^n c + 2nc + 1)\varepsilon_0 k_0 > (2^n - c)|W_0| > \frac{(2^n - c)k_0}{2(4c)^{2^n-1}}$$

and so

$$\varepsilon_0 > \frac{2^n - c}{2(3^n c + 2nc + 1)(4c)^{2^n-1}} = 2n\varepsilon_0 > \varepsilon_0,$$

which is absurd. Therefore, there exists a set $W = W_i \subseteq A$ such that $|W| \geq \varepsilon_0|A|$ and $|2W| \leq (c - \varepsilon_o)|W|$. This completes the proof.

Proof of Theorem 5.1. For $1 < c < 2^n$, let

$$\varepsilon_0 = \varepsilon_0(n, c) = \frac{2^n - c}{4n(3^n c + 2nc + 1)(4c)^{2^n - 1}}$$

be the positive real number defined in Lemma 5.12. Let $t = t(n, c)$ be the unique positive integer such that

$$t - 1 < \frac{c - 1}{\varepsilon_0} \leq t.$$

Let

$$\varepsilon_0^* = \varepsilon_0^t,$$

$$k^* = k^*(n, c) = (4c)^{2^n - 1},$$

and

$$k_0^* = k_0^*(n, c) = \varepsilon_0^{-t} k^*.$$

If $1 < c' < c$, then

$$\varepsilon_0(n, c) < \varepsilon_0(n, c')$$

and

$$k^*(n, c) > k^*(n, c').$$

Let A be a subset of V such that

$$|A| \geq k_0^* \geq k_0$$

and

$$|2A| \leq c|A|.$$

Suppose that

$$|A \cap H| \leq \varepsilon_0^*|A| \leq \varepsilon_0|A|$$

for every hyperplane H in V. By Lemma 5.12, there exists a set $W \subseteq A$ such that

$$|W| \geq \varepsilon_0|A| \geq \varepsilon_0 k_0^* = \varepsilon_0^{-(t-1)} k^* \geq k^*$$

and

$$|2W| \leq (c - \varepsilon_0)|W|.$$

Moreover, for every hyperplane H in V,

$$|W \cap H| \leq |A \cap H| \leq \varepsilon_0^t|A| \leq \varepsilon_0^{t-1}|W| \leq \varepsilon_0|W|.$$

Define $A_1' = W$ and $A_0' = A$. Let i $\leq j \leq t - 1$. Let $c' = c - j\varepsilon_0$. Then $j \leq t - 1$ implies that

$$1 < c' = c - j\varepsilon_0 < c.$$

Suppose that we have constructed sets

$$A'_j \subseteq A'_{j-1} \subseteq \ldots \subseteq A'_1 \subseteq A'_0 = A$$

such that

$$|A'_j| \geq \varepsilon_0^j |A| \geq \varepsilon_0^j k_0^* = \varepsilon_0^{-(t-j)} k^* \geq k^*$$

and

$$|2A'_j| \leq (c - j\varepsilon_0)|A_j| = c'|A_j|.$$

Moreover, for every hyperplane H in V,

$$\begin{aligned} |A'_j \cap H| &\leq |A \cap H| \\ &\leq \varepsilon_0^* |A| \\ &\leq \varepsilon_0^{t-j} |A'_j| \\ &\leq \varepsilon_0 |A'_j| \\ &= \varepsilon_0(n, c)|A'_j| \\ &< \varepsilon_0(n, c')|A'_j|. \end{aligned}$$

Since

$$|A'_j| \geq k^* = k^*(n, c) > k^*(n, c')$$

and

$$|2A'_j| \leq \big(c - j\varepsilon_0\big)|A'_j| = c'|A'_j|,$$

it follows from Lemma 5.12 that there exists a set $A'_{j+1} \subseteq A'_j$ such that

$$\begin{aligned} |A'_{j+1}| &\geq \varepsilon_0(n, c')|A'_j| \\ &> \varepsilon_0(n, c)|A'_j| \\ &\geq \varepsilon_0^{j+1}|A| \\ &\geq \varepsilon_0^{j+1} k_0^* \\ &= \varepsilon_0^{-(t-j-1)} k^* \\ &\geq k^* \end{aligned}$$

and

$$\begin{aligned} |2A'_{j+1}| &\leq (c' - \varepsilon_0(n, c'))|A'_{j+1}| \\ &\leq (c' - \varepsilon_0(n, c))|A'_{j+1}| \\ &= (c - (j+1)\varepsilon_0)\,|A'_{j+1}| \end{aligned}$$

In particular, when $j = t$, we obtain a set A'_t such that

$$|A'_t| \geq k^* > 1$$

and

$$|2A'_t| \leq (c - t\varepsilon_0)|A'_t| \leq |A'_t|.$$

This is impossible, because $|A'_t| > 1$ implies that

$$|2A'_t| > |A'_t|.$$

Therefore, there must exist a hyperplane H in V such that $|A \cap H| > \varepsilon_0^*|A|$. This completes the proof of Theorem 5.1.

Theorem 5.4 *Let $n \geq 2$, and let*

$$1 < c < 2^n.$$

There exist constants $k_0^* = k_0^*(n, c)$ and $\ell = \ell(n, c)$ such that if A is a finite subset of an n-dimensional Euclidean space V and if A satisfies

$$|A| \geq k_0^*$$

and

$$|2A| \leq c|A|,$$

then there exist ℓ parallel hyperplanes $H_1, \ldots, H_\ell$ in V such that

$$A \subseteq \bigcup_{i=1}^{\ell} H_i.$$

Proof. By Theorem 5.1, there exists a hyperplane H in V and a number $\varepsilon_0^* = \varepsilon_0^*(n, c) > 0$ such that $|A \cap H| > \varepsilon_0^*|A|$. Then $H = \{v \in V \mid (h, v) = \gamma\}$, where h is a nonzero vector in V and $\gamma \in \mathbf{R}$. If $a' \in A$, then a' lies in the hyperplane $H' = \{v \in V \mid (h, v) = \gamma'\}$, where $(h, a') = \gamma'$, and this hyperplane is parallel to H. Let $H = H_1, H_2, \ldots, H_\ell$ be a set of pairwise disjoint hyperplanes, each parallel to H, such that $A \cap H_i \neq \emptyset$ for $i = 1, \ldots, \ell$ and

$$A \subseteq \bigcup_{i=1}^{\ell} H_i.$$

Let

$$H_i = \{v \in V \mid (h, v) = \gamma_i\}.$$

Choose $a_i \in A \cap H_i$ for $i = 1, \ldots, \ell$. Then

$$\bigcup_{i=1}^{\ell}(a_i + (A \cap H)) \subseteq 2A$$

and

$$|a_i + (A \cap H)| = |A \cap H| > \varepsilon_0^*|A|.$$

If $b \in a_i + (A \cap H)$, then $(h, b) = \gamma_i + \gamma$, and so

$$(a_i + (A \cap H)) \cap \left(a_j + (A \cap H)\right) = \emptyset$$

for $i \neq j$. It follows that

$$\ell \varepsilon_0^* |A| < \sum_{i=1}^{\ell} |a_i + (A \cap H)| \leq |2A| \leq c|A|,$$

and so

$$\ell < \frac{c}{\varepsilon_0^*}.$$

This completes the proof.

5.5 Notes

It is not known if there exists a $\delta > 0$ such that Theorem 5.1 remains true if the condition $2|A| \leq c|A|$ is replaced by the condition $2|A| \leq c|A|^{1+\delta}$. Nor is it known how to generalize Theorem 5.1 to h-fold sumsets.

Theorem 5.1, in the case where A is a finite set of integer lattice points, appears as Lemma 2.12 in Freiman's monograph [54], but the proof in the book is difficult to follow. Fishburn [45] has published an independent proof for the special case $n = 2$, but this proof does not generalize to higher dimensions. In this chapter, the proof for all dimensions $n \geq 2$ and for arbitrary finite subsets of V comes from Nathanson [94] and includes complete proofs of various geometrical results on which Freiman's argument depends.

5.6 Exercises

1. Let H be a hyperplane, and let $v \in H$. Prove that $H - v$ is a subspace of dimension $n - 1$.

2. Let h_1 and h_2 be normal vectors to a hyperplane H. Prove that $h_1 = \theta h_2$ for some $\theta \neq 0$.

3. The set E is *parallel* to H in the vector space V if $E + v \subseteq H$ for some vector $v \in V$. Let H_1 and H_2 be hyperplanes with normal vectors h_1 and h_2, respectively. Prove that H_1 is parallel to H_2 if and only if $h_1 = \theta h_2$ for some $\theta \in \mathbf{R}, \theta \neq 0$.

4. Let $H = \{v \in V \mid (h, v) = \gamma\}$ be a hyperplane with normal vector h. Prove that h is perpendicular to every vector lying in H.

5. Let $f : V \to W$ be a linear map of vector spaces. If $K \subseteq V$ is convex, prove that $f(K)$ is convex. If $L \subseteq W$ is convex, prove that $f^{-1}(L)$ is convex.

6. Let $h \in V, h \neq 0$. Define $f : V \to \mathbf{R}$ by $f(v) = (h, v)$. Use the map f to show that the hyperplane $H = \{v \in V \mid (h, v) = 0\}$ and the open half-spaces $H^{(+1)}$ and $H^{(-1)}$ are convex.

7. Let V be an n-dimensional Euclidean space with inner product $(\ ,\ )$, and let $\{v_1, v_2, \ldots, v_m\} \subseteq V$. Prove that the vectors $v_1, v_2, \ldots, v_m$ are linearly independent if and only if there exist vectors $v_1^*, v_2^*, \ldots, v_m^*$ in V such that $(v_i, v_j^*) = \delta_{i,j}$ for $1 \leq i, j \leq m$.

8. Prove that a block is the convex hull of its vertices.

9. In the vector space $V = \mathbf{R}^2$, let $h_1 = (3, 1)$ and $h_2 = (1, 2)$. Let $H_1 = \{v \in V \mid (h_1, v) = 5\}$ and $H_2 = \{v \in V \mid (h_2, v) = 4\}$. Graph the hyperplanes H_1 and H_2, and label the four convex sets $H(\pm 1, \pm 1)$.

10. In the vector space $V = \mathbf{R}^2$, let $f_0 = (0, 0)$, $f_1 = (1, 2)$, and $f_2 = (3, 1)$. Let

$$\begin{aligned} S_2 &= \{f_2\}, \\ S_1 &= S_2 \cup (\{2f_1\} - S_2), \\ S_0 &= S_1 \cup (\{2f_0\} - S_1). \end{aligned}$$

Graph the block $B = \text{conv}(S_0)$, indicate the points in $\text{vert}(B)$, draw the four facial hyperplanes $F_{1,\pm 1}$ and $F_{2,\pm 1}$, and label the nine regions $D(\lambda_1, \lambda_2)$ for $(\lambda_1, \lambda_2) \in \{0, \pm 1\}^2$.

11. In the vector space $V = \mathbf{R}^3$, let $\mathbf{e}_1 = (4, 0, 0)$, $\mathbf{e}_2 = (0, 3, 0)$, and $\mathbf{e}_3 = (1, 1, 1)$. Sketch the block $B(0; \mathbf{e}_1, \mathbf{e}_2, \mathbf{e}_3)$. Find $(\lambda_1, \lambda_2, \lambda_3) \in \{0, \pm 1\}^3$ such that $(3, 4, -2) \in D(\lambda_1, \lambda_2, \lambda_3)$.

12. In the vector space $V = \mathbf{R}^3$, let $h_1 = (1, 0, 0)$, $h_2 = (-1, 1, 0)$, and $h_3 = (1, 2, 0)$. Consider the three hyperplanes

$$\begin{aligned} H_1 &= \{v \in V \mid (h_1, v) = 1\}, \\ H_2 &= \{v \in V \mid (h_2, v) = 2\}, \\ H_3 &= \{v \in V \mid (h_3, v) = -2\}. \end{aligned}$$

Prove that H_1, H_2, and H_3 are not linearly independent, and find all triples $(\mu_1, \mu_2, \mu_3) \in \{\pm 1\}^3$ such that $H(\mu_1, \mu_2, \mu_3) = \emptyset$.

13. Let $H_1, \ldots, H_n$ be n linearly independent hyperplanes in the n-dimensional vector space V. Prove that there exists a vector $v_0 \in V$ such that $\bigcap_{i=1}^n H_i = \{v_0\}$.

14. Let $H_1, \ldots, H_r$ be r linearly independent hyperplanes in the n-dimensional vector space V. Prove that there exists a vector $v_0 \in V$ and a subspace W of dimension $n - r$ such that $\bigcap_{i=1}^r H_i = \{v_0\} + W$.

15. Prove that Theorem 5.4 implies Theorem 5.1.

16. Let A be a finite subset of the vector space V. Prove that if $|A| > 1$, then $|2A| > |A|$.

17. Let A be a finite subset of the vector space V. Prove that $|2A| \geq 2|A| - 1$.

18. Let V be a vector space. For $\lambda \in \mathbf{R}$ and $A \subseteq V$, let

$$\lambda * A = \{\lambda a \mid a \in A\}.$$

Prove that $h\ (\lambda * A) = \lambda * (hA)$ and

$$\mathrm{conv}(hA) = h * \mathrm{conv}(A) = h\mathrm{conv}(A).$$

19. Let A be a finite subset of a real vector space, and let $|A| = k$. Prove that

$$h\mathrm{conv}(A) = k\mathrm{conv}(A) + (h - k)A$$

for all $h \geq k$.

20. Let V be a vector space, and let $\{a_1, \ldots, a_r\}$ be a set of r linearly independent vectors in V. Let $A = \{0, a_1, \ldots, a_r\}$. Prove that

$$|hA| = \frac{h^r}{r!} + O(h^{r-1}).$$

21. Let $\mathbf{Z}^n$ be the set of integer lattice points in $\mathbf{R}^n$, and let A be a finite subset of $\mathbf{Z}^n$ with $0 \in A$. Prove that there exists a constant $c = c(A)$ such that

$$|hA| \leq ch^n + O(h^{n-1}).$$

22. Let A be a finite subset of $\mathbf{Z}^n$ such that $0 \in A$ and A contains n linearly independent vectors. Prove that there exist constants $c_1 = c_1(A) > 0$ and $c_2 = c_2(A) > 0$ such that

$$c_1h^n + O(h^{n-1}) \leq |hA| \leq c_2h^n + O(h^{n-1}).$$

23. Let $A = \{a_0, a_1, \ldots, a_{k-1}\}$ be a finite subset of a vector space. The *affine dimension* of A is the maximum number of linearly independent vectors in the set $\{a_1 - a_0, \ldots, a_{k-1} - a_0\}$. Prove that if $A \subseteq \mathbf{Z}^n$ and the affine dimension of A is r, then there exist constants $c_1 = c_1(A) > 0$ and $c_2 = c_2(A) > 0$ such that

$$c_1h^r + O(h^{r-1}) \leq |hA| \leq c_2h^r + O(h^{r-1}).$$

6

Geometry of numbers

> But of [Minkowski] it might be said as of Saul that he went out to look after his father's asses and found a kingdom.
>
> H. Weyl [127]

6.1 Lattices and determinants

Minkowski's geometry of numbers is a beautiful and powerful tool that can be applied to many problems in number theory. In Section 6.3, for example, we shall give a geometric proof of the theorem of Lagrange that every nonnegative integer is the sum of four squares. The purpose of this chapter is to develop enough of the geometry of numbers to prove Theorem 6.12, which will be needed in the proof of Freiman's theorem in Chapter 8.

Let $\{\mathbf{a}_1, \ldots, \mathbf{a}_n\}$ be a basis for the Euclidean space $\mathbf{R}^n$. The abelian group generated by these n linearly independent vectors is the set of all sums of the form

$$u_1\mathbf{a}_1 + \cdots + u_n\mathbf{a}_n,$$

where $u_1, \ldots, u_n \in \mathbf{Z}$. A *lattice* in $\mathbf{R}^n$ is an abelian group Λ generated by a set of n linearly independent vectors. A set of n generators for Λ is called a *basis* for the lattice. For example, the *standard basis* for $\mathbf{R}^n$ is the set of vectors $\{\mathbf{e}_1, \ldots, \mathbf{e}_n\}$, where

$$\mathbf{e}_1 = (1, 0, 0, 0, \ldots, 0)$$

$$\begin{aligned}\mathbf{e}_2 &= (0, 1, 0, 0, \ldots, 0)\\ &\vdots\\ \mathbf{e}_n &= (0, 0, 0, \ldots, 0, 1).\end{aligned}$$

The integer lattice $\mathbf{Z}^n$ is the lattice with basis $\{\mathbf{e}_1, \ldots, \mathbf{e}_n\}$. The elements of $\mathbf{Z}^n$ are the vectors of the form $\mathbf{u} = (u_1, \ldots, u_n)$, where $u_i \in \mathbf{Z}$ for $i = 1, \ldots, n$.

The closure of the set X in $\mathbf{R}^n$ will be denoted $\overline{X}$. For $\mathbf{x} \in \mathbf{R}^n$, let $B(\mathbf{x}, \varepsilon)$ denote the open ball with center $\mathbf{x}$ and radius $\varepsilon > 0$. A group G in $\mathbf{R}^n$ is *discrete* if there exists $\varepsilon > 0$ such that $B(\mathbf{u}, \varepsilon) \cap G = \{\mathbf{u}\}$ for every $\mathbf{u} \in G$.

Theorem 6.1 *Let $\Lambda \subseteq \mathbf{R}^n$. Then Λ is a lattice if and only if Λ is a discrete subgroup that contains n linearly independent vectors.*

Theorem 6.2 *Let Λ be a lattice in $\mathbf{R}^n$, and let $\mathbf{b}_1, \ldots, \mathbf{b}_n$ be n linearly independent vectors contained in Λ. Then there exists a basis $\{\mathbf{a}_1, \ldots, \mathbf{a}_n\}$ for Λ such that each of the vectors $\mathbf{b}_j$ is of the form*

$$\mathbf{b}_j = \sum_{i=1}^{j} v_{i,j}\mathbf{a}_i,$$

where $v_{i,j} \in \mathbf{Z}$ for $j = 1, \ldots, n$ and $i = 1, \ldots, j$.

Proof. We shall prove both theorems at the same time.

Let Λ be a lattice in $\mathbf{R}^n$. Then Λ contains n linearly independent vectors. The integer lattice $\mathbf{Z}^n$ is discrete, since $B(\mathbf{g}, 1) \cap \mathbf{Z}^n = \{\mathbf{g}\}$ for all $\mathbf{g} \in \mathbf{Z}^n$. Let $\{\mathbf{a}_1, \ldots, \mathbf{a}_n\}$ be a basis for the lattice Λ, and let $T : \mathbf{R}^n \to \mathbf{R}^n$ be the linear transformation defined by $T(\mathbf{a}_i) = \mathbf{e}_i$ for $i = 1, \ldots, n$. Then T is an isomorphism, and $T(\Lambda) = \mathbf{Z}^n$. Let $U = T^{-1}(B(\mathbf{0}, 1))$. Since T is continuous, U is an open set in $\mathbf{R}^n$, and so there exists $\varepsilon > 0$ such that

$$\mathbf{0} \in B(\mathbf{0}, \varepsilon) \subseteq U.$$

If $\mathbf{u} \in B(\mathbf{0}, \varepsilon) \cap \Lambda$, then

$$T(\mathbf{u}) \in B(\mathbf{0}, 1) \cap \mathbf{Z}^n,$$

which implies that $T(\mathbf{u}) = \mathbf{0}$; hence $\mathbf{u} = \mathbf{0}$. Therefore, $B(\mathbf{0}, \varepsilon) \cap \Lambda = \{\mathbf{0}\}$.

Let $\mathbf{u}, \mathbf{u}' \in \Lambda$. Then $\mathbf{u} - \mathbf{u}' \in \Lambda$. If $\mathbf{u}' \in B(\mathbf{u}, \varepsilon)$, then $|\mathbf{u} - \mathbf{u}'| < \varepsilon$, and so

$$\mathbf{u} - \mathbf{u}' \in B(\mathbf{0}, \varepsilon) \cap \Lambda = \{\mathbf{0}\}$$

and so $\mathbf{u} - \mathbf{u}' = \mathbf{0}$. Thus,

$$B(\mathbf{u}, \varepsilon) \cap \Lambda = \{\mathbf{u}\}$$

for all $\mathbf{u} \in \Lambda$. This proves that the group Λ is discrete.

Conversely, let $\Lambda \neq \{\mathbf{0}\}$ be a discrete subgroup of $\mathbf{R}^n$ and let $\{\mathbf{b}_1, \ldots, \mathbf{b}_r\}$ be a maximal set of linearly independent vectors contained in Λ. Then $1 \leq r \leq n$. Fix $k \in [1, r]$, and let Λ_k be the set of all vectors $\mathbf{u} \in \Lambda$ of the form

$$\mathbf{u} = x_1\mathbf{b}_1 + \cdots + x_k\mathbf{b}_k,$$

where $x_k > 0$ and $0 \leq x_i < 1$ for $i = 1, \ldots, k-1$. Note that $\mathbf{b}_k \in \Lambda_k$, and so $\Lambda_k \neq \emptyset$. Let C_k be the set of all of the kth coordinates x_k of vectors in Λ_k, and let $c_{k,k} = \inf C_k$. Then there exists a sequence of vectors

$$\mathbf{u}_s = x_{1,s}\mathbf{b}_1 + \cdots + x_{k-1,s}\mathbf{b}_{k-1} + x_{k,s}\mathbf{b}_k \in \Lambda_k$$

such that $\lim_{s\to\infty} x_{k,s} = c_{k,k}$. Since $0 \leq x_{i,s} < 1$ for $i = 1, \ldots, k-1$ and $s = 1, 2, \ldots$, there exists a subsequence $\{\mathbf{u}_{s_j}\} \subseteq \Lambda_k$ that converges to a vector

$$\mathbf{a}_k = c_{1,k}\mathbf{b}_1 + \cdots + c_{k-1,k}\mathbf{b}_{k-1} + c_{k,k}\mathbf{b}_k \in \mathbf{R}^n.$$

Since Λ is a discrete subset of $\mathbf{R}^n$, the sequence $\{\mathbf{u}_{s_j}\}$ is eventually constant, and so $\mathbf{a}_k \in \Lambda_k$ and $c_{k,k} > 0$. Since the vectors $\mathbf{b}_1, \ldots, \mathbf{b}_r$ are linearly independent, it follows that the vectors $\mathbf{a}_1, \ldots, \mathbf{a}_r$ are also linearly independent.

We shall show that every element of Λ is an integral linear combination of $\mathbf{a}_1, \ldots, \mathbf{a}_r$. Let $\mathbf{u} \in \Lambda$. The vectors $\mathbf{a}_1, \ldots, \mathbf{a}_r$ span the vector subspace generated by Λ, and so there exist real numbers $u_1, \ldots, u_r$ such that

$$\mathbf{u} = u_1\mathbf{a}_1 + \cdots + u_r\mathbf{a}_r.$$

Suppose that $u_j \notin \mathbf{Z}$ for some $j \in [1, r]$. Let k be the greatest integer such that $u_k \notin \mathbf{Z}$, and let $u_k = g_k + x_k$, where $g_k \in \mathbf{Z}$ and $0 < x_k < 1$. The vector

$$g_k\mathbf{a}_k + u_{k+1}\mathbf{a}_{k+1} + \cdots + u_r\mathbf{a}_r$$

belongs to the group Λ since it is an integral linear combination of $\mathbf{a}_k, \ldots, \mathbf{a}_r$. Then

$$\begin{aligned} \mathbf{u}' &= \mathbf{u} - (g_k\mathbf{a}_k + u_{k+1}\mathbf{a}_{k+1} + \cdots + u_r\mathbf{a}_r) \\ &= u_1\mathbf{a}_1 + \cdots + u_{k-1}\mathbf{a}_{k-1} + x_k\mathbf{a}_k \end{aligned}$$

also belongs to Λ. Since $\mathbf{a}_j = \sum_{i=1}^{j} c_{i,j}\mathbf{b}_i$ for $j = 1, \ldots, k$, it follows that $\mathbf{u}'$ can be written as a linear combination of the vectors $\mathbf{b}_1, \ldots, \mathbf{b}_k$ in the form

$$\mathbf{u}' = u_1'\mathbf{b}_1 + \cdots + u_{k-1}'\mathbf{b}_{k-1} + x_k c_{k,k}\mathbf{b}_k.$$

For $i = 1, \ldots, k-1$, let $u_i' = g_i' + x_i$, where $g_i' \in \mathbf{Z}$ and $0 \leq x_i < 1$. Then $g_1'\mathbf{b}_1 + \cdots + g_{k-1}'\mathbf{b}_{k-1} \in \Lambda$, and so

$$\begin{aligned} \mathbf{u}'' &= \mathbf{u}' - (g_1'\mathbf{b}_1 + \cdots + g_{k-1}'\mathbf{b}_{k-1}) \\ &= x_1\mathbf{b}_1 + \cdots + x_{k-1}\mathbf{b}_{k-1} + x_k c_{k,k}\mathbf{b}_k \\ &\in \Lambda. \end{aligned}$$

Since $0 \leq x_i < 1$ and $x_k c_{k,k} > 0$, it follows that $\mathbf{u}'' \in \Lambda_k$, which is impossible, because the inequality

$$0 < x_k c_{k,k} < c_{k,k}$$

contradicts the minimality of $c_{k,k}$. Thus, every vector $\mathbf{u} \in \Lambda$ must be an integral linear combination of the linearly independent vectors $\mathbf{a}_1, \ldots, \mathbf{a}_r$. If Λ contains n linearly independent vectors, then $r = n$ and Λ is a lattice. This proves Theorem 6.1.

To prove Theorem 6.2, let $\mathbf{b}_1, \ldots, \mathbf{b}_n$ be n linearly independent vectors in the lattice Λ. Since Λ is discrete, the preceding argument shows that there exists a basis $\{\mathbf{a}_1, \ldots, \mathbf{a}_n\}$ for Λ such that each vector $\mathbf{a}_j$ is of the form

$$\mathbf{a}_j = \sum_{i=1}^{j} c_{i,j}\mathbf{b}_i,$$

where $c_{i,j} \in \mathbf{R}$ for $j = 1, \ldots, n$, $i = 1, \ldots, j$, and $c_{j,j} > 0$. Solving these equations for $\mathbf{b}_1, \ldots, \mathbf{b}_n$, we obtain real numbers $v_{i,j}$ such that

$$\mathbf{b}_j = \sum_{i=1}^{j} v_{i,j}\mathbf{a}_i$$

for $j = 1, \ldots, n$. Since $\{\mathbf{a}_1, \ldots, \mathbf{a}_n\}$ is a basis for Λ, it follows that $v_{i,j} \in \mathbf{Z}$ for $j = 1, \ldots, n$ and $i = 1, \ldots, j$. Also, $v_{j,j} = 1/c_{j,j} \geq 1$ for $j = 1, \ldots, n$. This completes the proof.

The basis of a lattice is not uniquely determined by the lattice. For example, let Λ be the lattice in $\mathbf{Z}^2$ generated by the vectors $\mathbf{a}_1 = (7, 5)$ and $\mathbf{a}_2 = (4, 3)$. Since $\mathbf{a}_1, \mathbf{a}_2 \in \mathbf{Z}^2$, it follows that $\Lambda \subseteq \mathbf{Z}^2$. Conversely, since

$$\mathbf{e}_1 = 3\mathbf{a}_1 - 5\mathbf{a}_2 \in \Lambda$$

and

$$\mathbf{e}_2 = -4\mathbf{a}_1 + 7\mathbf{a}_2 \in \Lambda$$

it follows that $\mathbf{Z}^2 \subseteq \Lambda$. Thus, $\mathbf{Z}^2 = \Lambda$, and the sets $\{(1, 0), (0, 1)\}$ and $\{(7, 5), (4, 3)\}$ are distinct bases for $\mathbf{Z}^2$. Observe that $\mathbf{a}_1 = 7\mathbf{e}_1 + 5\mathbf{e}_2$, $\mathbf{a}_2 = 4\mathbf{e}_1 + 3\mathbf{e}_2$, and the determinant

$$\begin{vmatrix} 7 & 4 \\ 5 & 3 \end{vmatrix} = 1.$$

This example can be generalized. Let $U = (u_{ij})$ be an $n \times n$ *unimodular matrix*, that is, a matrix with integer entries and determinant $\det(U) = \pm 1$. Then the inverse matrix $U^{-1} = V = (v_{ij})$ also has integer entries and $\det(V) = \det(U) = \pm 1$. Since $UV = VU = I$, where I is the $n \times n$ identity matrix,

$$\sum_{k=1}^{n} u_{ik}v_{kj} = \sum_{k=1}^{n} v_{ik}u_{kj} = \delta_{ij} = \begin{cases} 1 & \text{if } i = j \\ 0 & \text{if } i \neq j. \end{cases}$$

Let $\mathbf{a}_1, \ldots, \mathbf{a}_n$ be a basis for the lattice Λ in $\mathbf{R}^n$. We shall use the matrix U to construct another basis for Λ. For $j = 1, \ldots, n$, let

$$\mathbf{a}'_j = \sum_{i=1}^{n} u_{ij}\mathbf{a}_i \in \Lambda.$$

Let Λ' be the group in $\mathbf{R}^n$ generated by the n vectors $\{\mathbf{a}'_1, \ldots, \mathbf{a}'_n\}$. Since $\mathbf{a}'_j \in \Lambda$ for $j = 1, \ldots, n$, it follows that $\Lambda' \subseteq \Lambda$. Since V is the inverse of the matrix U,

$$
\begin{aligned}
\sum_{k=1}^{n} v_{ki}\mathbf{a}'_k &= \sum_{k=1}^{n} v_{ki} \sum_{j=1}^{n} u_{jk}\mathbf{a}_j \\
&= \sum_{j=1}^{n} \left(\sum_{k=1}^{n} u_{jk}v_{ki} \right) \mathbf{a}_j \\
&= \sum_{j=1}^{n} \delta_{ji}\mathbf{a}_j \\
&= \mathbf{a}_i \in \Lambda',
\end{aligned}
$$

and so $\Lambda \subseteq \Lambda'$. Thus, $\Lambda = \Lambda'$, and the vectors $\mathbf{a}_1, \ldots, \mathbf{a}_n$ and $\mathbf{a}'_1, \ldots, \mathbf{a}'_n$ are both bases for the lattice Λ.

Conversely, let $\{\mathbf{a}_1, \ldots, \mathbf{a}_n\}$ and $\{\mathbf{a}'_1, \ldots, \mathbf{a}'_n\}$ be two bases for the lattice Λ. For $i, j = 1, \ldots, n$, there exist integers u_{ij} and v_{ij} such that

$$
\mathbf{a}'_j = \sum_{i=1}^{n} u_{ij}\mathbf{a}_i
$$

and

$$
\mathbf{a}_j = \sum_{i=1}^{n} v_{ij}\mathbf{a}'_i.
$$

Then

$$
\begin{aligned}
\mathbf{a}_j &= \sum_{k=1}^{n} v_{kj}\mathbf{a}'_k \\
&= \sum_{k=1}^{n} v_{kj} \sum_{i=1}^{n} u_{ik}\mathbf{a}_i \\
&= \sum_{i=1}^{n} \left(\sum_{k=1}^{n} u_{ik}v_{kj} \right) \mathbf{a}_i.
\end{aligned}
$$

Since the n vectors $\mathbf{a}_1, \ldots, \mathbf{a}_n$ are linearly independent, it follows that

$$
\sum_{k=1}^{n} u_{ik}v_{kj} = \delta_{ij}
$$

for $i, j = 1, \ldots, n$. Similarly,

$$
\sum_{k=1}^{n} v_{ik}u_{kj} = \delta_{ij}
$$

for $i, j = 1, \ldots, n$. Let U and V be the matrices $U = (u_{ij})$ and $V = (v_{ij})$. Then $V = U^{-1}$ and, since the matrix elements u_{ij} and v_{ij} are integers, $\det(U) = \det(V) = \pm 1$. Thus, any two bases for a lattice Λ in $\mathbf{R}^n$ are related by a unimodular matrix.

Let $\mathbf{a}_1, \ldots, \mathbf{a}_n \in \mathbf{R}^n$, and let

$$\mathbf{a}_j = \sum_{i=1}^{n} a_{ij}\mathbf{e}_i$$

for $j = 1, \ldots, n$, where a_{ij} are the coordinates of $\mathbf{a}_i$ with respect to the standard basis vectors $\mathbf{e}_1, \ldots, \mathbf{e}_n$. The $n \times n$ matrix $A = (a_{ij})$ is called the *matrix of the vectors* $\mathbf{a}_1, \ldots, \mathbf{a}_n$. The vectors $\mathbf{a}_1, \ldots, \mathbf{a}_n$ are linearly independent if and only if $\det(A) \neq 0$.

The *determinant of the lattice* Λ, denoted $\det(\Lambda)$, plays a fundamental role in the geometry of numbers. This determinant is defined by

$$\det(\Lambda) = |\det(A)|,$$

where A is the matrix of a basis $\{\mathbf{a}_1, \ldots, \mathbf{a}_n\}$ for Λ. Then $\det(\Lambda) \neq 0$, since a basis for a lattice is a set of n linearly independent vectors in $\mathbf{R}^n$. We shall prove that $\det(\Lambda)$ is independent of the choice of basis for the lattice Λ.

Let $\{\mathbf{a}_1, \ldots, \mathbf{a}_n\}$ and $\{\mathbf{a}'_1, \ldots, \mathbf{a}'_n\}$ be two bases for the lattice Λ, and let $U = (u_{ij})$ be the unimodular matrix such that

$$\mathbf{a}'_j = \sum_{i=1}^{n} u_{ij}\mathbf{a}_i.$$

Let

$$\mathbf{a}_j = \sum_{i=1}^{n} a_{ij}\mathbf{e}_i$$

and

$$\mathbf{a}'_j = \sum_{i=1}^{n} a'_{ij}\mathbf{e}_i,$$

where a_{ij} and a'_{ij} are the coordinates of $\mathbf{a}_j$ and $\mathbf{a}'_j$ with respect to the standard basis $\{\mathbf{e}_1, \ldots, \mathbf{e}_n\}$. Then $\det(U) = \pm 1$. Let $A = (a_{ij})$ and $A' = \left(a'_{ij}\right)$ be the matrices of the bases $\{\mathbf{a}_1, \ldots, \mathbf{a}_n\}$ and $\{\mathbf{a}'_1, \ldots, \mathbf{a}'_n\}$, respectively. We shall show that $A' = AU$. Observe that

$$\begin{aligned}
\sum_{i=1}^{n} a'_{ij}\mathbf{e}_i &= \mathbf{a}'_j \\
&= \sum_{k=1}^{n} u_{kj}\mathbf{a}_k \\
&= \sum_{k=1}^{n} u_{kj} \sum_{i=1}^{n} a_{ik}\mathbf{e}_i \\
&= \sum_{i=1}^{n} \left(\sum_{k=1}^{n} a_{ik}u_{kj} \right) \mathbf{e}_i,
\end{aligned}$$

and so

$$a'_{ij} = \sum_{k=1}^{n} a_{ik} u_{kj}$$

for $i, j = 1, \ldots, n$. This is equivalent to the matrix equation

$$A' = AU.$$

It follows that

$$|\det(A')| = |\det(AU)| = |\det(A)||\det(U)| = |\det(A)|. \tag{6.1}$$

This proves that $\det(\Lambda)$ is well defined.

The *fundamental parallelepiped* of the lattice Λ with respect to the basis $\{\mathbf{a}_1, \ldots, \mathbf{a}_n\}$ is the set

$$\begin{aligned} F(\Lambda) &= F(\Lambda; \mathbf{a}_1, \ldots, \mathbf{a}_n) \\ &= \left\{ \sum_{i=1}^{n} x_i \mathbf{a}_i : 0 \leq x_i < 1 \text{ for } i = 1, \ldots, n \right\} \subseteq \mathbf{R}^n. \end{aligned}$$

If $\mathbf{a}_j = \sum_{i=1}^{n} a_{ij} \mathbf{e}_i$, then the volume of the fundamental parallelepiped is

$$\begin{aligned} \operatorname{vol}(F(\Lambda; \mathbf{a}_1, \ldots, \mathbf{a}_n)) &= \int \cdots \int_{F(\Lambda; \mathbf{a}_1, \ldots, \mathbf{a}_n)} dV \\ &= \int_0^1 \cdots \int_0^1 |\det(a_{ij})| dx_1 \cdots dx_n \\ &= |\det(a_{ij})| \\ &= \det(\Lambda). \end{aligned}$$

Thus, while the fundamental parallelepiped of a lattice is a set in $\mathbf{R}^n$ that depends on the choice of basis for the lattice, the volume of a fundamental parallelepiped is independent of the choice of basis.

Theorem 6.3 *Let Λ be a lattice in $\mathbf{R}^n$, and let $F(\Lambda)$ be the fundamental parallelepiped of Λ with respect to the basis $\{\mathbf{a}_1, \ldots, \mathbf{a}_n\}$. Then every vector in $\mathbf{R}^n$ has a unique representation as the sum of an element in the lattice and an element in the fundamental parallelepiped.*

Proof. Let $\mathbf{v}$ be any vector in $\mathbf{R}^n$. Since $\{\mathbf{a}_1, \ldots, \mathbf{a}_n\}$ is a basis for $\mathbf{R}^n$, there exist real numbers $v_1, \ldots, v_n$ such that

$$\mathbf{v} = \sum_{i=1}^{n} v_i \mathbf{a}_i.$$

Let $v_i = u_i + x_i$, where $u_i \in \mathbf{Z}$ and $x_i \in [0, 1)$. Then

$$\begin{aligned} \sum_{i=1}^{n} u_i \mathbf{a}_i &\in \Lambda, \\ \sum_{i=1}^{n} x_i \mathbf{a}_i &\in F(\Lambda), \end{aligned}$$

and so

$$\mathbf{v} = \sum_{i=1}^{n} v_i \mathbf{a}_i = \sum_{i=1}^{n} u_i \mathbf{a}_i + \sum_{i=1}^{n} x_i \mathbf{a}_i \in \Lambda + F(\Lambda).$$

Thus, $\mathbf{R}^n = \Lambda + F(\Lambda)$. If

$$\mathbf{v} = \sum_{i=1}^{n} u_i \mathbf{a}_i + \sum_{i=1}^{n} x_i \mathbf{a}_i = \sum_{i=1}^{n} u_i' \mathbf{a}_i + \sum_{i=1}^{n} x_i' \mathbf{a}_i$$

with $u_i, u_i' \in \mathbf{Z}$ and $x_i, x_i' \in [0, 1)$, then the linear independence of the vectors $\mathbf{a}_1, \ldots, \mathbf{a}_n$ implies that $u_i + x_i = u_i' + x_i'$, and so $u_i - u_i' = x_i' - x_i \in (-1, 1)$ for $i = 1, \ldots, n$. Since $u_i - u_i' \in \mathbf{Z}$, it follows that $u_i = u_i'$ and $x_i = x_i'$. This completes the proof.

6.2 Convex bodies and Minkowski's First Theorem

If $\mathbf{a}$ and $\mathbf{b}$ are vectors in $\mathbf{R}^n$, the *line segment* from $\mathbf{a}$ to $\mathbf{b}$ is the set of all vectors of the form $(1 - t)\mathbf{a} + t\mathbf{b}$, where $t \in [0, 1]$. The set K in $\mathbf{R}^n$ is *convex* if, for every pair of points $\mathbf{a}, \mathbf{b} \in K$, the line segment from $\mathbf{a}$ to $\mathbf{b}$ also belongs to K. A *body* in $\mathbf{R}^n$ is a bounded open set. We shall consider only nonempty convex bodies K with finite Jordan volume, denoted $\mathrm{vol}(K)$. For $\lambda \in \mathbf{R}$, $\lambda \geq 0$, we define

$$\lambda * K = \{\lambda \mathbf{a} : \mathbf{a} \in K\}.$$

Then $\mathrm{vol}(\lambda * K) = \lambda^n \mathrm{vol}(K)$.

Let $r, r_1, \ldots, r_n$ be positive real numbers. The following are simple examples of convex bodies in $\mathbf{R}^n$:

(i) The *ball* $B(\mathbf{0}, r)$ consisting of all vectors $(x_1, \ldots, x_n)$ such that

$$x_1^2 + \cdots + x_n^2 < r^2.$$

(ii) The *ellipsoid* consisting of all vectors $(x_1, \ldots, x_n)$ such that

$$\frac{x_1^2}{r_1^2} + \cdots + \frac{x_n^2}{r_n^2} < 1.$$

(iii) The *cube* consisting of all vectors $(x_1, \ldots, x_n)$ such that

$$\max(|x_1|, \ldots, |x_n|) < r.$$

(iv) The *box* consisting of all vectors $(x_1, \ldots, x_n)$ such that

$$|x_i| < r_i$$

for $i = 1, \ldots, n$.

(v) The *interior of the block* B, denoted $\text{int}(B)$, where $\mathbf{a}_1, \ldots, \mathbf{a}_n$ are linearly independent vectors and

$$\text{int}(B) = \{x_1\mathbf{a}_1 + \cdots + x_n\mathbf{a}_n : -1 < x_i < 1 \text{ for } i = 1, \ldots, n\}.$$

(vi) The *octahedron* consisting of all vectors $(x_1, \ldots, x_n)$ such that

$$|x_1| + \cdots + |x_n| < r.$$

(vii) The *simplex* consisting of all vectors $(x_1, \ldots, x_n)$ such that $0 < x_i < r$ for $i = 1, \ldots, n$ and

$$x_1 + \cdots + x_n < r.$$

The set K is *symmetric* if $\mathbf{a} \in K$ implies that $-\mathbf{a} \in K$. If K is a symmetric convex body and $\mathbf{a} \in K$, then $-\mathbf{a} \in K$, and so $\mathbf{0} = (1/2)\mathbf{a} + (1/2)(-\mathbf{a}) \in K$. The simplex is not symmetric. Examples (i)–(vi) are symmetric convex bodies.

Lemma 6.1 (Blichfeldt) *Let* Λ *be a lattice in* $\mathbf{R}^n$ *and* K *be a body in* $\mathbf{R}^n$ *with volume greater than* $\det(\Lambda)$. *Then there exist vectors* $\mathbf{a}$ *and* $\mathbf{b}$ *in* K *such that* $\mathbf{a} - \mathbf{b} \in \Lambda \setminus \{\mathbf{0}\}$.

Proof. Fix a basis for the lattice Λ, and let $F = F(\Lambda)$ be the fundamental parallelepiped of the lattice Λ with respect to this basis. Then $\text{vol}(F) = \det(\Lambda)$. Since K is bounded and Λ is discrete, there exist only finitely many lattice points $\mathbf{u} \in \Lambda$ such that $K \cap (\mathbf{u} + F) \neq \emptyset$. Since

$$\mathbf{R}^n = \bigcup_{u \in \Lambda} (\mathbf{u} + F),$$

it follows that

$$K = \bigcup_{u \in \Lambda} (K \cap (\mathbf{u} + F))$$

and

$$\begin{aligned} \text{vol}(K) &= \sum_{u \in \Lambda} \text{vol}(K \cap (\mathbf{u} + F)) \\ &= \sum_{u \in \Lambda} \text{vol}((K - \mathbf{u}) \cap F) \\ &> \text{vol}(F). \end{aligned}$$

Since $(K - \mathbf{u}) \cap F \subseteq F$ for all lattice points $\mathbf{u}$, it follows that the sets $(K - \mathbf{u}) \cap F$ cannot be pairwise disjoint, and so there exist distinct lattice points $\mathbf{u}_1, \mathbf{u}_2 \in \Lambda$ such that $(K - \mathbf{u}_1) \cap (K - \mathbf{u}_2) \neq \emptyset$. This means that there exist distinct vectors $\mathbf{a}, \mathbf{b} \in K$ such that $\mathbf{a} - \mathbf{u}_1 = \mathbf{b} - \mathbf{u}_2$, and so $\mathbf{a} - \mathbf{b} = \mathbf{u}_1 - \mathbf{u}_2$ is a nonzero element of Λ.

Theorem 6.4 (Minkowski's first theorem) *Let* Λ *be a lattice in* $\mathbf{R}^n$, *and let* K *be a symmetric convex body in* $\mathbf{R}^n$ *with volume greater than* $2^n \det(\Lambda)$. *Then* K *contains a nonzero element of the lattice* Λ.

Proof. Let $K' = (1/2)K$. Then K' is a symmetric convex body with

$$\text{vol}(K') = \frac{\text{vol}(K)}{2^n} > \det(\Lambda).$$

It follows from Lemma 6.1 that there exist vectors $\mathbf{a}', \mathbf{b}' \in K'$ such that $\mathbf{a}' - \mathbf{b}'$ is a nonzero element of Λ, and $\mathbf{a}' = \mathbf{a}/2$ and $\mathbf{b}' = \mathbf{b}/2$, where $\mathbf{a}, \mathbf{b} \in K$. Since K is symmetric, $-\mathbf{b} \in K$, and since K is convex,

$$(1/2)\mathbf{a} + (1/2)(-\mathbf{b}) = \mathbf{a}' - \mathbf{b}' \in K.$$

Thus, K contains the nonzero lattice point $\mathbf{a}' - \mathbf{b}'$.

Corollary 6.1 *Let K be a symmetric convex body in $\mathbf{R}^n$ with volume greater than 2^n. Then K contains a nonzero element of the lattice $\mathbf{Z}^n$.*

Proof. This follows immediately from the theorem since $\det(\mathbf{Z}^n) = 1$.

Corollary 6.2 *Let Λ be a lattice in $\mathbf{R}^n$, and let K be a symmetric convex body in $\mathbf{R}^n$. Let*

$$\lambda_1 = \inf\{\lambda > 0 : (\lambda * K) \cap (\Lambda \setminus \{\mathbf{0}\}) \neq \emptyset\}. \tag{6.2}$$

Then

$$\lambda_1^n \, vol(K) \leq 2^n \det(\Lambda). \tag{6.3}$$

Proof. Since the lattice Λ is discrete, there exists $\varepsilon > 0$ such that the ball $B(0, \varepsilon)$ contains no nonzero lattice point. Since the convex body K is bounded, there exists $\mu > 0$ such that $\mu * K \subseteq B(0, \varepsilon)$. Then

$$(\mu * K) \cap \Lambda \subseteq B(0, \varepsilon) \cap \Lambda = \{\mathbf{0}\},$$

and so $\lambda_1 \geq \mu > 0$. Suppose that there exists a nonzero lattice point $\mathbf{u}$ such that $\mathbf{u} \in \lambda_1 * K$. Then $\mathbf{u} = \lambda_1 * \mathbf{x}$ for some $\mathbf{u} \in K$. Since K is open, there exists $\delta \in (0, 1)$ such that $B(\mathbf{x}, \delta) \subseteq K$. Let $\delta' = \delta/(2|\mathbf{x}|)$. Then

$$\mathbf{x}' = \mathbf{x} + \delta'\mathbf{x} = (1 + \delta')\mathbf{x} \in K.$$

This implies that

$$\mathbf{u} = \lambda_1\mathbf{x} = \left(\frac{\lambda_1}{1+\delta'}\right)\mathbf{x}' \in \left(\left(\frac{\lambda_1}{1+\delta'}\right) * K\right) \cap (\Lambda \setminus \{\mathbf{0}\}),$$

which is impossible, because

$$0 < \frac{\lambda_1}{1+\delta'} < \lambda_1.$$

Therefore,

$$(\lambda_1 * K) \cap (\Lambda \setminus \{\mathbf{0}\}) = \emptyset,$$

and Minkowski's theorem implies that

$$\lambda_1^n \operatorname{vol}(K) = \operatorname{vol}(\lambda_1 * K) \leq 2^n \det(\Lambda).$$

This completes the proof.

It is easy to see that inequality (6.3) implies Minkowski's first theorem. Suppose that $\operatorname{vol}(K) > 2^n \det(\Lambda)$. Then (6.3) implies that $\lambda_1 < 1$. Choose $\lambda \in \mathbf{R}$ such that $\lambda_1 < \lambda < 1$. Then $\lambda * K$ contains a nonzero element of Λ, and $\lambda * K \subseteq K$.

6.3 Application: Sums of four squares

We shall use Minkowski's First Theorem to give a simple proof of the famous theorem of Lagrange that every nonnegative integer can be represented as the sum of four squares. We need three simple lemmas.

Lemma 6.2 *Let m be an odd, positive integer. There exist integers a and b such that*

$$a^2 + b^2 + 1 \equiv 0 \pmod{m}.$$

Proof. The proof is in three steps.

Step 1. Let $m = p$ be an odd prime, and let

$$A = \{a^2 : a = 0, 1, \ldots, (p-1)/2\}$$

and

$$B = \{-b^2 - 1 : b = 0, 1, \ldots, (p-1)/2\}.$$

Since $|A| = |B| = (p+1)/2$ and the elements of the set A (resp. B) are pairwise incongruent modulo p, it follows from the pigeonhole principle that there exist integers $a, b \in [0, (p-1)/2]$ such that

$$a^2 \equiv -b^2 - 1 \pmod{p}.$$

Step 2. Let $m = p^k$, where p is an odd prime and $k \geq 1$. We shall prove by induction on k that the congruence

$$a^2 + b^2 + 1 \equiv 0 \pmod{p^k}$$

is solvable. The case $k = 1$ has just been proven. Suppose that the congruence holds for some $k \geq 1$. Then at least one of the integers a, b is not divisible by p, say, $a \not\equiv 0 \pmod{p}$. There is an integer s such that

$$a^2 = -b^2 - 1 + sp^k.$$

Since $(2a, p) = 1$, there exists an integer t such that

$$s + 2at \equiv 0 \pmod{p}.$$

Let $a_1 = a + tp^k$. Then

$$\begin{aligned} a_1^2 &= (a+tp^k)^2 \\ &= a^2 + 2atp^k + t^2p^{2k} \\ &\equiv -b^2 - 1 + sp^k + 2atp^k \pmod{p^{k+1}} \\ &\equiv -b^2 - 1 + (s+2at)p^k \pmod{p^{k+1}} \\ &\equiv -b^2 - 1 \pmod{p^{k+1}}. \end{aligned}$$

This completes the induction.

Step 3. Let m be a positive, odd integer. The result is trivial for $m = 1$, so we can assume that $m \geq 3$. Then

$$m = \prod_{i=1}^{r} p_i^{k_i},$$

where $p_1, \ldots, p_r$ are distinct odd primes and $k_i \geq 1$ for $i = 1, \ldots, r$. For each of the r prime powers $p_i^{k_i}$, there are integers a_i, b_i such that

$$a_i^2 + b_i^2 + 1 \equiv 0 \pmod{p_i^{k_i}}.$$

By the Chinese remainder theorem, there exist integers a, b such that

$$a \equiv a_i \pmod{p_i^{k_i}}$$

and

$$b \equiv b_i \pmod{p_i^{k_i}}$$

for all $i = 1, \ldots, r$. Then

$$a^2 + b^2 + 1 \equiv 0 \pmod{m}.$$

This completes the proof of the lemma.

Lemma 6.3 *If every odd, positive integer is the sum of four squares, then every positive integer is the sum of four squares.*

Proof. If n is the sum of four squares, say,

$$n = a^2 + b^2 + c^2 + d^2,$$

then

$$2n = (a+b)^2 + (a-b)^2 + (c+d)^2 + (c-d)^2,$$

and so $2n$ is also the sum of four squares. Iterating this argument, we obtain that $2^k n$ is a sum of four squares for every $k \geq 0$. The lemma follows from the observation that every positive integer is the form $2^k n$ for some odd number n.

Lemma 6.4 *Let $B(\mathbf{0}, r)$ be the ball of radius r in $\mathbf{R}^4$. Then vol$(B(\mathbf{0}, r)) = \pi^2 r^4/2$.*

Proof. The volume of $B(\mathbf{0}, r)$ is the value of the iterated integral

$$\int_{-r}^{r}\int_{-\sqrt{r^2-x_1^2}}^{\sqrt{r^2-x_1^2}}\int_{-\sqrt{r^2-x_1^2-x_2^2}}^{\sqrt{r^2-x_1^2-x_2^2}}\int_{-\sqrt{r^2-x_1^2-x_2^2-x_3^2}}^{\sqrt{r^2-x_1^2-x_2^2-x_3^2}} dx_4dx_3dx_2dx_1.$$

The computation is left to the reader (Exercise 15).

Theorem 6.5 (Lagrange) *Every positive integer is the sum of four squares.*

Proof. By Lemma 6.3, it suffices to prove the theorem for odd integers. Let m be an odd positive integer. By Lemma 6.2, there exist integers a and b such that $a^2 + b^2 + 1 \equiv 0 \pmod{m}$. Let Λ be the lattice in $\mathbf{R}^4$ with basis vectors

$$\begin{aligned}
\mathbf{a}_1 &= (m, 0, 0, 0),\\
\mathbf{a}_2 &= (0, m, 0, 0),\\
\mathbf{a}_3 &= (a, b, 1, 0),\\
\mathbf{a}_4 &= (b, -a, 0, 1).
\end{aligned}$$

Then $\Lambda \subseteq \mathbf{Z}^4$, $\det(\Lambda) = m^2$, and the lattice consists of all vectors of the form

$$\begin{aligned}
\mathbf{u} &= u_1\mathbf{a}_1 + u_2\mathbf{a}_2 + u_3\mathbf{a}_3 + u_4\mathbf{a}_4\\
&= (u_1m + u_3a + u_4b, u_2m + u_3b - u_4a, u_3, u_4),
\end{aligned}$$

where $u_1, u_2, u_3, u_4 \in \mathbf{Z}$. The congruence condition on a and b implies that

$$\begin{aligned}
|\mathbf{u}|^2 &= (u_1m + u_3a + u_4b)^2 + (u_2m + u_3b - u_4a)^2 + u_3^2 + u_4^2\\
&\equiv (u_3^2 + u_4^2)(a^2 + b^2 + 1) \pmod{m}\\
&\equiv 0 \pmod{m}
\end{aligned}$$

for all lattice points $\mathbf{u} \in \Lambda$. Let $K = B(\mathbf{0}, \sqrt{2m})$ be the ball of radius $\sqrt{2m}$ in $\mathbf{R}^4$. Then K is a symmetric convex body and, by Lemma 6.4,

$$\mathrm{vol}(K) = \frac{\pi^2(\sqrt{2m})^4}{2} = 2\pi^2m^2 > 16m^2 = 2^4\det(\Lambda).$$

It follows from Minkowski's first theorem (Theorem 6.4) that the ball K contains a nonzero lattice point

$$\begin{aligned}
\mathbf{u} &= u_1\mathbf{a}_1 + u_2\mathbf{a}_2 + u_3\mathbf{a}_3 + u_4\mathbf{a}_4\\
&= v_1\mathbf{e}_1 + v_2\mathbf{e}_e + v_3\mathbf{e}_3 + v_4\mathbf{e}_4,
\end{aligned}$$

where $u_1, u_2, u_3, u_4, v_1, v_2, v_3, v_4 \in \mathbf{Z}$. Since

$$|\mathbf{u}|^2 = v_1^2 + v_2^2 + v_3^2 + v_4^2 \equiv 0 \pmod{m}$$

and

$$0 < |\mathbf{u}|^2 = v_1^2 + v_2^2 + v_3^2 + v_4^2 < (\sqrt{2m})^2 = 2m,$$

it follows that

$$v_1^2 + v_2^2 + v_3^2 + v_4^2 = m.$$

This completes the proof.

6.4 Successive minima and Minkowski's second theorem

For vectors $\mathbf{x}, \mathbf{y} \in \mathbf{R}^n$, we let $d(\mathbf{x}, \mathbf{y}) = |\mathbf{x} - \mathbf{y}|$ denote the usual Euclidean distance from $\mathbf{x}$ to $\mathbf{y}$. For any nonempty compact set L in $\mathbf{R}^n$ and vector $\mathbf{x} \in \mathbf{R}^n$, the *distance from* $\mathbf{x}$ *to* L is defined by

$$d(\mathbf{x}, L) = \inf\{d(\mathbf{x}, \mathbf{y}) : \mathbf{y} \in L\}.$$

Let L_1 and L_2 be nonempty compact subsets of $\mathbf{R}^n$, and let

$$d(L_1, L_2) = \sup\{d(\mathbf{x}_1, L_2) : \mathbf{x}_1 \in L_1\} + \sup\{d(\mathbf{x}_2, L_1) : \mathbf{x}_2 \in L_2\}. \tag{6.4}$$

Let X be a fixed compact set in $\mathbf{R}^n$, and let $\Omega(X)$ be the set of all nonempty compact subsets of X. For $L_1, L_2 \in \Omega(X)$, we define the distance between L_1 and L_2 by (6.4). Then $\Omega(X)$ is a metric space (Exercise 22). A sequence $\{L_i\}_{i=1}^{\infty}$ of compact sets in $\Omega(X)$ converges to $L \in \Omega(X)$ if and only if $\lim_{i\to\infty} d(L_i, L) = 0$.

The metric topology of $\Omega(X)$ is used to construct some continuous functions that will be used in the proof of Minkowski's second theorem. Let $\mathbf{R}^n = V \bigoplus W$, where V and W are vector spaces, $\dim V = r$, and $\dim W = n - r$. Let $\pi : \mathbf{R}^n \to W$ be the canonical projection onto W. Let X be any nonempty compact set in $\mathbf{R}^n$, and let $X' = \pi(X) \subseteq W$. The map from X' to $\Omega(X)$ defined by

$$\mathbf{x}' \longmapsto \pi^{-1}(\mathbf{x}') \cap X$$

is continuous, and so the map

$$\sigma : X \to \Omega(X)$$

defined by

$$\sigma(\mathbf{x}) = (\pi^{-1}(\pi(\mathbf{x}))) \cap X = (V + \{\mathbf{x}\}) \cap X$$

is a continuous map from X into $\Omega(X)$. Moreover, $\sigma(\mathbf{x})$ lies in the r-dimensional affine subspace $V + \{\pi\mathbf{x})\} = V + \{\mathbf{x}\}$.

Let $X = \overline{K}$, where K is a convex body in $\mathbf{R}^n$. Then $\overline{K}$ is a compact convex set with nonempty interior. If $\mathbf{x} \in \overline{K}$, then $\sigma(\mathbf{x})$ is a compact convex subset of $V + \{\pi(\mathbf{x})\}$. Let $\mathrm{vol}_r(\mathbf{x})$ and $\mathbf{c}(\mathbf{x})$ denote the r-dimensional volume and center of mass, respectively, of $\sigma(\mathbf{x})$. Then $\mathrm{vol}_r : \overline{K} \to \mathbf{R}$ and $\mathbf{c} : \overline{K} \to \overline{K}$ are continuous functions. (see Exercises 25 and 27).

Let K be a convex body in $\mathbf{R}^n$. For $\lambda \in \mathbf{R}$, $\lambda \geq 0$, let

$$\lambda * K = \{\lambda\mathbf{u} : \mathbf{u} \in K\}.$$

Then $\lambda * \overline{K} = \overline{\lambda * K}$. If $\mathbf{0} \in K$ and $\lambda \leq \mu$, then $\lambda * K \subseteq \mu * K$.

Let Λ be a lattice in $\mathbf{R}^n$, and let $\{\mathbf{a}_1, \ldots, \mathbf{a}_n\}$ be a basis for Λ. Since K is open and $0 \in K$, there exists $\varepsilon > 0$ such that

$$0 \in B(0, \varepsilon) \subseteq K$$

and so

$$B(0, \lambda\varepsilon) \subseteq \lambda * K$$

for all $\lambda \geq 0$. In particular, $\{\mathbf{a}_1, \ldots, \mathbf{a}_n\} \subseteq \lambda * K$ for λ sufficiently large. Since K is bounded and Λ is discrete, there exists $\lambda > 0$ sufficiently small that $(\lambda * K) \cap \Lambda = \{\mathbf{0}\}$.

The *successive minima* of the convex body K with respect to the lattice Λ are the real numbers $\lambda_1, \lambda_2, \ldots, \lambda_n$ defined as follows:

$$\lambda_k = \inf\{\lambda > 0 : \lambda * K \text{ contains } k \text{ linearly independent elements of } \Lambda\}.$$

It is easy to see that $0 < \lambda_1 \leq \lambda_2 \leq \cdots \leq \lambda_n$, and that the definition of λ_1 is equivalent to (6.2).

Because K is an open set, it follows that $\lambda_k * K$ contains at most $k - 1$ linearly independent vectors in Λ and that $\lambda_k * \overline{K}$ contains at least k linearly independent vectors.

There is an equivalent way to define the successive minima and at the same time identify a linearly independent set $\{\mathbf{b}_1, \ldots, \mathbf{b}_n\}$ of vectors in the lattice Λ such that every vector $\mathbf{u} \in (\lambda_k * K) \cap \Lambda$ is a linear combination of $\mathbf{b}_1, \ldots, \mathbf{b}_{k-1}$. Let

$$\lambda_1 = \inf\{\lambda > 0 : \lambda * \overline{K} \text{ contains a nonzero vector } \mathbf{b}_1 \in \Lambda\}$$

and, for $2 \leq k \leq n$,

$$\lambda_k = \inf\{\lambda > 0 : \lambda * \overline{K} \text{ contains a vector } \mathbf{b}_k \in \Lambda \text{ linearly independent of } \mathbf{b}_1, \mathbf{b}_2, \ldots, \mathbf{b}_{k-1}\}.$$

For example, let $\Lambda = \mathbf{Z}^n$ and let K be the box

$$K = \{(x_1, \ldots, x_n) \in \mathbf{R}^n : |x_i| < r_i \text{ for } i = 1, \ldots, n\},$$

where $0 < r_n \leq r_{n-1} \leq \cdots \leq r_2 \leq r_1 \leq 1$. Then $((1/r_1) * K) \cap \mathbf{Z}^n = \{0\}$ and $\pm\mathbf{e}_1 \in (\lambda * K) \cap \mathbf{Z}^n$ for all $\lambda > 1/r_1$; hence $\lambda_1 = 1/r_1$. Similarly, $\lambda_i = 1/r_i$ for $i = 2, \ldots, n$. Since $\mathrm{vol}(K) = 2^n r_1 r_2 \cdots r_n$ and $\det(\Lambda) = 1$, we see that

$$\lambda_1 \cdots \lambda_n \,\mathrm{vol}(K) = 2^n \det(\Lambda).$$

This simple example shows that the following theorem is best possible.

Theorem 6.6 (Minkowski's second theorem) *Let K be a symmetric, convex body in $\mathbf{R}^n$, and let Λ be a lattice in $\mathbf{R}^n$. Let $\lambda_1, \ldots, \lambda_n$ be the successive minima of K with respect to the lattice Λ. Then*

$$\lambda_1 \cdots \lambda_n \, vol(K) \leq 2^n det(\Lambda).$$

Proof. Corresponding to the successive minima $\lambda_1, \ldots, \lambda_n$ are n linearly independent vectors $\mathbf{b}_1, \ldots, \mathbf{b}_n$ in the lattice Λ such that, for $k = 1, \ldots, n$, every vector in $(\lambda_k * K) \cap \Lambda$ is a linear combination of $\mathbf{b}_1, \ldots, \mathbf{b}_{k-1}$ and

$$\{\mathbf{b}_1, \ldots, \mathbf{b}_k\} \subseteq \lambda_k * \overline{K}.$$

We shall use the basis $\{\mathbf{b}_1, \mathbf{b}_2, \ldots, \mathbf{b}_n\}$ to construct a continuous map

$$\varphi : \overline{K} \to \lambda_n * \overline{K}$$

such that

$$\varphi(K) \subseteq \lambda_n * K.$$

For $j = 1, \ldots, n$, let V_j be the subspace of $\mathbf{R}^n$ spanned by $\{\mathbf{b}_1, \mathbf{b}_2, \ldots, \mathbf{b}_{j-1}\}$ and let W_j be the subspace of $\mathbf{R}^n$ spanned by $\{\mathbf{b}_j, \mathbf{b}_{j+1}, \ldots, \mathbf{b}_n\}$. Then

$$\mathbf{R}^n = V_j \bigoplus W_j.$$

Let

$$\pi_j : \mathbf{R}^n \to W_j$$

be the projection onto W_j. For every vector $\mathbf{y} \in W_j$,

$$\pi_j^{-1}(\mathbf{y}) = V_j + \{\mathbf{y}\}$$

is an affine subspace (or *plane*) of dimension $j - 1$. Let $K'_j = \pi_j(K)$. Then K'_j is a convex body in W_j, and $\overline{K'_j} = \pi_j(\overline{K})$ is a compact convex set in W_j.

Let $\mathbf{c}_j : \overline{K} \to \overline{K}$ be the continuous function that maps $\mathbf{x} \in \overline{K}$ to the center of mass of $\sigma_j(\mathbf{x}) = \pi_j^{-1}\left(\pi_j(\mathbf{x})\right) \cap \overline{K}$. We define the coordinate functions $c_{ij}(\mathbf{x})$ by

$$\mathbf{c}_j(\mathbf{x}) = \sum_{i=1}^{n} c_{ij}(\mathbf{x})\mathbf{b}_i.$$

If $\mathbf{x} = \sum_{i=1}^{n} x_i \mathbf{b}_i$, then

$$c_{ij}(\mathbf{x}) = x_i \text{ for } i = j, j+1, \ldots, n$$

and $c_{ij}(\mathbf{x})$ is a continuous function of $x_j, \ldots, x_n$ for $i = 1, \ldots, j-1$.

Let $\lambda_1, \ldots, \lambda_n$ be the successive minima of K, and let $\lambda_0 = 0$. For $\mathbf{x} \in \overline{K}$, we define

$$\varphi(\mathbf{x}) = \sum_{j=1}^{n} (\lambda_j - \lambda_{j-1})\mathbf{c}_j(\mathbf{x}).$$

Let $t_j = (\lambda_j - \lambda_{j-1})/\lambda_n$ for $j = 1, \ldots, n$. Then $t_j \geq 0$ for all j and $t_1 + \cdots + t_n = 1$. Since $\mathbf{c}_j(\mathbf{x}) \in \overline{K}$ for all $\mathbf{x} \in \overline{K}$, it follows that

$$\varphi(\mathbf{x}) = \lambda_n \sum_{j=1}^{n} t_j \mathbf{c}_j(\mathbf{x}) \in \lambda_n * \overline{K},$$

and so $\varphi : \overline{K} \to \lambda_n * \overline{K}$ is a continuous function that satisfies $\varphi(K) \subseteq \lambda_n * K$. Moreover,

$$\varphi(\mathbf{x}) = \sum_{j=1}^{n} (\lambda_j - \lambda_{j-1})\mathbf{c}_j(\mathbf{x})$$

$$
\begin{aligned}
&= \sum_{j=1}^{n}(\lambda_j - \lambda_{j-1})\sum_{i=1}^{n} c_{i,j}(\mathbf{x})\mathbf{b}_i \\
&= \sum_{i=1}^{n}\left(\sum_{j=1}^{n}(\lambda_j - \lambda_{j-1})c_{i,j}\right)(\mathbf{x})\mathbf{b}_i \\
&= \sum_{i=1}^{n}\left(\sum_{j=1}^{i}(\lambda_j - \lambda_{j-1})x_i + \sum_{j=i+1}^{n}(\lambda_j - \lambda_{j-1})c_{i,j}(\mathbf{x})\right)\mathbf{b}_i \\
&= \sum_{i=1}^{n}\left(\lambda_i x_i + \sum_{j=i+1}^{n}(\lambda_j - \lambda_{j-1})c_{i,j}(\mathbf{x})\right)\mathbf{b}_i .
\end{aligned}
$$

Let

$$\varphi_i(\mathbf{x}) = \sum_{j=i+1}^{n}(\lambda_j - \lambda_{j-1})c_{i,j}(\mathbf{x}).$$

Then $\varphi_i(\mathbf{x}) = \varphi_i(x_{i+1}, \ldots, x_n)$ is a continuous, real-valued function of the $n - i$ coefficients $x_{i+1}, \ldots, x_n$ of the vector $\mathbf{x} \in K$, and

$$\varphi(\mathbf{x}) = \sum_{i=1}^{n-1}(\lambda_i x_i + \varphi_i(x_{i+1}, \ldots, x_n))\,\mathbf{b}_i + \lambda_n x_n \mathbf{b}_n. \tag{6.5}$$

This has two important consequences.

First, the function φ is one-to-one. Let $\mathbf{x}, \mathbf{x}' \in K$, where $\mathbf{x} = x_1\mathbf{b}_1 + \cdots + x_n\mathbf{b}_n$ and $\mathbf{x}' = x_1'\mathbf{b}_1 + \cdots + x_n'\mathbf{b}_n$. If $\varphi(\mathbf{x}) = \varphi(\mathbf{x}')$, then

$$
\begin{aligned}
&\sum_{i=1}^{n-1}(\lambda_i x_i + \varphi_i(x_{i+1}, \ldots, x_n))\,\mathbf{b}_i + \lambda_n x_n \mathbf{b}_n \\
&\quad = \sum_{i=1}^{n-1}\left(\lambda_i x_i' + \varphi_i(x_{i+1}', \ldots, x_n')\right)\mathbf{b}_i + \lambda_n x_n' \mathbf{b}_n .
\end{aligned}
$$

Equating the coefficients of $\mathbf{b}_n$, we see that $\lambda_n x_n = \lambda_n x_n'$, and so $x_n = x_n'$. Equating the coefficients of $\mathbf{b}_{n-1}$, we see that

$$\lambda_{n-1}x_{n-1} + \varphi_{n-1}(x_n) = \lambda_{n-1}x_{n-1}' + \varphi_{n-1}(x_n') = \lambda_{n-1}x_{n-1}' + \varphi_{n-1}(x_n),$$

and so $x_{n-1} = x_{n-1}'$. Applying this argument inductively, we obtain $x_i = x_i'$ for $i = 1, \ldots, n$, and so $\mathbf{x} = \mathbf{x}'$.

The second consequence is

$$\text{vol}(\varphi(\overline{K})) = \lambda_1 \cdots \lambda_n \text{ vol}(\overline{K}).$$

In the special case when the functions $\varphi_i(x_i, \ldots, x_n)$ are identically equal to zero for $i = 1, \ldots, n$, the function φ is given by the simple formula

$$\varphi(x_1, \ldots, x_n) = \lambda_1 x_1 \mathbf{b}_1 + \cdots + \lambda_n x_n \mathbf{b}_n,$$

and

$$\mathrm{vol}(\varphi(K)) = \lambda_1 \cdots \lambda_n \, \mathrm{vol}(K). \tag{6.6}$$

Since the functions $\varphi_i(x_{i+1}, \ldots, x_n)$ are continuous for $i = 1, \ldots, n$, this formula for the volume of $\varphi(K)$ holds in general.

The set $\varphi(K)$ is not necessarily convex, but it is a bounded open subset of $\mathbf{R}^n$ to which we can apply Lemma 6.1 of Blichfeldt. Let $K' = \varphi(K)$. If

$$\mathrm{vol}(K') = \lambda_1 \cdots \lambda_n \, \mathrm{vol}(K) > 2^n \det(\Lambda),$$

then

$$\mathrm{vol}((1/2) * K') > \det(\Lambda),$$

and Lemma 6.1 implies that there exist vectors $\mathbf{x}'_1, \mathbf{x}'_2 \in K'$ such that

$$\frac{\mathbf{x}'_1 - \mathbf{x}'_2}{2} \in \Lambda \setminus \{\mathbf{0}\}.$$

Since $K' = \varphi(K)$, there exist vectors $\mathbf{x}_1, \mathbf{x}_2 \in K$ such that $\varphi(\mathbf{x}_1) = \mathbf{x}'_1$ and $\varphi(\mathbf{x}_2) = \mathbf{x}'_2$. Let

$$\mathbf{x}_1 = \sum_{i=1}^{n} x_{i,1} \mathbf{b}_i$$

and

$$\mathbf{x}_2 = \sum_{i=1}^{n} x_{i,2} \mathbf{b}_i .$$

Since $\mathbf{x}_1 \neq \mathbf{x}_2$, there exists $k \geq 1$ such that $x_{k,1} \neq x_{k,2}$ and $x_{i,1} = x_{i,2}$ for $i = k+1, \ldots, n$. Recall that if $\mathbf{x} = \sum_{i=1}^{n} x_i \mathbf{b}_i$, then the center of mass $\mathbf{c}_j(x_j, \ldots, x_n) \in K$ for $j = 1, \ldots, n$. Since K is a symmetric convex body,

$$\frac{\mathbf{c}_j(x_{j+1,1}, \ldots, x_{n,1}) - \mathbf{c}_j(x_{j+1,2}, \ldots, x_{n,2})}{2} \in K,$$

and so

$$\frac{\lambda_k \left(\mathbf{c}_j(x_{j+1,1}, \ldots, x_{n,1}) - \mathbf{c}_j(x_{j+1,2}, \ldots, x_{n,2}) \right)}{2} \in \lambda_k * K$$

for $j = 1, \ldots, n$. Let $t_j = (\lambda_j - \lambda_{j-1}/\lambda_k$ for $j = 1, \ldots, k$. Then

$$\begin{aligned}
\frac{\mathbf{x}'_1 - \mathbf{x}'_2}{2} &= \frac{\varphi(\mathbf{x}_1) - \varphi(\mathbf{x}_2)}{2} \\
&= \sum_{j=1}^{n} (\lambda_j - \lambda_{j-1}) \left(\frac{\mathbf{c}_j(\mathbf{x}_1) - \mathbf{c}_j(\mathbf{x}_2)}{2} \right) \\
&= \sum_{j=1}^{k} (\lambda_j - \lambda_{j-1}) \left(\frac{\mathbf{c}_j(\mathbf{x}_1) - \mathbf{c}_j(\mathbf{x}_2)}{2} \right) \\
&= \sum_{j=1}^{k} t_j \lambda_k \left(\frac{\mathbf{c}_j(\mathbf{x}_1) - \mathbf{c}_j(\mathbf{x}_2)}{2} \right) \\
&\in (\lambda_k * K) \cap (\Lambda \setminus \{\mathbf{0}\}).
\end{aligned}$$

Representing the map φ in the form (6.5), we see that

$$\begin{aligned}
\frac{\mathbf{x}_1' - \mathbf{x}_2')}{2} &= \frac{\varphi(\mathbf{x}_1) - \varphi(\mathbf{x}_2)}{2} \\
&= \sum_{i=1}^{n} \left(\lambda_i \left(\frac{x_{i,1} - x_{i,2}}{2} \right) + \left(\frac{\varphi_i(\mathbf{x}_1) - \varphi_i(\mathbf{x}_2)}{2} \right) \right) \mathbf{b}_i \\
&= \sum_{i=1}^{k} \left(\lambda_i \left(\frac{x_{i,1} - x_{i,2}}{2} \right) + \left(\frac{\varphi_i(\mathbf{x}_1) - \varphi_i(\mathbf{x}_2)}{2} \right) \right) \mathbf{b}_i \\
&= \sum_{i=1}^{k-1} \left(\lambda_i \left(\frac{x_{i,1} - x_{i,2}}{2} \right) + \left(\frac{\varphi_i(\mathbf{x}_1) - \varphi_i(\mathbf{x}_2)}{2} \right) \right) \mathbf{b}_i \\
&\quad + \lambda_k \left(\frac{x_{k,1} - x_{k,2}}{2} \right) \mathbf{b}_k,
\end{aligned}$$

and

$$\lambda_k \left(\frac{x_{k,1} - x_{k,2}}{2} \right) \neq 0.$$

Since

$$\frac{\mathbf{x}_1' - \mathbf{x}_2'}{2} \in (\lambda_k * K) \cap \Lambda,$$

it follows from the definition of the successive minima $\lambda_1, \ldots, \lambda_n$ that $(\mathbf{x}_1' - \mathbf{x}_2')/2$ can be represented as a linear combination of the vectors $\mathbf{b}_1, \ldots, \mathbf{b}_{k-1}$. On the other hand, we have just shown that $(\mathbf{x}_1' - \mathbf{x}_2')/2$ is also a linear combination with real coefficients of the vectors $\mathbf{b}_1, \ldots, \mathbf{b}_{k-1}, \mathbf{b}_k$ and that the coefficient of $\mathbf{b}_k$ in this representation is $\lambda_k(x_{k,1} - x_{k,2})/2 \neq 0$. This implies that $\mathbf{b}_k$ is a linear combination of the vectors $\mathbf{b}_1, \ldots, \mathbf{b}_{k-1}$, which contradicts the linear independence of the vectors $\mathbf{b}_1, \ldots, \mathbf{b}_n$. This completes the proof of Minkowski's second theorem.

6.5 Bases for sublattices

Let Λ and M be lattices in $\mathbf{R}^n$ with bases $\{\mathbf{a}_1, \ldots, \mathbf{a}_n\}$ and $\{\mathbf{b}_1, \ldots, \mathbf{b}_n\}$, respectively. If $M \subseteq \Lambda$, then M is called a *sublattice* of Λ, and $\mathbf{b}_j \in \Lambda$ for $j = 1, \ldots, n$. Therefore, there exist integers v_{ij} such that

$$\mathbf{b}_j = \sum_{i=1}^{n} v_{ij} \mathbf{a}_i.$$

Let

$$\mathbf{a}_j = \sum_{i=1}^{n} a_{ij} \mathbf{e}_i$$

and

$$\mathbf{b}_j = \sum_{i=1}^{n} b_{ij} \mathbf{e}_i.$$

Then

$$\begin{aligned}
\mathbf{b}_j &= \sum_{k=1}^{n} v_{k,j}\mathbf{a}_k \\
&= \sum_{k=1}^{n} v_{k,j} \sum_{i=1}^{n} a_{i,k}\mathbf{e}_i \\
&= \sum_{i=1}^{n} \left(\sum_{k=1}^{n} a_{i,k} v_{k,j} \right) \mathbf{e}_i \\
&= \sum_{i=1}^{n} b_{ij}\mathbf{e}_i,
\end{aligned}$$

and so

$$b_{ij} = \sum_{k=1}^{n} a_{ik} v_{kj}.$$

This implies that

$$\det(M) = |\det(b_{ij})| = |\det(a_{ik})||\det(v_{kj})| = |\det(v_{ik})| \det(\Lambda).$$

Let

$$d = |\det(v_{ik})| = \frac{\det(M)}{\det(\Lambda)}.$$

Then d is an integer, since $v_{ik} \in \mathbf{Z}$ for all $i, k = 1, \ldots, n$. Also, d is nonzero, since both $\det(M)$ and $\det(\Lambda)$ are nonzero. Moreover, d is independent of the choice of bases for the lattices Λ and M. The number d, called the *geometrical index* of the lattice M in Λ, is the ratio of the volumes of the fundamental parallelepipeds of the lattices M and Λ.

Since $\det(v_{i,j}) = \pm d$, it follows that the matrix $V = (v_{i,j}) \in M_n(\mathbf{Z})$ has an inverse V^{-1} of the form

$$V^{-1} = (v_{ij}^{-1}) = (w_{i,j}/d),$$

where $w_{i,j} \in \mathbf{Z}$ for $i, j = 1, \ldots, n$. This means that

$$\sum_{k=1}^{n} w_{i,k} v_{k,j} = \sum_{k=1}^{n} v_{i,k} w_{k,j} = d\delta_{i,j}$$

for $i, j = 1, \ldots, n$. Then

$$\begin{aligned}
\sum_{k=1}^{n} w_{k,j}\mathbf{b}_k &= \sum_{k=1}^{n} w_{k,j} \sum_{i=1}^{n} v_{i,k}\mathbf{a}_i \\
&= \sum_{i=1}^{n} \left(\sum_{k=1}^{n} v_{i,k} w_{k,j} \right) \mathbf{a}_i \\
&= \sum_{i=1}^{n} d\delta_{ij}\mathbf{a}_i \\
&= d\mathbf{a}_j \in M
\end{aligned}$$

for $j = 1, \ldots, n$, and so

$$d * \Lambda \subseteq M \subseteq \Lambda. \tag{6.7}$$

Theorem 6.7 *Let M be a sublattice of the lattice Λ in $\mathbf{R}^n$, and let $\{\mathbf{b}_1, \ldots, \mathbf{b}_n\}$ be a basis for M. There exists a basis $\{\mathbf{a}_1, \ldots, \mathbf{a}_n\}$ of Λ such that*

$$\begin{aligned}
\mathbf{b}_1 &= v_{1,1}\mathbf{a}_1 \\
\mathbf{b}_2 &= v_{1,2}\mathbf{a}_1 + v_{2,2}\mathbf{a}_2 \\
\mathbf{b}_3 &= v_{1,3}\mathbf{a}_1 + v_{2,3}\mathbf{a}_2 + v_{3,3}\mathbf{a}_3 \\
&\vdots \\
\mathbf{b}_n &= v_{1,n}\mathbf{a}_1 + v_{2,n}\mathbf{a}_2 + \ldots + v_{n,n}\mathbf{a}_n,
\end{aligned}$$

where $v_{i,j} \in \mathbf{Z}$ for $j = 1, \ldots, n$ and $i = 1, \ldots, j$, and $v_{j,j} \geq 1$ for $j = 1, \ldots, n$.

Proof. This is simply a restatement of Theorem 6.2.

Theorem 6.8 *Let Λ be a lattice in $\mathbf{R}^n$ with basis $\{\mathbf{a}_1, \ldots, \mathbf{a}_n\}$, and let M be a sublattice of Λ. There exists a basis $\{\mathbf{b}_1, \ldots, \mathbf{b}_n\}$ of M such that*

$$\begin{aligned}
\mathbf{b}_1 &= v_{1,1}\mathbf{a}_1 \\
\mathbf{b}_2 &= v_{1,2}\mathbf{a}_1 + v_{2,2}\mathbf{a}_2 \\
\mathbf{b}_3 &= v_{1,3}\mathbf{a}_1 + v_{2,3}\mathbf{a}_2 + v_{3,3}\mathbf{a}_3 \\
&\vdots \\
\mathbf{b}_n &= v_{1,n}\mathbf{a}_1 + v_{2,n}\mathbf{a}_2 + \ldots + v_{n,n}\mathbf{a}_n,
\end{aligned}$$

where $v_{i,j} \in \mathbf{Z}$ for $j = 1, \ldots, n$ and $i = 1, \ldots, j$, and $v_{j,j} \geq 1$ for $j = 1, \ldots, n$.

Proof. Let d be the geometric index of M in Λ. By (6.7),

$$d * \Lambda \subseteq M.$$

Thus, $d * \Lambda$ is a sublattice of M, and $\{d\mathbf{a}_1, \ldots, d\mathbf{a}_n\}$ is a basis for $d * \Lambda$. It follows from Theorem 6.7 that there exists a basis $\{\mathbf{b}_1, \ldots, \mathbf{b}_n\}$ for M such that

$$d\mathbf{a}_j = \sum_{i=1}^{j} u_{i,j}\mathbf{b}_i$$

for $j = 1, \ldots, n$, where the coefficients $u_{i,j}$ are integers. Solving these equations for the vectors $\mathbf{b}_1, \ldots, \mathbf{b}_n$, we obtain expressions of the form

$$\mathbf{b}_j = \sum_{i=1}^{j} v_{i,j}\mathbf{a}_i,$$

where the coefficients $v_{i,j}$ are integers because $\{\mathbf{b}_1, \ldots, \mathbf{b}_n\} \subseteq \Lambda$ and the vectors $\mathbf{a}_1, \ldots, \mathbf{a}_n$ form a basis for Λ.

Corollary 6.3 *Let M be a sublattice of the lattice Λ, and let $\{\mathbf{a}_1, \ldots, \mathbf{a}_n\}$ and $\{\mathbf{b}_1, \ldots, \mathbf{b}_n\}$ be bases for Λ and M, respectively, such that $\mathbf{b}_j = \sum_{i=1}^{j} v_{i,j}\mathbf{a}_i$. Let d be the geometric index of M in Λ. Then*

$$d = v_{1,1}v_{2,2}\cdots v_{n,n}.$$

Proof. This follows immediately from the fact that $d = \det(v_{i,j})$, and the matrix $(v_{i,j})$ is upper triangular.

Corollary 6.4 *Let Λ be a sublattice of the integer lattice $\mathbf{Z}^n$. Then there exists a basis $\{\mathbf{a}_1, \ldots, \mathbf{a}_n\}$ for Λ such that*

$$\begin{aligned} \mathbf{a}_1 &= a_{1,1}\mathbf{e}_1 \\ \mathbf{a}_2 &= a_{1,2}\mathbf{e}_1 + a_{2,2}\mathbf{e}_2 \\ &\vdots \\ \mathbf{a}_n &= a_{1,n}\mathbf{e}_1 + a_{2,n}\mathbf{e}_2 \cdots + a_{n,n}\mathbf{e}_n, \end{aligned}$$

where $a_{i,j} \in \mathbf{Z}$ for $j = 1, \ldots, n$ and $i = 1, \ldots, j$, and $a_{j,j} \geq 1$ for $j = 1, \ldots, n$. Moreover, $d(\Lambda) = a_{1,1}a_{2,2}\cdots a_{n,n}$.

Proof. The standard basis $\mathbf{e}_1, \ldots, \mathbf{e}_n$ for $\mathbf{R}^n$ is a basis for the integer lattice $\mathbf{Z}^n$, and Λ is a sublattice of $\mathbf{Z}^n$. The result follows immediately from Theorem 6.8.

Let M be a sublattice of the lattice Λ. Then M is a subgroup of the abelian group Λ. The quotient group Λ/M is the set of all cosets

$$\mathbf{a} + M = \{\mathbf{a} + \mathbf{b} \mid \mathbf{b} \in M\},$$

where $\mathbf{a} \in \Lambda$, and the addition of two cosets is defined by $(\mathbf{a}_1 + M) + (\mathbf{a}_2 + M) = (\mathbf{a}_1 + \mathbf{a}_2) + M$ for all $\mathbf{a}_1, \mathbf{a}_2 \in \Lambda$. Then $\mathbf{a}_1 + M = \mathbf{a}_2 + M$ if and only if $\mathbf{a}_1 - \mathbf{a}_2 \in M$. The *algebraic index* of M in Λ is the order of the quotient group Λ/M and is denoted $[\Lambda : M]$.

Theorem 6.9 *Let M be a sublattice of the lattice Λ in $\mathbf{R}^n$. Then*

$$[\Lambda : M] = \frac{\det(M)}{\det(\Lambda)},$$

that is, the algebraic index of M in Λ is equal to the geometric index of M in Λ.

Proof. Let $\{\mathbf{a}_1, \ldots, \mathbf{a}_n\}$ be a basis for Λ, and let $\{\mathbf{b}_1, \ldots, \mathbf{b}_n\}$ be a basis for the sublattice M of the form

$$\mathbf{b}_j = \sum_{i=1}^{j} v_{i,j}\mathbf{a}_i,$$

where $v_{j,j}$ is a positive integer for $j = 1, \ldots, n$. Then

$$v_{1,1}\cdots v_{n,n} = \frac{\det(M)}{\det(\Lambda)}.$$

Let $\mathbf{b} \in M$ and $\mathbf{b} \neq \mathbf{0}$. Then $\mathbf{b}$ has a unique representation in the form

$$\mathbf{b} = \sum_{i=1}^{n} g_i \mathbf{a}_i = \sum_{i=1}^{k} g_i \mathbf{a}_i,$$

where k is the largest integer such that $g_k \neq 0$. We shall prove that

$$|g_k| \geq v_{k,k}.$$

Since $\mathbf{b} \in M$, there exist unique integers $h_1, \ldots, h_n$ such that

$$\begin{aligned} \mathbf{b} &= \sum_{j=1}^{n} h_j \mathbf{b}_j \\ &= \sum_{j=1}^{n} h_j \sum_{i=1}^{j} v_{i,j} \mathbf{a}_i \\ &= \sum_{i=1}^{n} \left(\sum_{j=i}^{n} v_{i,j} h_j \right) \mathbf{a}_i, \end{aligned}$$

and so

$$g_i = \sum_{j=i}^{n} v_{i,j} h_j$$

for $i = 1, \ldots, n$. Since $v_{j,j} \neq 0$ for all j and $\sum_{j=i}^{n} v_{i,j} h_j = 0$ for $i = k+1, \ldots, n$, it follows that $h_i = 0$ for $i = k+1, \ldots, n$ and $g_k = v_{k,k} h_k \neq 0$. Therefore, $h_k \neq 0$ and

$$|g_k| = |v_{k,k} h_k| = v_{k,k} |h_k| \geq v_{k,k}.$$

We shall prove that the set

$$S = \left\{ \sum_{i=1}^{n} g_i \mathbf{a}_i : g_i \in \mathbf{Z}, 0 \leq g_i < v_{i,i} \right\}$$

is a complete set of coset representatives for Λ / M.

Let $\mathbf{s} = \sum_{i=1}^{n} g_i \mathbf{a}_i \in S$ and $\mathbf{s}' = \sum_{i=1}^{n} g_i' \mathbf{a}_i \in S$. If $\mathbf{s} + M = \mathbf{s}' + M$, then

$$\mathbf{s} - \mathbf{s}' = \sum_{i=1}^{n} (g_i - g_i') \mathbf{a}_i \in M.$$

If $\mathbf{s} \neq \mathbf{s}'$, there exists a largest integer $k \geq 1$ such that $g_k \neq g_k'$. Then

$$\mathbf{s} - \mathbf{s}' = \sum_{i=1}^{k} (g_i - g_i') \mathbf{a}_i \in M,$$

and so $|g_k - g_k'| \geq v_{k,k}$. But $0 \leq g_k, g_k' < v_{k,k}$ implies that $|g_k - g_k'| < v_{k,k}$, which is a contradiction. Therefore, $\mathbf{s} + M \neq \mathbf{s}' + M$, and so the elements of S represent distinct cosets in Λ / M.

Let $\mathbf{u} = \sum_{i=1}^{k} g_i \mathbf{a}_i \in \Lambda$, where $g_k \neq 0$ and $k = 0, 1, \ldots, n$. We shall prove by induction on k that there exists $\mathbf{s} \in S$ such that $\mathbf{u} \in \mathbf{s} + M$. If $k = 0$, then $\mathbf{u} = \mathbf{0} \in S$, and $\mathbf{u} \in \mathbf{0} + M$. Let $k \geq 1$, and assume that the statement holds for $k' = 0, 1, \ldots, k-1$. Let

$$\mathbf{u} = \sum_{i=1}^{k} g_i \mathbf{a}_i \in \Lambda,$$

where $g_k \neq 0$. Let

$$g_k = q_k v_{k,k} + r_k,$$

where $q_k, r_k \in \mathbf{Z}$ and $0 \leq r_k < v_{k,k}$. Then

$$\begin{aligned}
\mathbf{u} &= \sum_{i=1}^{k-1} g_i \mathbf{a}_i + g_k \mathbf{a}_k \\
&= \sum_{i=1}^{k-1} g_i \mathbf{a}_i + q_k v_{k,k} \mathbf{a}_k + r_k \mathbf{a}_k \\
&= \sum_{i=1}^{k-1} g_i \mathbf{a}_i + q_k \left(\mathbf{b}_k - \sum_{i=1}^{k-1} v_{i,k} \mathbf{a}_i \right) + r_k \mathbf{a}_k \\
&= \sum_{i=1}^{k-1} (g_i - q_k v_{i,k}) \mathbf{a}_i + q_k \mathbf{b}_k + r_k \mathbf{a}_k.
\end{aligned}$$

It follows from the induction hypothesis that

$$\sum_{i=1}^{k-1} (g_i - q_k v_{i,k}) \mathbf{a}_i = \mathbf{s}' + \mathbf{b},$$

where $\mathbf{b} \in M$ and

$$\mathbf{s}' = \sum_{i=1}^{k-1} g_i' \mathbf{a}_i \in S.$$

Then $\mathbf{s} = \mathbf{s}' + r_k \mathbf{a}_k \in S$ and $\mathbf{b} + q_k \mathbf{b}_k \in M$, and so

$$\mathbf{u} = \mathbf{s}' + \mathbf{b} + q_k \mathbf{b}_k + r_k \mathbf{a}_k \in \mathbf{s} + M.$$

Thus, S is a complete set of representatives for the quotient lattice Λ / M, and

$$[\Lambda : M] = |S| = v_{1,1} \cdots v_{n,n} = \frac{\det(M)}{\det(\Lambda)}.$$

6.6 Torsion-free abelian groups

The abelian group G is *torsion-free* if every nonzero element of G has infinite order, that is, if $g \in G$, $g \neq 0$ implies that $mg \neq 0$ for all $m \in \mathbf{Z}$, $m \neq 0$. The set

$\{g_i\}_{i \in I} \subseteq G$ is a *set of generators* for G if every element g of G can be represented in the form

$$g = \sum_{i \in I} m_i g_i,$$

where $m_i \in \mathbf{Z}$ and $m_i = 0$ for all but finitely many $i \in I$. The abelian group G is *finitely generated* if it contains a finite set of generators. The abelian group G is *free* if it contains a subset $\{g_i\}_{i \in I}$ such that every element $g \in G$ has a *unique* representation in the form

$$g = \sum_{i \in I} m_i g_i,$$

where $m_i \in \mathbf{Z}$ and $m_i = 0$ for all but finitely many $i \in I$. In this case, the set $\{g_i\}_{i \in I}$ is called a *basis* for G. Every free abelian group is torsion-free. The group $G = \{0\}$ is the free abelian group whose basis is the empty set.

Let G be an abelian group, and let

$$m * G = \{mg \ : g \in G\}.$$

Since G is abelian, $m * G$ is a subgroup of G for every $m \geq 2$. Let $[G : m * G]$ denote the index of $m * G$ in G.

Lemma 6.5 *Let G be a free abelian group. If $[G : 2 * G]$ is infinite, then every basis for G is infinite. If $[G : 2 * G]$ is finite, then every basis for G has cardinality*

$$\frac{\log[G : 2 * G]}{\log 2}.$$

Proof. Let $\{g_i\}_{i \in I}$ be a basis for G. The map $\varphi : G \to \bigoplus_{i \in I} \mathbf{Z}/2\mathbf{Z}$ given by

$$\varphi\left(\sum_{i \in I} m_i g_i\right) = (m_i + 2\mathbf{Z})_{i \in I}$$

is a well-defined surjective homomorphism with kernel $2 * G$. Therefore,

$$G/(2 * G) \cong \bigoplus_{i \in I} \mathbf{Z}/2\mathbf{Z}.$$

If the quotient group $G/(2 * G)$ is infinite, then I must be infinite, and so every basis for G is infinite. If the quotient group $G/(2 * G)$ is finite, then I is finite and

$$[G : 2 * G] = |G/(2 * G)| = \left|\bigoplus_{i \in I} \mathbf{Z}/2\mathbf{Z}\right| = 2^{|I|};$$

hence every basis for G is has cardinality $\log[G : 2 * G]/\log 2$. This completes the proof.

Let $G \neq \{0\}$ be a free abelian group with a finite basis. The *rank* of G is the cardinality of a basis for G. By the preceding lemma, the rank of a free abelian group is well-defined. If $G = \{0\}$, we say that G has rank 0. If G_1 is a free abelian group of rank n_1 and G_2 is a free abelian group of rank n_2, then $G_1 \bigoplus G_2$ is a free abelian group of rank $n_1 + n_2$.

Lemma 6.6 *Let $G \neq \{0\}$ be a free abelian group of finite rank. Then $G \cong \mathbf{Z}^n$ for some $n \geq 1$.*

Proof. Let $\{g_1, \ldots, g_n\}$ be a basis for G. The map $\varphi : G \to \mathbf{Z}^n$ given by

$$\varphi\left(\sum_{i=1}^{n} m_i g_i\right) = (m_1, \ldots, m_n)$$

is well-defined and an isomorphism. .

Lemma 6.7 *Let G be an abelian group, and let A be a free abelian group. Let $\varphi : G \to A$ be a surjective homomorphism. Then $G = K \bigoplus H$, where K is the kernel of φ and H is a subgroup G such that $\varphi : H \to A$ is an isomorphism.*

Proof. Let $\{a_i\}_{i\in I}$ be a basis for the free abelian group A. Since the map φ is onto, there exist elements $h_i \in G$ such that $\varphi(h_i) = a_i$. Let H be the subgroup of G generated by $\{h_i\}_{i\in I}$. The homomorphism φ restricted to H maps H onto A. It follows that for every $g \in G$ there exists $h \in H$ such that $\varphi(g) = \varphi(h)$, and so $g - h \in K$. Therefore, $G = K + H$.

Let $h = \sum_{i\in I} m_i h_i \in H$. Then

$$\varphi(h) = \sum_{i\in I} m_i \varphi(h_i) = \sum_{i\in I} m_i a_i = 0$$

if and only if $m_i = 0$ for all $i \in I$ or, equivalently, $\varphi(h) = 0$ if and only if $h = 0$. Thus, $\varphi : H \to A$ is an isomorphism. Let $g \in K \cap H$. Then $g \in K$ implies that $\varphi(g) = 0$ and $g \in H$ implies that $g = 0$. Therefore, $K \cap H = \{0\}$, and so $G = K \bigoplus H$. This completes the proof.

Lemma 6.8 *A subgroup of a free abelian group of rank n is a free abelian group of rank at most n.*

Proof. Let G be a free abelian group of rank n, and let $\{g_1, \ldots, g_n\}$ be a basis for G. Let G' be a subgroup of G. If $G' = \{0\}$, then G' has rank 0. Therefore, we can assume that $G' \neq \{0\}$.

The proof will be by induction on n. If $n = 1$, then $G = \mathbf{Z}g_1$ for some $g_1 \in G$. Let

$$H = \{r \in \mathbf{Z} : rg_1 \in G'\}.$$

Then H is a subgroup of $\mathbf{Z}$, $H \neq \{0\}$, and so $H = d\mathbf{Z}$ for some $d \in \mathbf{Z}$, $d \geq 1$. It follows that $G' = \mathbf{Z}dg_1$ is a free abelian group of rank 1.

Let $n \geq 2$, and suppose that the Lemma holds for any free abelian group of rank at most $n - 1$. Let G be a free abelian group with basis $\{g_1, \ldots, g_n\}$, and let K be the subgroup of G with basis $\{g_1, \ldots, g_{n-1}\}$. Then K is a free abelian group of rank $n - 1$. If $G' \subseteq K$, the induction hypothesis implies that G' is a free abelian group of rank at most $n - 1$.

Suppose that $G' \not\subseteq K$. The homomorphism $\varphi : G \to \mathbf{Z}g_n$ defined by

$$\varphi\left(\sum_{i=1}^{n} m_i g_i\right) = m_n g_n.$$

has kernel K. Let $\psi : G' \to \mathbf{Z}g_n$ be the restriction of the homomorphism φ to G'. The condition $G' \not\subseteq K$ implies that $\psi(G') = \mathbf{Z}dg_n$ for some $d \in \mathbf{Z}$, $d \geq 1$. Let K' be the kernel of ψ. Since

$$K' = K \cap G' \subseteq K$$

and K is free of rank $n-1$, it follows from the induction hypothesis that K' is a free abelian group of rank at most $n-1$. The map ψ maps G' onto the free abelian group $\mathbf{Z}dg_n$. By Lemma 6.7,

$$G' \cong H' \bigoplus K',$$

where H' is a subgroup of G' such that ψ restricted to H' is an isomorphism. This means that H' is a free abelian group of rank 1, and so G' is a free abelian group of rank at most $(n-1)+1 = n$.

Theorem 6.10 *Let $G \neq \{0\}$ be a finitely generated torsion-free abelian group. Then G is a free abelian group of finite rank, and so G is isomorphic to the integer lattice $\mathbf{Z}^n$ for some $n \geq 1$.*

Proof. Let $\Gamma = \{g_1, \ldots, g_k\}$ be a finite set of generators for G, and let $\Gamma' = \{g'_1, \ldots, g'_r\}$ be a maximal subset of Γ such that $\sum_{i=1}^{r} m_i g'_i = 0$ with $m_i \in \mathbf{Z}$ if and only if $m_i = 0$ for all $i = 1, \ldots, r$. Let G' be the subgroup of G generated by Γ'. Then G' is a free abelian group of rank r. Let $g_i \in \Gamma$. By the maximality of Γ', there exist integers $u_i, m_{i,1}, \ldots, m_{i,r}$ not all zero such that

$$u_i g_i + m_{i,1} g'_1 + \cdots + m_{i,r} g'_r = 0.$$

If $u_i = 0$, then $m_{i,j} = 0$ for $j = 1, \ldots, r$, which is impossible. Thus, $u_i \neq 0$, and so $u_i g_i \in G'$. Let m be the least common multiple of the integers $|u_1|, |u_2|, \ldots, |u_k|$. Then $mg_i \in G'$ for all $g_i \in \Gamma$. Since Γ generates G, it follows that

$$m * G = \{mg : g \in G\} \subseteq G'.$$

Since G' is a free abelian group of finite rank, it follows from Lemma 6.8 that the subgroup $m * G$ is also a free abelian group of finite rank. Since G is torsion-free, the map $\varphi : G \to m * G$ defined by $\varphi(g) = mg$ is an isomorphism, and so G is a free abelian group of finite rank. The theorem follows from Lemma 6.6.

Theorem 6.11 *Let M be a lattice in $\mathbf{R}^n$, and let Λ be a subgroup of $\mathbf{R}^n$ such that $M \subseteq \Lambda$ and $[\Lambda : M] < \infty$. Then Λ is a lattice.*

Proof. The group Λ is torsion-free and abelian because $\mathbf{R}^n$ is torsion-free and abelian. Also, $\Lambda \neq \{\mathbf{0}\}$ because $M \subseteq \Lambda$. Let $\{\mathbf{b}_1, \ldots, \mathbf{b}_n\}$ be a basis for M. Let

$[\Lambda : M] = r$ and let $\mathbf{u}_1, \ldots, \mathbf{u}_r \in \Lambda$ be a complete set of coset representatives for the finite quotient group Λ/M. Since every element of Λ belongs to some coset $\mathbf{u}_i + M$, it follows that $\{\mathbf{u}_1, \ldots, \mathbf{u}_r, \mathbf{b}_1, \ldots, \mathbf{b}_n\}$ is a finite set of generators for the torsion-free group Λ. By Theorem 6.10, Λ is a free abelian group of finite rank m. Since M is a subgroup of Λ and M is a free abelian group of rank n, it follows from Lemma 6.8 that $n \leq m$. Since Λ is free, the map $\mathbf{u} \mapsto r\mathbf{u}$ is an isomorphism of Λ onto $r * \Lambda$, and so $r * \Lambda$ is a free abelian group of rank m. Since the quotient group Λ/M has order r, it follows that $r(\mathbf{u} + M) = M$, and so $r\mathbf{u} \in M$ for every $\mathbf{u} \in \Lambda$. Therefore,

$$r * \Lambda \subseteq M.$$

Lemma 6.8 implies that $m \leq n$. Thus, $m = n$ and Λ is a free abelian group of rank n.

Let $\mathbf{a}_1, \ldots, \mathbf{a}_n$ be a set of generators for Λ. Since M is a lattice, it contains a set of n linearly independent vectors. Since Λ contains M, the generators $\mathbf{a}_1, \ldots, \mathbf{a}_n$ are linearly independent. Therefore, Λ is a lattice.

6.7 An important example

The results in this section will be applied in Chapter 8 to prove Theorem 8.7, which is part of the proof of Freiman's theorem.

Let $m \geq 2$ and let $\mathbf{u} = (u_1, \ldots, u_n), \mathbf{v} = (v_1, \ldots, v_n) \in \mathbf{Z}^n$. We write

$$\mathbf{u} \equiv \mathbf{v} \pmod{m}$$

if $u_i \equiv v_i \pmod{m}$ for $i = 1, \ldots, n$.

Theorem 6.12 *Let $m \geq 2$ and let $r_1, \ldots, r_n$ be integers such that*

$$(r_1, \ldots, r_n, m) = 1. \tag{6.8}$$

Let

$$\mathbf{r} = (r_1, \ldots, r_n) \in \mathbf{Z}^n,$$

and let

$$\Lambda = \{\mathbf{u} \in \mathbf{Z}^n : \mathbf{u} \equiv q\mathbf{r} \pmod{m} \text{ for some } q \in \mathbf{Z}\}.$$

Then Λ is a lattice, and $\det(\Lambda) = m^{n-1}$. Moreover, there exist positive real numbers $\lambda_1, \ldots, \lambda_n$ such that

$$\lambda_1 \cdots \lambda_n \leq 4^n m^{n-1}$$

and there exist linearly independent vectors $\mathbf{b}_1, \ldots, \mathbf{b}_n \in \Lambda$ such that

$$\mathbf{b}_j = (b_{1,j}, \ldots, b_{n,j})$$

and

$$|b_{i,j}| \leq \frac{\lambda_j}{4},$$

for $i, j = 1, \ldots, n$.

Proof. Let

$$M = (m * \mathbf{Z})^n = \{\mathbf{u} \in \mathbf{Z}^n : \mathbf{u} \equiv \mathbf{0} \pmod{m}\}$$

be the lattice in $\mathbf{R}^n$ with basis $(m\mathbf{e}_1, \ldots, m\mathbf{e}_n)$. Then M is a subgroup of the group Λ, and the determinant of M is $\det(M) = m^n$. For every integer q, we have

$$\{\mathbf{u} \in \mathbf{Z}^n : \mathbf{u} \equiv q\mathbf{r} \pmod{m}\} = q\mathbf{r} + M \in \Lambda/M.$$

If $q \equiv q' \pmod{m}$, then $q\mathbf{r}+M = q'\mathbf{r}+M$. If $q\mathbf{r}+M = q'\mathbf{r}+M$, then $(q-q')\mathbf{r} \in M$, and so $(q-q')r_i \equiv 0 \pmod{m}$ for $i = 1, \ldots, n$. Let $((q-q'), m) = d$. Then

$$\left(\frac{q-q'}{d}\right) r_i \equiv 0 \pmod{m/d},$$

and so

$$r_i \equiv 0 \pmod{m/d}$$

for $i = 1, \ldots, n$. It follows from (6.8) that $d = m$ and $q \equiv q' \pmod{m}$. This implies that

$$\Lambda = \bigcup_{q=0}^{m-1} (q\mathbf{r} + M),$$

and so $[\Lambda : M] = m < \infty$. Theorem 6.11 implies that Λ is a lattice. By Theorem 6.9,

$$\det(\Lambda) = \frac{\det(M)}{[\Lambda : M]} = m^{n-1}.$$

Let

$$K = \{(x_1, \ldots, x_n) \in \mathbf{R}^n : |x_i| < 1/4 \text{ for } i = 1, \ldots, n\}.$$

The set K is a symmetric, convex body of volume $\mathrm{vol}(K) = 2^{-n}$. Applying Minkowski's second theorem to the set K and the lattice Λ, we see that the successive minima $\lambda_1, \ldots, \lambda_n$ satisfy

$$\lambda_1 \cdots \lambda_n \leq \frac{2^n \det(\Lambda)}{\mathrm{vol}(K)} = 4^n m^{n-1},$$

and so there exist linearly independent vectors $\mathbf{b}_1, \ldots, \mathbf{b}_n \in \Lambda$ such that

$$\mathbf{b}_j = (b_{1.j}, b_{2.j}, \ldots, b_{n.j}) \in \overline{\lambda_j * K} = \lambda_j * \overline{K}$$

for $j = 1, \ldots, n$. Therefore,

$$|b_{i.j}| \leq \frac{\lambda_j}{4}$$

for $i, j = 1, \ldots, n$. This completes the proof.

6.8 Notes

The material in this chapter is classical. Standard references for the geometry of numbers are Cassels [14], Gruber and Lekkerkerker [60], and Siegel [119]. An excellent book on convexity is Eggleston [32]. The proof of Minkowski's second theorem follows Siegel [119]. Alternate proofs have been given by Bambah, Woods, and Zassenhaus [7], Danicic [21], Davenport [23], and Weyl [127]. The proof of Lagrange's theorem by means of the geometry of numbers is due to Davenport [24]. The proof of Theorem 6.10 follows Lang [78].

6.9 Exercises

1. Let Λ be the lattice in $\mathbf{R}^2$ with basis $\mathbf{a}_1 = (1, 2)$ and $\mathbf{a}_2 = (2, 1)$. Draw the lattice in the plane, and identify its fundamental parallelepiped $F(\Lambda)$. Show that Λ consists of all vectors of the form $(u + 2v, 2u + v)$, where $u, v \in \mathbf{Z}$. Express the vector $(8, 7)$ as the sum of a vector in Λ and a vector in $F(\Lambda)$. What is the volume of the fundamental parallelepiped of Λ?

2. Let Λ be the lattice in $\mathbf{R}^3$ with basis $\mathbf{a}_1 = (1, 2, 3), \mathbf{a}_2 = (3, 1, 2), \mathbf{a}_3 = (2, 3, 1)$. Compute $\det(\Lambda)$.

3. Let Λ be the subgroup of $\mathbf{R}^2$ generated by the vectors $(1, 0)$, $(0, 1)$, and $(1/2, 1/2)$. Prove that Λ is a lattice, and find a basis for Λ.

4. Let Λ be a lattice in $\mathbf{R}^n$, and let $\mathbf{a}_1, \ldots, \mathbf{a}_n$ be a basis for Λ, where $\mathbf{a}_j = \sum_{i=1}^{n} a_{i,j}\mathbf{e}_i$. Let $A = (a_{i,j})$ be the matrix of the basis $\mathbf{a}_1, \ldots, \mathbf{a}_n$. Prove that Λ consists of all vectors of the form $A\mathbf{u}$, where $\mathbf{u}$ is a column vector in $\mathbf{Z}^n$.

5. Let Λ be a lattice in $\mathbf{R}^n$ such that $\Lambda \subseteq \mathbf{Z}^n$. Prove that $\Lambda = \mathbf{Z}^n$ if and only if $\det(\Lambda) = 1$.

6. Construct a lattice $\Lambda \subseteq \mathbf{R}^n$ such that $\det(\Lambda) = 1$ but $\Lambda \neq \mathbf{Z}^n$.

7. Let K be a convex set, and let $t_1, \ldots, t_r$ be nonnegative real numbers such that $t_1 + \cdots + t_r = 1$. Prove that if $\mathbf{u}_1, \ldots, \mathbf{u}_r \in K$, then $t_1\mathbf{u}_1 + \cdots + t_r\mathbf{u}_r \in K$.

8. Let $A : \mathbf{R}^n \to \mathbf{R}^n$ be an isomorphism. Prove that if K is a convex body, then $A(K)$ is a convex body.

9. Let K be a convex set in $\mathbf{R}^n$, and let $\overline{K}$ be the closure of K. Prove that $\overline{K}$ is convex.

10. Let K be a convex body in $\mathbf{R}^n$ with $\mathbf{0} \in K$. Prove that if $\lambda \leq \mu$, then $\lambda * K \subseteq \mu * K$. Construct a convex body K with $\mathbf{0} \notin K$ such that $K \cap (2 * K) = \emptyset$.

11. Let K be a convex body in $\mathbf{R}^n$ with $\mathbf{0} \in K$. If $\lambda_1 > \lambda_2 > \lambda_3 > \cdots > 0$ and $\lim_{i\to\infty} \lambda_i = \lambda$, prove that

$$\bigcap_{i=1}^{\infty} \lambda_i * K = \lambda * \overline{K}.$$

12. Let f be a continuous real-valued function on $\mathbf{R}^n$ such that

    (a) $f(\mathbf{u}) > 0$ for all $\mathbf{u} \in \mathbf{R}^n$, $\mathbf{u} \neq \mathbf{0}$,
    (b) $f(t\mathbf{u}) = tf(\mathbf{u})$ for all $t \in \mathbf{R}$, $t \geq 0$, and $\mathbf{u} \in \mathbf{R}^n$,
    (c) $f(\mathbf{u}_1 + \mathbf{u}_2) \leq f(\mathbf{u}_1) + f(\mathbf{u}_2)$ for all $\mathbf{u}_1, \mathbf{u}_2 \in \mathbf{R}^n$.

    Let $K = \{\mathbf{u} \in \mathbf{R}^n : f(\mathbf{u}) < 1\}$. Prove that K is a convex body.

13. Let K be a convex body in $\mathbf{R}^n$ with $\mathbf{0} \in K$.. For $\mathbf{u} \in \mathbf{R}^n$, $\mathbf{u} \neq \mathbf{0}$, let

$$t_0 = \inf\{t \in \mathbf{R} : t\mathbf{u} \in K\}.$$

    Show that $t_0 > 0$, $t\mathbf{u} \in K$ for $t < t_0$, and $t_0\mathbf{u} \in \partial K$, where $\partial K = \overline{K} \setminus K$ is the boundary of K.

14. Let Λ be a lattice in $\mathbf{R}^n$, and let K be a symmetric convex body in $\mathbf{R}^n$ such that $\text{vol}(K) = 2^n \det(\Lambda)$. Prove that $\overline{K}$ contains a nonzero element of Λ.

15. Prove that the volume of the 4-dimensional ball of radius r is $\pi^2 r^4/2$.

16. Let K be a convex body containing $\mathbf{0}$. For $\lambda \in \mathbf{R}$, $\lambda \geq 0$, let

$$\lambda * K = \{\lambda\mathbf{u} : \mathbf{u} \in K\}.$$

    Prove that

    (a) $\lambda * \overline{K} = \overline{\lambda * K}$
    (b) $\partial(\lambda * K) = \lambda * \partial(K)$, where $\partial(K) = \overline{K} \setminus K$.

17. Let K be a convex body in $\mathbf{R}^n$ with $\mathbf{0} \in K$. The *gauge function* f of the convex body K is defined by

$$f(\mathbf{x}) = \inf\{\lambda \in \mathbf{R} : \lambda \geq 0, \mathbf{x} \in \lambda K\}$$

    for all $\mathbf{x} \in \mathbf{R}^n$. Prove that

    (a) $f(\mathbf{0}) = 0$ and $f(\mathbf{x}) > 0$ for all $\mathbf{x} \neq \mathbf{0}$.
    (b) $f(t\mathbf{x}) = tf(\mathbf{x})$ for all $\mathbf{x} \in \mathbf{R}^n$ and $t \in \mathbf{R}$, $t \geq 0$.
    (c) $f(\mathbf{x}_1 + \mathbf{x}_2) \leq f(\mathbf{x}_1) + f(\mathbf{x}_2)$ for all $\mathbf{x}_1, \mathbf{x}_2 \in \mathbf{R}^n$.
    (d) $K = \{\mathbf{x} \in \mathbf{R}^n : f(\mathbf{x}) < 1\}$.
    (e) $\lambda * K = \{\mathbf{x} \in \mathbf{R}^n : f(\mathbf{x}) < \lambda\}$.

(f) $\partial(\lambda * K) = \{\mathbf{x} \in \mathbf{R}^n : f(\mathbf{x}) = \lambda\}$.

(g) K is symmetric if and only if $f(\mathbf{x}) = f(-\mathbf{x})$ for all $\mathbf{x} \in \mathbf{R}^n$.

18. Compute the gauge function for the unit ball $B(\mathbf{0}, \mathbf{1})$ in $\mathbf{R}^n$.

19. Compute the gauge function for the unit cube in $\mathbf{R}^n$.

20. Let $K = \{(x, y) \in \mathbf{R}^2 : x^2/3 + y^2/12 < 1\}$. Prove that K is convex. Compute the gauge function for K.

21. Prove that the gauge function of the convex body K in $\mathbf{R}^n$ is convex in the sense that
$$f((1-t)\mathbf{x}_1 + t\mathbf{x}_2) \leq (1-t)f(\mathbf{x}_1) + tf(\mathbf{x}_2)$$
for all $\mathbf{x}_1, \mathbf{x}_2 \in \mathbf{R}^n$ and $0 \leq t \leq 1$.

22. Let L_1, L_2 be compact sets in $\mathbf{R}^n$, and define $d(L_1, L_2)$ by (6.4). Prove that $d(L_1, L_2)$ is a metric on the set of all compact subsets of $\mathbf{R}^n$, that is, prove that for any compact sets L_1, L_2, and L_3,

    (a) $d(L_1, L_2) \geq 0$,

    (b) $d(L_1, L_2) = 0$ if and only if $L_1 = L_2$,

    (c) $d(L_1, L_2) = d(L_2, L_1)$,

    (d) $d(L_1, L_2) \leq d(L_1, L_3) + d(L_3, L_2)$.

23. For any set $X \subseteq \mathbf{R}^n$ and $\varepsilon > 0$, let
$$X(\varepsilon) = \{\mathbf{v} \in \mathbf{R}^n : |\mathbf{v} - \mathbf{x}| < \varepsilon \text{ for some } \mathbf{x} \in X\}.$$
Let L_1 and L_2 be compact subsets of $\mathbf{R}^n$. Let
$$\delta_1 = \inf\{\varepsilon > 0 : L_1 \subseteq L_2(\varepsilon)\}$$
and let
$$\delta_2 = \inf\{\varepsilon > 0 : L_2 \subseteq L_1(\varepsilon)\}.$$
Define
$$d(L_1, L_2) = \delta_1 + \delta_2.$$
Prove that this definition of the distance function $d(L_1, L_2)$ is equivalent to (6.4).

24. Let X be a compact subset of $\mathbf{R}^n$ and let $\{L_i\}$ be a sequence of compact convex subsets of X that converges to the compact set L, that is, $\lim_{i\to\infty} d(L_i, L) = 0$. Prove that L is convex.

25. Let $\Omega(X)$ be the metric space of all compact subsets of a compact set X in $\mathbf{R}^n$, and let $L \in \Omega(X)$. Let $\mathrm{vol}(L)$ denote the volume of L. Prove that $\mathrm{vol}(L)$ is a continuous function on $\Omega(X)$.

26. Prove that the center of mass of a convex body lies in the convex body. Construct a nonconvex set X such that the center of mass of X does not lie inside X.

27. Let $\Omega(X)$ be the metric space of all compact subsets of a compact set X in $\mathbf{R}^n$, and let $L \in \Omega(X)$. Let $\mathbf{c}(L)$ denote the center of mass of L. Prove that $\mathbf{c}(L)$ is a continuous function on $\Omega(X)$.

7

Plünnecke's inequality

7.1 Plünnecke graphs

A *directed graph* $G = (V(G), E(G))$ consists of a finite set $V(G)$ of *vertices* and a set $E(G)$ of *edges*, where each edge $e \in E(G)$ is an ordered pair (v, v') of distinct elements of $V(G)$. Let $h \geq 1$. A directed graph $G = (V(G), E(G))$ is a *graph of level h* if the vertex set $V(G)$ is the union of $h + 1$ pairwise disjoint nonempty sets $V_0, V_1, \ldots, V_h$ and if every edge of G is of the form (v, v'), where $v \in V_{i-1}$ and $v' \in V_i$ for some $i = 1, \ldots, h$; thus,

$$E(G) \subseteq \bigcup_{i=1}^{h} (V_{i-1} \times V_i).$$

A directed graph $G = (V(G), E(G))$ of level h is a *Plünnecke graph of level h* if it satisfies the following two conditions:

(i) Let $1 \leq i \leq h - 1$ and $k \geq 2$. Let $u \in V_{i-1}$, $v \in V_i$, and $w_1, \ldots, w_k \in V_{i+1}$ be $k + 2$ distinct vertices of G such that $(u, v) \in E(G)$ and $(v, w_j) \in E(G)$ for $j = 1, \ldots, k$. Then there exist distinct vertices $v_1, \ldots, v_k \in V_i$ such that $(u, v_j) \in E(G)$ and $(v_j, w_j) \in E(G)$ for $j = 1, \ldots, k$. This can be represented by the following diagram:

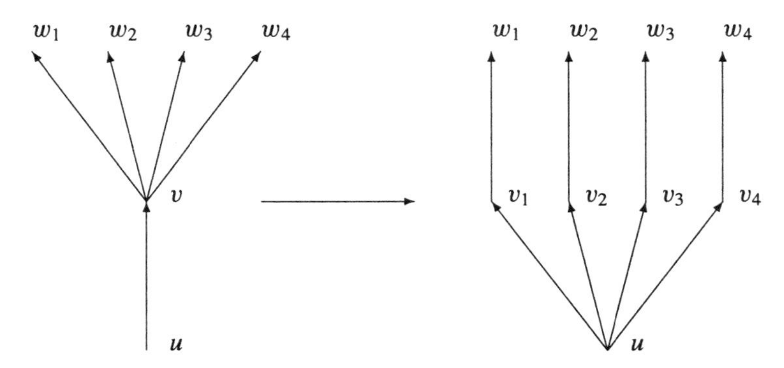

(ii) Let $1 \leq i \leq h-1$ and $k \geq 2$. Let $u_1, \ldots, u_k \in V_{i-1}$, $v \in V_i$, and $w \in V_{i+1}$ be $k+2$ distinct vertices of G such that $(u_j, v) \in E(G)$ for $j = 1, \ldots, k$, and $(v, w) \in E(G)$. Then there exist distinct vertices $v_1, \ldots, v_k \in V_i$ such that $(u_j, v_j) \in E(G)$ and $(v_j, w) \in E(G)$ for $j = 1, \ldots, k$. This can be represented by the following diagram:

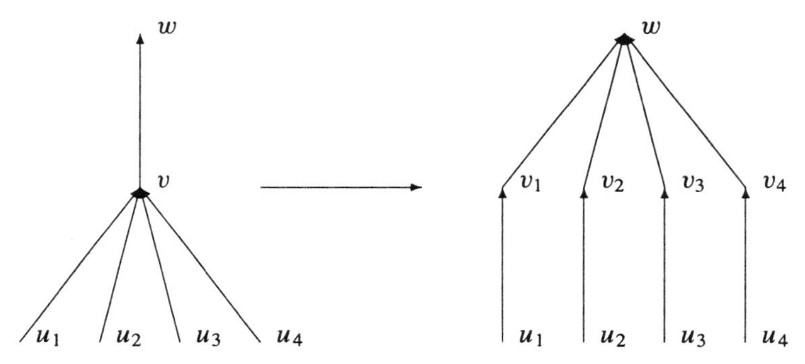

Let $G = (V(G), E(G))$ be a directed graph. A *path in G from vertex a to vertex b* is a finite sequence of vertices $a = v_0, v_1, \ldots, v_{k-1}, v_k = b$ such that $(v_{i-1}, v_i) \in E(G)$ for $i = 1, \ldots, k$ and $v_i \neq a, b$ for $i = 1, 2, \ldots, k-1$. The path can also be identified with the sequence of edges $(v_0, v_1), (v_1, v_2), \ldots, (v_{k-1}, v_k)$. The vertices $v_1, \ldots, v_{k-1}$ are called the *intermediate vertices* of the path.

Let X and Y be nonempty subsets of $V(G)$. The *image* of X in Y, denoted $\mathrm{im}_G(X, Y)$ or simply $\mathrm{im}(X, Y)$, is the set of all $y \in Y$ such that there exists a path from x to y for some $x \in X$. The *magnification ratio* of X in Y is

$$D(X, Y) = \min \left\{ \frac{|\mathrm{im}(Z, Y)|}{|Z|} : \emptyset \neq Z \subseteq X \right\}.$$

Let G be a graph of level h with vertex set $V(G) = \bigcup_{i=0}^{h} V_i$. For $i = 1, \ldots, h$, the ith *magnification ratio* of G is

$$D_i = D_i(G) = D(V_0, V_i).$$

Plünnecke proved that if G is a Plünnecke graph of level h, then the sequence $D_i^{1/i}$ is decreasing, that is,

$$D_1 \geq D_2^{1/2} \geq D_3^{1/3} \geq \cdots \geq D_h^{1/h}.$$

These simple inequalities have powerful consequences in additive number theory.

7.2 Examples of Plünnecke graphs

Addition graphs. Let A and B be nonempty, finite subsets of an abelian group. We want to construct a Plünnecke graph of level h whose ith vertex set is the sumset $A + iB$ and whose edges are the ordered pairs of group elements of the form $(v, v + b)$, where $b \in B$ and $v \in A + (i - 1)B$ for some $i = 1, \ldots, h$. Since the sumsets $\{A + iB\}_{i=0}^{h}$ are not necessarily pairwise disjoint, we have to be careful about the construction of this graph.

Let $h \geq 1$, and let A and B be finite, nonempty subsets of an abelian group. The *addition graph G of level h* constructed from the pair A, B is the graph whose vertex set $V(G)$ and edge set $E(G)$ are defined as follows:

$$V(G) = \bigcup_{i=0}^{h} V_i,$$

where

$$V_i = (A + iB) \times \{i\}$$

and

$$E(G) = \bigcup_{i=1}^{h} \{((v, i - 1), (v + b, i)) : v \in A + (i - 1)B, b \in B\}.$$

It is clear that the sets $V_0, V_1, \ldots, V_h$ are pairwise disjoint and that the pair $((v, i), (v', i')) \in E(G)$ if and only if $i' = i + 1$ and $v' - v \in B$. Thus, the addition graph is a directed graph of level h.

We shall prove that the addition graph is a Plünnecke graph. Let $1 \leq i \leq h - 1$, and let $(u, i - 1) \in V_{i-1}$, $(v, i) \in V_i$, and $(w_1, i + 1), \ldots, (w_k, i + 1) \in V_{i+1}$ be $k + 2$ distinct vertices in $V(G)$ such that

$$((u, i - 1), (v, i)) \in E(G)$$

and

$$((v, i), (w_j, i + 1)) \in E(G)$$

for $j = 1, \ldots, k$. Then $v - u = b \in B$ and $w_j - v = b_j \in B$ for $j = 1, \ldots, k$. Let $v_j = b_j + u$. Since the k elements w_j are distinct, the k elements b_j are distinct, and so the k elements v_j are distinct. Since $u \in A + (i - 1)B$, it follows that $v_j \in A + iB$

and so $((u, i-1), (v_j, i)) \in E(G)$ for $j = 1, \ldots, k$. Since A and B are subsets of an abelian group, it follows that

$$w_j = b_j + v = b_j + (b + u) = b + (b_j + u) = b + v_j,$$

and so $((v_j, i), (w_j, i+1)) \in E(G)$ for $j = 1, \ldots, k$. This shows that the addition graph satisfies property (i) of Plünnecke graphs. It can be shown in the same way that property (ii) is satisfied.

Addition graphs are an important class of Plünnecke graphs. To simplify notation, we shall henceforth denote the ith vertex set in the addition graph by $A + iB$.

Truncated addition graphs. Let $n \geq 1$, and let A and B be nonempty sets of nonnegative integers such that $A \cap [1, n] \neq \emptyset$ and $0 \in B$. Let

$$V_i = (A + iB) \cap [1, n]$$

for $i = 0, 1, 2, \ldots$. The *truncated addition graph* G of level h constructed from the pair of sets A and B is the graph with vertex set $V(G) = \bigcup_{i=0}^{h} V_i$ and edge set

$$E(G) = \left\{(v, v') \in V_{i-1} \times V_i : v' - v \in B\right\}.$$

This is a Plünnecke graph of level h.

Independent addition graphs. Let $B = \{b_1, \ldots, b_n\}$ be a set of n elements of an abelian group such that the $\binom{n+h-1}{h}$ h-fold sums of the form $b_{j_1} + \cdots + b_{j_h}$ with $1 \leq j_1 \leq \cdots \leq j_h \leq n$ are distinct. Let $I_{n,h}$ denote the addition graph of level h constructed from the sets $A = \{0\}$ and B. Then $V_0 = \{0\}$, $V_i = iB$, and

$$|V_i| = \binom{n+i-1}{i} = \frac{n(n+1)\cdots(n+i-1)}{i!}$$

for $i = 1, \ldots, h$. Moreover,

$$\frac{n^i}{i!} \leq |V_i| \leq n^i$$

since

$$\begin{aligned}
\frac{n^i}{i!} &\leq \frac{n(n+1)\cdots(n+i-1)}{i!} \\
&= \left(1+\frac{1}{n}\right)\left(1+\frac{2}{n}\right)\cdots\left(1+\frac{i-1}{n}\right)\left(\frac{n^i}{i!}\right) \\
&\leq \frac{i!n^i}{i!} \\
&= n^i.
\end{aligned}$$

The graph $I_{n,h}$ is called the *independent addition graph* of level h on n elements.

It is easy to construct examples of independent addition graphs. Let $A = \{0\}$ and $B = \{b_1, \ldots, b_n\}$ be a set of n positive integers such that $b_i > hb_{i-1}$ for $i = 2, \ldots, n$. Then the h-fold sums of elements of $B \cup \{0\}$ are distinct, and the $h+1$ sets $\{V_i\}_{i=0}^{h}$ are pairwise disjoint.

Contracted graphs. Let $G = (V(G), E(G))$ be a Plünnecke graph of level h, where $V(G) = \bigcup_{i=0}^{h} V_i$. Let $a, b \in V(G)$. A *path in G from vertex a to vertex b* is a finite sequence of vertices $a = v_0, v_1, \ldots, v_{k-1}, v_k = b$ such that $(v_{i-1}, v_i) \in E(G)$ for $i = 1, \ldots, k$ and $v_i \neq a, b$ for $i = 1, \ldots, k-1$. Let $0 \leq j < k \leq h$, and let X and Y be nonempty subsets of V_j and V_k, respectively, such that for some $a \in X$ and $b \in Y$ there is a path in G from a to b. Let $V(X, Y)$ be the set of all vertices $v \in V(G)$ that lie on some path from X to Y. Let

$$V_i(X, Y) = V(X, Y) \cap V_{i+j}$$

for $i = 0, 1, \ldots, k-j$. Then

$$V(X, Y) = \bigcup_{i=0}^{k-j} V_i(X, Y).$$

Since there exists a path from the vertex $a \in X$ to the vertex $b \in Y$, it follows that $V_i(X, Y) \neq \emptyset$ for $i = 0, 1, \ldots, k-j$. Let $G(X, Y)$ be the graph with vertex set $V(X, Y)$ and edge set $E(X, Y)$ consisting of all edges $(v, v') \in E(G)$ with $v, v' \in V(X, Y)$. Then the *contracted graph* $G(X, Y)$ is a Plünnecke graph of level $k - j$.

Product graphs. Let $h \geq 1$, $r \geq 2$, and let $G_1, \ldots, G_r$ be Plünnecke graphs of level h. We construct the *product graph* $G = G_1 \times \cdots \times G_r$ as follows: Let $V(G_j) = \bigcup_{i=0}^{h} V_{i,j}$ for $j = 1, \ldots, r$ and define

$$V_i = V_{i,1} \times \cdots \times V_{i,r}$$

for $i = 0, 1, \ldots, h$. Let $E(G)$ consist of all ordered pairs of r-tuples

$$\left(\left(v_{i-1,1}, \ldots, v_{i-1,r}\right), \left(v_{i,1}, \ldots, v_{i,r}\right)\right) \in V_{i-1} \times V_i$$

such that $(v_{i-1,j}, v_{i,j}) \in E(G_j)$ for $j = 1, \ldots, r$. It is easy to check that the product graph G is a Plünnecke graph of level h.

Inverse graphs. Let G be a Plünnecke graph of level h with vertex set $V(G) = \bigcup_{i=0}^{h} V_i$ and edge set $E(G)$. Let $V_i^{-1} = V_{h-i}$ for $i = 0, 1, \ldots, h$. The *inverse graph* G^{-1} is the graph whose vertex set is $V(G^{-1}) = \bigcup_{i=0}^{h} V_i^{-1}$ and whose edge set $E(G^{-1})$ is determined by the condition that $(v, v') \in E(G^{-1})$ if and only if $(v', v) \in E(G)$. Then G^{-1} is a Plünnecke graph of level h.

7.3 Multiplicativity of magnification ratios

In this section, we shall prove that the magnification ratios of graphs of level h are multiplicative.

Theorem 7.1 *Let G' and G'' be directed graphs of level h. Then*

$$D_i(G' \times G'') = D_i(G')D_i(G'')$$

for $i = 1, \ldots, h$.

Proof. Let $V(G') = \bigcup_{i=0}^{h} V_i'$ and $V(G'') = \bigcup_{i=0}^{h} V_i''$. Let Z' and Z'' be nonempty subsets of V_0' and V_0'', respectively, such that

$$D_i(G') = \frac{|\mathrm{im}(Z', V_i')|}{|Z'|}$$

and

$$D_i(G'') = \frac{D|\mathrm{im}(Z'', V_i'')|}{|Z''|}.$$

Then

$$Z' \times Z'' \subseteq V_0' \times V_0'' = V_0(G' \times G'')$$

and

$$\mathrm{im}(Z' \times Z'', V_0' \times V_0'') = \mathrm{im}(Z', V_0') \times \mathrm{im}(Z'', V_0'').$$

Since $|Z' \times Z''| = |Z'||Z''| \neq 0$ and

$$|\mathrm{im}(Z' \times Z'', V_i' \times V_i'')| = |\mathrm{im}(Z', V_i')||\mathrm{im}(Z'', V_i'')|,$$

it follows that

$$\begin{aligned} D_i(G' \times G'') &\leq \frac{|\mathrm{im}(Z' \times Z'', V_i' \times V_i'')|}{|Z' \times Z''|} \\ &= \frac{|\mathrm{im}(Z', V_i')|}{|Z'|} \frac{|\mathrm{im}(Z'', V_i'')|}{|Z''|} \\ &= D_i(G')D_i(G''). \end{aligned}$$

To complete the proof of the theorem, we must prove the reverse inequality

$$D_i(G' \times G'') \geq D_i(G')D_i(G'').$$

We do this first in the case where G' and G'' are graphs of level 1, G'' is a graph of a very special type, namely, $V(G'') = V_0'' \cup V_1''$, where the two vertex sets V_0'' and V_1'' are both copies of a nonempty set T, and $E(G'')$ consists of the $|T|$ edges (t, t) for $t \in T$. If $Z'' \subseteq T = V_0''$, then $\mathrm{im}(Z'', V_1'') = Z''$, and so

$$D_1(G'') = 1.$$

(Note: Since the vertex sets V_0'' and V_1'' of a graph of level 1 must be disjoint, we should, just as in the construction of the addition graph, formally define the vertex sets $V_0'' = \{(t, 0) : t \in T\}$ and $V_1'' = \{(t, 1) : t \in T\}$ and the edge set $E(G'') = \{((t, 0), (t, 1)) : t \in T\}$.)

Let $G = G' \times G''$, and let Z be a nonempty subset of $V_0(G) = V_0' \times V_0'' = V_0' \times T$. Let

$$Z_t' = \{v' \in V_0' : (v', t) \in Z\}.$$

Then

$$Z = \bigcup_{t \in T} (Z_t' \times \{t\})$$

and

$$\mathrm{im}(Z, V_1' \times V_1'') = \bigcup_{t \in T} \left(\mathrm{im}(Z_t', V_1'') \times \{t\}\right).$$

Therefore,

$$\begin{aligned} |\mathrm{im}(Z, V_1' \times V_1'')| &= \sum_{t \in T} |\mathrm{im}(Z_t', V_1')| \\ &\geq D_1(G') \sum_{t \in T} |Z_t'| \\ &= D_1(G')|Z|, \end{aligned}$$

and so

$$D_1(G') \leq \frac{|\mathrm{im}(Z, V_1' \times V_1'')|}{|Z|}.$$

Since $D_1(G'') = 1$, it follows that

$$D_1(G')D_1(G'') = D_1(G') \leq D_1(G' \times G'').$$

Therefore,

$$D_1(G')D_1(G'') = D_1(G' \times G'')$$

for graphs of this special type.

We now consider the general case. Let G' and G'' be graphs of level h, and let $1 \leq i \leq h$. We construct a graph $H = (V(H), E(H))$ of level 2 with vertices $V(H) = W_0 \cup W_1 \cup W_2$, where

$$\begin{aligned} W_2 &= V_i' \times V_i'' = V_i(G' \times G'') \\ W_1 &= V_i' \times V_0'' \\ W_0 &= V_0' \times V_0'' = V_0(G' \times G''). \end{aligned}$$

The edges from W_0 to W_1 will consist of all pairs $((a, c), (b, c))$ such that

$$a \in V_0', b \in V_i', c \in V_0'',$$

and there exists a path in G' from $a \in V_0'$ to $b \in V_i'$. The edges from W_1 to W_2 will consist of all pairs $((a, c), (a, d))$ such that

$$a \in V_i', c \in V_0'', d \in V_i'',$$

and there exists a path in G'' from $c \in V_0''$ to $d \in V_i''$. It follows that there is a path in H from $(v_0', v_0'') \in W_0$ to $(v_i', v_i'') \in W_2$ if and only if there is a path in $G' \times G''$ from $(v_0', v_0'') \in V_0(G' \times G'')$ to $(v_i', v_i'') \in V_i(G' \times G'')$. Therefore,

$$D_2(H) = D(W_0, W_2) = D_i(G' \times G'').$$

We shall prove that

$$D_1(H) = D(W_0, W_1) \geq D_i(G').$$

Let Z be a nonempty subset of $W_0 = V_0' \times V_0''$. For $t \in V_0''$, let

$$Z_t' = \{v' \in V_0' : (v', t) \in Z\}.$$

Then

$$Z = \bigcup_{t \in V_0''} (Z_t' \times \{t\})$$

and

$$\mathrm{im}(Z, W_1) = \bigcup_{t \in V_0''} \left(\mathrm{im}(Z_t', V_i') \times \{t\}\right).$$

Therefore,

$$\begin{aligned} |\mathrm{im}(Z, W_1)| &= \sum_{t \in V_0''} |\mathrm{im}(Z_t', V_i')| \\ &\geq D_i(G') \sum_{t \in V_0''} |Z_t'| \\ &= D_i(G')|Z|. \end{aligned}$$

Therefore,

$$D(W_0, W_1) = \min\left\{ \frac{|\mathrm{im}(Z, W_1)|}{|Z|} : \emptyset \neq Z \subseteq W_0 \right\} \geq D_i(G').$$

A similar argument proves that

$$D(W_1, W_2) \geq D_i(G'').$$

If $|\mathrm{im}(Z, W_1)| = 0$ for some nonempty set $Z \subseteq W_0$, then $D_i(G') = D_i(G' \times G'') = 0$ and we are done. Therefore, we can assume that $|\mathrm{im}(Z, W_1)| \neq 0$ for all $\emptyset \neq Z \subseteq W_0$. It follows that

$$\begin{aligned} D_i(G' \times G'') &= D_2(H) \\ &= D(W_0, W_2) \\ &= \min\left\{ \frac{|\mathrm{im}(Z, W_2)|}{|Z|} : \emptyset \neq Z \subseteq W_0 \right\} \\ &= \min\left\{ \frac{|\mathrm{im}(Z, W_1)|}{|Z|} \frac{|\mathrm{im}(Z, W_2)|}{|\mathrm{im}(Z, W_1)|} : \emptyset \neq Z \subseteq W_0 \right\} \\ &= \min\left\{ \frac{|\mathrm{im}(Z, W_1)|}{|Z|} \frac{|\mathrm{im}(\mathrm{im}(Z, W_1), W_2)|}{|\mathrm{im}(Z, W_1)|} : \emptyset \neq Z \subseteq W_0 \right\} \\ &\geq \min\left\{ \frac{|\mathrm{im}(Z, W_1)|}{|Z|} : \emptyset \neq Z \subseteq W_0 \right\} \\ &\quad \times \min\left\{ \frac{|\mathrm{im}(\mathrm{im}(Z, W_1), W_2)|}{|\mathrm{im}(Z, W_1)|} : \emptyset \neq Z \subseteq W_0 \right\} \end{aligned}$$

$$\begin{aligned}
&\geq \min\left\{\frac{|\mathrm{im}(Z, W_1)|}{|Z|} : \emptyset \neq Z \subseteq W_0\right\} \\
&\quad \times \min\left\{\frac{|\mathrm{im}(Z', W_2)|}{|Z'|} : \emptyset \neq Z' \subseteq W_1\right\} \\
&= D(W_0, W_1)D(W_1, W_2) \\
&\geq D_i(G')D_i(G'').
\end{aligned}$$

This completes the proof.

Corollary 7.1 *Let $h \geq 1$ and $r \geq 2$, and let $G_1, \ldots, G_r$ be graphs of level h. Let G be the product graph $G_1 \times \cdots \times G_r$. Then*

$$D_i(G) = D_i(G_1) \cdots D_i(G_r)$$

for $i = 1, \ldots, h$.

7.4 Menger's theorem

Let $G = (V(G), E(G))$ be a directed graph. Let $a = v_0, v_1, \ldots, v_k = b$ and $a = w_0, w_1, \ldots, w_l = b$ be two paths in G from vertex a to vertex b. These paths are *disjoint* if $v_i \neq w_j$ for $i = 1, \ldots, k-1$ and $j = 1, \ldots, l-1$. A set S of vertices *separates vertex a from vertex b* if every path from a to b contains at least one element of S. Let S be a set that separates vertex a from vertex b, but contains neither a nor b. Let m be the maximum number of pairwise disjoint paths from a to b. Since S contains at least one vertex from each of these paths, it follows that $|S| \geq m$. Menger's theorem states that there exists a separating set S such that $|S| = m$.

Notation. Let $G = (V(G), E(G))$ be a directed graph. The directed graph $G' = (V(G'), E(G'))$ is a *subgraph* of G if $V(G') \subseteq V(G)$ and $E(G') \subseteq (V(G') \times V(G')) \cap E(G)$. Let $V' \subseteq V(G)$. The *complete subgraph* of G generated by V' is the graph with vertex set V' and edge set

$$E' = \{(v, v') \in E(G) : v, v' \in V'\} = (V' \times V') \cap E(G).$$

Let $W \subseteq V$ and $F \subseteq E(G)$. Let $G \setminus \{W, F\}$ denote the graph with vertex set $V(G) \setminus W$ and edge set consisting of all edges $(v, v') \in E(G) \setminus F$ such that $v, v' \in V(G) \setminus W$.

Theorem 7.2 (Menger) *Let a and b be vertices of a directed graph $G = (V(G), E(G))$, and suppose that $(a, b) \notin E(G)$. Let m be the maximum number of pairwise disjoint paths from the vertex a to the vertex b. Let ℓ be the cardinality of the smallest set S of vertices that separates the vertex a from the vertex b and that contains neither a nor b. Then $\ell = m$.*

Proof. The proof will be by induction on ℓ. If $\ell = 0$, then S is empty, so there are no paths from a to b and $m = 0$. If $\ell = 1$, then $S = \{v\}$, and there exists at least one path from a to b. Moreover, every path from a to b must contain the intermediate vertex v. It follows that there cannot exist two pairwise disjoint paths from a to b, and so $m = 1$.

Let $\ell \geq 2$. Suppose that the theorem is true for any two vertices a, b of a directed graph G such that $(a, b) \notin E(G)$ and there exists a separating set of cardinality at most $\ell - 1$ that does not contain a or b. If the theorem is false for ℓ, then there is a graph G with vertices a, b such that $(a, b) \notin E(G)$, there are at most $\ell - 1$ pairwise disjoint paths from a to b, but ℓ is the cardinality of the smallest set that separates a and b and contains neither a nor b.

Consider such graphs in which the number of vertices is minimal, and from this set of graphs choose one for which the number of edges is minimal. Call this graph $G = (V(G), E(G))$. Let e be any edge in $E(V)$, and let

$$G' = G \setminus \{e\} = (V(G), E(G) \setminus \{e\}).$$

Since G' is a subgraph of G, there are at most $\ell - 1$ pairwise disjoint paths from a to b in G'. Let $S(e)$ be a set of vertices of minimum cardinality that separates a from b in G' and contains neither a nor b. By the minimality of G, the set $S(e)$ contains at most $\ell - 1$ elements. If $e = (v_1, v_2)$ and $v_1 \neq a$, then $S(e) \cup \{v_1\}$ separates a from b in G, since every path from a to b either contains some vertex in $S(e)$, or includes the edge e. Similarly, if $v_2 \neq b$, then $S(e) \cup \{v_2\}$ separates a from b in G. This implies that

$$|S(e)| = \ell - 1. \tag{7.1}$$

Let $v \in V(G)$. If a, v, b is a path in G, then $v \in S$ for every separating set S such that $S \cap \{a, b\} = \emptyset$. Therefore, $S' = S \setminus \{v\}$ separates a from b in the graph $G' = G \setminus \{v\}$. Similarly, if S' separates a from b in the graph G', then $S' \cup \{v\}$ separates a from b in the graph G. Therefore, $|S'| \geq \ell - 1$ and $|S'| = \ell - 1$ for some separating set in G'. By the induction hypothesis, the graph G' contains $\ell - 1$ pairwise disjoint paths from a to b, and these paths, together with the path a, v, b, give ℓ pairwise disjoint paths from a to b in G, which is false. Therefore, if v is any vertex in $V(G)$, then either $(a, v) \notin E(G)$ or $(v, b) \notin E(G)$.

Let S separate a from b in G, and suppose that $|S| = \ell$ and $a \notin S$ and $b \notin S$. We shall prove that either $(a, s) \in E(G)$ for all $s \in S$ or $(s, b) \in E(G)$ for all $s \in S$.

Let $P(a, S)$ denote the set of all paths in G that start at the vertex a and end at some vertex $s \in S$, with no intermediate vertex in S. Let $P(S, b)$ denote the set of all paths in G that start at some vertex $s \in S$ and end at b and have no intermediate vertex in S. Since S is a separating set, every path in G from a to b contains at least one point that lies in S, and so the path has an initial segment that belongs to $P(a, S)$ and a terminal segment that belongs to $P(S, b)$. Moreover, since the separating set S is minimal, for every $s \in S$ there is a path from a to s and from s to b.

Let $I(a, S)$ be the set of intermediate vertices of paths in $P(a, S)$, and let $I(S, b)$ be the set of intermediate vertices of paths in $P(S, b)$. If there exists $c \in I(a, S) \cap I(S, b)$, then there is a path in G from a to c that contains no vertex of S, and there is a path in G from c to b that contains no vertex of S. Concatenating these two paths produces a path in G from a to b that contains no element of S, which is impossible. Therefore,

$$I(a, S) \cap I(S, b) = \emptyset.$$

We shall prove that either $I(a, S) = \emptyset$ or $I(S, b) = \emptyset$. Suppose that both sets $I(a, S)$ and $I(S, b)$ are nonempty. Let $H = (V(H), E(H))$ be the graph with vertex set

$$\begin{aligned} V(H) &= \{a\} \cup S \cup I(S, b) \cup \{b\} \\ &= \{a\} \cup \{v \in V(G) : v \text{ lies on a path in } P(S, b)\} \end{aligned}$$

and edge set

$$\begin{aligned} E(H) &= \{(a, s) : s \in S\} \\ & \cup \left\{(v, v') \in E(G) : (v, v') \text{ lies on a path in } P(S, b)\right\}. \end{aligned}$$

Since $I(a, S) \cap I(S, b) = \emptyset$ and $I(a, S) \neq \emptyset$, it follows that $I(a, S) \cap V(H) = \emptyset$, and so $|V(H)| < |V(G)|$.

Let T be a set that separates a from b in the graph H, and suppose that $a, b \notin T$. Let π be a path from a to b in G, and let π' be the terminal segment of this path that belongs to $P(S, b)$. Then π' is a path from some $s \in S$ to b. Since (a, s) is an edge in H, it follows that a, s concatenated with the path π' is a path in H from a to b. Since T separates a from b in H, it follows that T contains an intermediate vertex of this path, and so either $s \in T$ or some intermediate vertex of π' belongs to T. Thus, T also separates a from b in G. Therefore, $|T| \geq \ell$. Since the graph H contains strictly fewer vertices than the graph G, it follows from the induction hypothesis that there are ℓ pairwise disjoint paths in H from a to b. In particular, for each $s \in S$ there exists a path $\pi_2(s) \in P(S, b)$ from s to b such that the ℓ paths $\pi_2(s)$ are pairwise disjoint.

Similarly, for each $s \in S$ there exists a path $\pi_1(s) \in P(a, S)$ from a to s such that the ℓ paths $\pi_1(s)$ are pairwise disjoint. Concatenating the path $\pi_1(s)$ from a to s with the path $\pi_2(s)$ from s to b produces a path $\pi(s)$ from a to b, and these $|S| = \ell$ paths are pairwise disjoint. This is impossible in the graph G. Therefore, either $I(a, S) = \emptyset$ and $(a, s) \in E(G)$ for all $s \in S$, or $I(S, b) = \emptyset$ and $(s, b) \in E(G)$ for all $s \in S$. Moreover, since G contains no path of the form a, s, b, it follows that these two possibilities are mutually exclusive.

Let a, v, v' be the initial segment of a shortest path in G from a to b. Then $v' \neq b$ and $e = (v, v') \in E(G)$. Let $S(e)$ be a minimal separating set for the graph $G \setminus \{e\}$. By (7.1), we have $|S(e)| = \ell - 1 \geq 1$, and so $S(e) \cup \{v\}$ is a separating set for G of minimum cardinality ℓ. Since $(a, v) \in E(G)$, it follows that $(a, s) \in E(G)$ for all $s \in S(e)$. Similarly, $S(e) \cup \left\{v'\right\}$ is a separating set for G. If $(a, v') \in E(G)$, then there is a shorter path from a to b than the one that starts a, v, v', which is impossible. Therefore, $(a, v') \notin E(G)$. It follows that $(v', b) \in E(H)$, and so $(s, b) \in E(G)$ for all $s \in S(e)$. Thus, if $s \in S(e)$, then $(a, s) \in E(G)$ and $(s, b) \in E(G)$, and so a, s, b is a path in G, which is impossible. This completes the proof of Menger's theorem.

Let $G = (V(G), E(G))$ be a directed graph, and let X and Y be nonempty, disjoint sets of vertices of G. A *path in G from the set X to the set Y* is a finite sequence of vertices $v_0, v_1, \ldots, v_{k-1}, v_k$ such that $v_0 \in X$, $v_k \in Y$, and $(v_{i-1}, v_i) \in E(G)$ for $i = 1, \ldots, k$. Let $v_0, v_1, \ldots, v_k$ and $w_0, w_1, \ldots, w_l$ be two paths in the graph G from X to Y. These paths are *totally disjoint* if $v_i \neq w_j$ for $i = 0, 1, \ldots, k$ and $j = 0, 1, \ldots, l$. A set S of vertices *separates the set* X *from the set* Y if every path from X to Y contains at least one element of S.

Theorem 7.3 *Let X and Y be nonempty, disjoint sets of vertices of a directed graph G. Let m be the maximum number of pairwise totally disjoint paths from X to Y. Let ℓ be the cardinality of the smallest set S of vertices that separates the set X from the set Y. Then $\ell = m$.*

Proof. Let $G = (V(G), E(G))$ be a directed graph. Let X and Y be nonempty, disjoint subsets of $V(G)$, and let m be the maximum number of pairwise totally disjoint paths from X to Y. Let a and b be elements not belonging to $V(G)$. We construct a new graph $G^* = (V(G^*), E(G^*))$ by adjoining two new vertices a and b to $V(G)$ as follows: Let

$$V(G^*) = V(G) \cup \{a, b\}$$

and

$$E(G^*) = E(G) \cup \{(a, x) : x \in X\} \cup \{(y, b) : y \in Y\}.$$

Then $a = v_0, v_1, \ldots, v_{k-1}, v_k = b$ is a path in G^* from vertex a to vertex b if and only if $v_1, \ldots, v_{k-1}$ is a path in G from the set X to the set Y. Two paths $a = v_0, v_1, \ldots, v_{k-1}, v_k = b$ and $a = w_0, w_1, \ldots, w_{l-1}, w_l = b$ are disjoint paths from a to b in G^* if and only if $v_1, \ldots, v_{k-1}$ and $w_1, \ldots, w_{l-1}$ are totally disjoint paths from X to Y in G. It follows that m is also the maximum number of pairwise disjoint paths from a to b in the graph G^*. By Theorem 7.2, there exists a set $S \subseteq V(G^*)$ such that S separates a from b, $a, b \notin S$, and $|S| = m$. Then $S \subseteq V(G)$, and S separates X from Y in G. This completes the proof.

7.5 Plünnecke's inequality

Let $G = (V(G), E(G))$ be a directed graph, and let $v \in V(G)$. Let $d^+(v) = d^+(v, G)$ denote the number of vertices $v' \in V(G)$ such that $(v, v') \in E(G)$. Let $d^-(v) = d^-(v, G)$ denote the number of vertices $v' \in V(G)$ such that $(v', v) \in E(G)$.

Lemma 7.1 *Let G be a Plünnecke graph, and let $(u, v) \in E(G)$. Then*

$$d^+(u) \geq d^+(v)$$

and

$$d^-(u) \leq d^-(v).$$

Proof. These inequalities follow immediately from properties (i) and (ii) of the definition of a Plünnecke graph.

Lemma 7.2 *Let G be a Plünnecke graph of level h, and let $V(G) = \bigcup_{i=0}^{h} V_i$. If $D_h \geq 1$, then there are $|V_0|$ totally disjoint paths from V_0 to V_h.*

Proof. Let m be the maximum number of pairwise totally disjoint paths from V_0 to V_h. By Theorem 7.3, there exists a set S of cardinality m that separates V_0 from V_h. We shall prove that $|S| = |V_0|$.

For $v \in V(G)$, let $i(v)$ denote the unique integer i such that $v \in V_i$. Choose a separating set S such that $|S| = m$ and

$$\sum_{s \in S} i(s)$$

is minimal. We shall prove that

$$S \subseteq V_0 \cup V_h.$$

Suppose not. Then $S \cap V_j \neq \emptyset$ for some $j \in [1, h-1]$. Let

$$S \cap V_j = \{s_1, \ldots, s_q\},$$

where $q = |S \cap V_j| \geq 1$. Let $\pi_1, \ldots, \pi_m$ be m pairwise totally disjoint paths from V_0 to V_h, and suppose that s_i is a vertex on path π_i for $i = 1, \ldots, q$. Since $0 < j < h$, it follows that, for each $i \in [1, q]$, the vertex s_i has a predecessor $r_i \in V_{j-1}$ on the path π_i and a successor $t_i \in V_{j+1}$ on the path π_i. It follows from the minimality of $\sum_{s \in S} i(s)$ that the set

$$S^* = \{r_1, \ldots, r_q, s_{q+1}, \ldots, s_m\}$$

does not separate V_0 from V_h, and so there exists a path π^* from V_0 to V_h that does not intersect S^*. However, since π^* cannot avoid the separating set S, it follows that s_i lies on π^* for some $i \in [1, q]$, say, $i = 1$. Let r^* be the predecessor of s_1 on the path π^*. Then $r^* \notin \{r_1, \ldots, r_q\}$.

We shall consider the following sets of vertices:

$$\begin{aligned} S_q^- &= \{r_1, \ldots, r_q\} \subseteq V_{j-1}, \\ S_q^* &= \{r^*, r_1, \ldots, r_q\} \subseteq V_{j-1}, \\ S_q &= \{s_1, \ldots, s_q\} \subseteq V_j, \\ S_q^+ &= \{t_1, \ldots, t_q\} \subseteq V_{j+1}. \end{aligned}$$

The contracted graph (defined in Section 7.2)

$$G^* = G(S_q^*, S_q^+)$$

is a Plünnecke graph of level 2. Let

$$V(G^*) = V_0^* \cup V_1^* \cup V_2^*.$$

Since r_i, s_i, t_i are successive vertices on the path π_i for $i = 1, \ldots, q$ and r^*, s_1, t_1 are successive vertices on the path π^*, it follows that

$$\begin{aligned} V_2^* &= S_q^+ \\ V_1^* &\supseteq S_q \\ V_0^* &= S_q^*. \end{aligned}$$

We shall prove that $V_1^* = S_q$. If there exists a vertex $s' \in V_1^* \setminus S_q$, then $(r', s') \in E(G)$ and $(s', t') \in E(G)$ for some $r' \in S_q^* = \{r^*\} \cup S_q^-$ and $t' \in S_q^+$. Therefore, $t' = t_l$ for some $l \in [1, q]$. The path π_l has a terminal segment that starts at t_l, ends in V_h, and does not intersect S. If $r' = r_{l'} \in S_q^-$, then the path $\pi_{l'}$ has an initial segment that starts in V_0, ends at r', and does not intersect S. Combining this initial segment with the path r', s', t', and then with the terminal segment of π_l from r_l into V_h, we obtain a path from V_0 into V_h that does not intersect the separating set S, which is impossible.

If $r' = r^*$, then the path π^* passes through r' and does not intersect $\{s_{q+1}, \ldots, s_m\}$, which are the elements of the separating set S that do not belong to V_j. The initial segment of the path π^* from V_0 to r', followed by the segment r', s', t' and then the terminal segment of the path π_l, is again a path from V_0 into V_h that does not intersect the separating set S, which is impossible. Therefore,

$$V_1^* = S_q.$$

Since (r_i, s_i) and (s_i, t_i) are edges in the contracted graph G^*, it follows from Lemma 7.1 that

$$d^+(r_i, G^*) \geq d^+(s_i, G^*)$$

and

$$d^-(t_i, G^*) \geq d^-(s_i, G^*).$$

Since the number of edges leaving V_0^* is exactly equal to the number of edges going into V_1^*, and the number of edges leaving V_1^* is exactly equal to the number of edges going into V_2^*, it follows that

$$\begin{aligned} \sum_{i=1}^{q} d^+(r_i, G^*) &\geq \sum_{i=1}^{q} d^+(s_i, G^*) \\ &= \sum_{i=1}^{q} d^-(t_i, G^*) \\ &\geq \sum_{i=1}^{q} d^-(s_i, G^*) \\ &= d^+(r^*, G^*) + \sum_{i=1}^{q} d^+(r_i, G^*) \\ &\geq 1 + \sum_{i=1}^{q} d^+(r_i, G^*), \end{aligned}$$

which is absurd. Therefore, $S \subseteq V_0 \cup V_h$.

Since $|S|$ is the maximum number of pairwise totally disjoint paths from V_0 to V_h, it follows that

$$|S| \leq |V_0|.$$

If $V_0 \subseteq S$, then $|V_0| \leq |S|$, and so $|V_0| = |S|$. If $V_0 \not\subseteq S$, then $V_0 \setminus S$ is nonempty. Since S is a separating set and $S \subseteq V_0 \cup V_h$, every path in G from $V_0 \setminus S$ must end in $V_h \cap S$. It follows that

$$1 \leq D_h \leq \frac{|\mathrm{im}(V_0 \setminus S, V_h)|}{|V_0 \setminus S|} \leq \frac{|V_h \cap S|}{|V_0 \setminus S|},$$

and so

$$|V_0 \setminus S| \leq |V_h \cap S|.$$

Therefore,

$$\begin{aligned} |S| &= |V_0 \cap S| + |V_h \cap S| \\ &\geq |V_0 \cap S| + |V_0 \setminus S| \\ &= |V_0| \end{aligned}$$

and $|V_0| = |S|$. This completes the proof.

Lemma 7.3 *Let G be a Plünnecke graph of level $h \geq 2$. If $D_h \geq 1$, then $D_i \geq 1$ for $i = 1, \ldots, h$.*

Proof. Since $D_h \geq 1$, it follows from the previous lemma that there are $|V_0|$ totally pairwise disjoint paths from V_0 to V_h, and each of these paths contains a vertex from the set V_i. The vertices in V_i belonging to different totally disjoint paths are distinct. Let Z be a nonempty subset of V_0. Since there are $|Z|$ pairwise totally disjoint paths emanating from Z, it follows that

$$|Z| \leq |\mathrm{im}(Z, V_i)|,$$

and so

$$D_i = D(V_0, V_i) = \min\left\{\frac{|\mathrm{im}(Z, V_i)|}{|Z|}\right\} \geq 1.$$

This completes the proof.

Theorem 7.4 (Plünnecke) *Let G be a Plünnecke graph of level $h \geq 2$, and let $D_1, \ldots, D_h$ be the magnification ratios of G. Then*

$$D_1 \geq D_2^{1/2} \geq \cdots \geq D_h^{1/h}.$$

Proof. It suffices to prove that $D_i \geq D_h^{i/h}$ for $i = 1, \ldots, h$. If $D_h = 1$, then Lemma 7.3 implies that $D_i \geq 1 = D_h^{i/h}$. If $D_h = 0$, the result is obvious.

Let $A = \{0\}$, and let B be a set of n integers such that the addition graph of level h constructed from A and B is the independent addition graph $I_{n,h}$, as defined in Section 7.2. Then

$$\frac{n^i}{i!} \leq |iB| = D(\{0\}, iB) = D_i(I_{n,h}) \leq n^i$$

for $i = 1, \ldots, h$.

Suppose that $0 < D_h < 1$. Let r be any positive integer, and let

$$n = \left[1 + (h!D_h(G)^{-r})^{1/h}\right].$$

Then

$$D_h(G)^r n^h \geq h!.$$

Let G^r be the product of r copies of the graph G, and consider the product graph $G^r \times I_{n,h}$. This is a Plünnecke graph of level h and, because of the multiplicativity of magnification ratios,

$$D_h(G^r \times I_{n,h}) = D_h(G)^r D_h(I_{n,h}) \geq \frac{D_h(G)^r n^h}{h!} \geq 1.$$

It follows from Lemma 7.3 that for $i = 1, \ldots, h$,

$$1 \leq D_i(G^r \times I_{n,h}) = D_i(G)^r D_i(I_{n,h}) \leq D_i(G)^r n^i,$$

and so

$$D_i(G) \geq n^{-i/r}.$$

Since $D_h = D_h(G) < 1$, we have

$$(h!D_h(G)^{-r})^{1/h} < n \leq 1 + (h!D_h(G)^{-r})^{1/h} \leq 2(h!D_h(G)^{-r})^{1/h}.$$

Therefore,

$$\begin{aligned} D_i(G) &\geq n^{-i/r} \\ &\geq \left(2(h!D_h(G)^{-r})^{1/h}\right)^{-i/r} \\ &= \left(2(h!)^{1/h}\right)^{-i/r} D_h(G)^{i/h}. \end{aligned}$$

Since this inequality holds for all $r \geq 1$, and

$$\lim_{r\to\infty} \left(2(h!)^{1/h}\right)^{-i/r} = 1,$$

we conclude that $D_i(G) \geq D_h(G)^{i/h}$ for $i = 1, \ldots, h$.

Finally, we consider the case $D_h > 1$. Let r be a positive integer such that

$$n = [D_h(G)^{r/h}] > 1.$$

Then

$$2 \leq n \leq D_h(G)^{r/h} < n + 1 < 2n$$

and

$$D_h(G)^r n^{-h} \geq 1.$$

Let $I_{n,h}^{-1}$ be the inverse of the independent addition graph. Its magnification ratios satisfy the conditions

$$D_h(I_{n,h}^{-1}) = |hB|^{-1} = \binom{n+h-1}{h-1}^{-1} \geq n^{-h}$$

and

$$D_i(I_{n,h}^{-1}) \leq \frac{|(h-i)B|}{|hB|} \leq \frac{n^{h-i}}{n^h/h!} = h!n^{-i}.$$

Consider the $(r+1)$-fold product graph $G^r \times I_{n,h}$. It follows from the multiplicativity of magnification ratios that

$$D_h(G^r \times I_{n,h}^{-1}) = D_h(G)^r D_h(I_{n,h}^{-1}) \geq D_h(G)^r n^{-h} \geq 1,$$

and so, by Lemma 7.3,

$$1 \leq D_i(G^r \times I_{n,h}^{-1}) \leq D_i(G)^r h! n^{-i}.$$

This implies that

$$\begin{aligned} D_i(G) &\geq (h!)^{-1/r} n^{i/r} \\ &> (h!)^{-1/r} \left(\frac{D_h(G)^{r/h}}{2} \right)^{i/r} \\ &= (2^i h!)^{-1/r} D_h(G)^{i/h}. \end{aligned}$$

Since this inequality holds for all $r \geq 1$, it follows that $D_i(G) \geq D_h(G)^{i/h}$ for $i = 1, \ldots, h$. This completes the proof of Plünnecke's inequality.

7.6 Application: Estimates for sumsets in groups

Theorem 7.5 *Let B be a finite subset of an abelian group. Then*

$$|hB| \leq |iB|^{h/i}$$

for $i = 1, \ldots, h$.

Proof. Let G be the addition graph of level h constructed from the sets $A = \{0\}$ and B. Since $D_i = |iB|$ for $i = 1, \ldots, h$, it follows from Plünnecke's inequality that

$$|hB| = D_h \leq D_i^{h/i} = |iB|^{h/i}.$$

Theorem 7.6 *Let A and B be finite subsets of an abelian group, and let $1 \leq i \leq h$. If $|A| = n$ and $|A + iB| \leq cn$, then there is a nonempty subset $A' \subseteq A$ such that*

$$\left|A' + hB\right| \leq c^{h/i}|A'|.$$

Proof. Let G be the addition graph of level h constructed from the sets A and B. There exists a nonempty set $A' \subseteq A$ such that

$$D_h = \frac{|A' + hB|}{|A'|}.$$

Then $D_h \leq D_i^{h/i}$ implies that

$$\begin{aligned} |A' + hB| &= |A'|D_h \\ &\leq |A'|D_i^{h/i} \\ &\leq |A'|\left(\frac{|A + iB|}{|A|}\right)^{h/i} \\ &\leq c^{h/i}|A'|. \end{aligned}$$

Theorem 7.7 *Let B be a finite subset of an abelian group. If $|B| = k$ and $|2B| \leq ck$, then*

$$|hB| \leq c^h k$$

for all $h \geq 2$.

Proof. We apply Theorem 7.6 with $A = B$ and $i = 1$. Then there is a nonempty set $B' \subseteq B$ such that

$$|hB| \leq |B' + hB| \leq c^h|B'| \leq c^h|B| = c^h k.$$

Lemma 7.4 *Let U, V, and W be nonempty, finite subsets of an abelian group. Then*

$$|U||V - W| \leq |U + V||U + W|.$$

Proof. For each $d \in V - W$, we choose elements $v(d) \in V$ and $w(d) \in W$ such that $d = v(d) - w(d)$. Define the function

$$\phi : U \times (V - W) \to (U + V) \times (U + W)$$

by

$$\phi(u, d) = (u + v(d), u + w(d)).$$

We shall prove that the map ϕ is one-to-one. If

$$\phi(u_1, d_1) = \phi(u_2, d_2),$$

then

$$(u_1 + v(d_1), u_1 + w(d_1)) = (u_2 + v(d_2), u_2 + w(d_2)),$$

and so

$$u_1 + v(d_1) = u_2 + v(d_2)$$

and

$$u_1 + w(d_1) = u_2 + w(d_2).$$

Therefore,

$$d_1 = v(d_1) - w(d_1) = v(d_2) - w(d_2) = d_2.$$

Since each $d \in V - W$ uniquely determines $v(d) \in V$, it follows that $v(d_1) = v(d_2)$ and so $u_1 = u_2$. Therefore, ϕ is a one-to-one mapping of finite sets, and

$$\begin{aligned} |U||V - W| &= |U \times (V - W)| \\ &\leq |(U + V) \times (U + W)| \\ &= |U + V||U + W|. \end{aligned}$$

This completes the proof.

Theorem 7.8 *Let A and B be finite subsets of an abelian group such that $|A| = n$ and $|A + B| \leq cn$. Let $k \geq 1$ and $l \geq 1$. Then*

$$|kB - lB| \leq c^{k+l} n.$$

Proof. Since $|kB - lB| = |lB - kB|$, we can assume without loss of generality that $k \leq l$. Applying Theorem 7.6 with $i = 1$, we obtain a nonempty set $A' \subseteq A$ such that

$$|A' + kB| \leq c^k |A'| = c'n', \tag{7.2}$$

where $c' = c^k$ and $n' = |A'|$. Applying Theorem 7.6 with A, i, h, n, and c replaced by A', k, l, n', and c', respectively, we obtain from inequality (7.2) a nonempty subset $A'' \subseteq A'$ such that

$$|A'' + lB| \leq (c')^{l/k} |A''| = c^l |A''|. \tag{7.3}$$

It follows from Lemma 7.4 with $U = A''$, $V = kB$, and $W = lB$ and from inequalities (7.2) and (7.3) that

$$\begin{aligned} |A''||kB - lB| &\leq |A'' + kB||A'' + lB| \\ &\leq |A' + kB||A'' + lB| \\ &\leq c^k |A'| c^l |A''| \\ &\leq c^{k+l} n |A''|. \end{aligned}$$

We complete the proof by dividing this inequality by $|A''|$.

Theorem 7.9 *Let G be an abelian group such that every element of G has order at most r. Let B be a finite subset of G, and let $H(B)$ be the subgroup generated by B. If there exists a subset A of G such that $|A| = |B| = n$ and*

$$|A + B| \leq cn,$$

then

$$|H(B)| \leq f(r, c)n,$$

where

$$f(r, c) = r^{c^4} c^2.$$

Proof. It follows from Theorem 7.8 that

$$|kB - lB| \leq c^{k+l} n$$

for all nonnegative integers k and l. In particular,

$$|B - B| \leq c^2 n$$

and

$$|2B - 2B| \leq c^4 n.$$

If $w \in 2B - B$, then $w - B \subseteq 2B - 2B$. Let $W = \{w_1, \ldots, w_k\}$ be a maximal subset of $2B - B$ such that the sets $w_i - B$ are pairwise disjoint. Then

$$\bigcup_{i=1}^{k} (w_i - B) \subseteq 2B - 2B$$

and

$$kn = \sum_{i=1}^{k} |w_i - B| = \left| \bigcup_{i=1}^{k} (w_i - B) \right| \leq |2B - 2B| \leq c^4 n.$$

Therefore,

$$|W| = k \leq c^4.$$

We shall prove that

$$lB - B \subseteq (l - 1)W + B - B \tag{7.4}$$

for all $l \geq 1$. The proof will be by induction on l. This is clear for $l = 1$, since $0 \cdot W = \{0\}$. Let $w \in 2B - B$. Since the set W is maximal, there exists $w_i \in W$ such that

$$(w - B) \cap (w_i - B) \neq \emptyset.$$

Therefore, there exist group elements $b, b' \in B$ such that

$$w - b = w_i - b',$$

and so

$$w = w_i + b - b' \in W + B - B.$$

Thus,

$$2B - B \subseteq W + B - B.$$

This proves (7.4) in the case $l = 2$. Suppose that (7.4) holds for some $l \geq 2$. Then

$$\begin{aligned}
(l+1)B - B &= (2B - B) + (l-1)B \\
&\subseteq (W + B - B) + (l-1)B \\
&= W + lB - B \\
&\subseteq W + (l-1)W + B - B \\
&= lW + B - B.
\end{aligned}$$

This completes the induction.

Let $H(W)$ be the subgroup of G generated by W. Since $|W| = k \leq c^4$ and every element of the abelian group G has order at most r, it follows that

$$|H(W)| \leq r^k \leq r^{c^4}.$$

Then

$$lB - B \subseteq (l-1)W + B - B \subseteq H(W) + B - B$$

for all $l \geq 1$. Since B is finite and every element of B has order at most r, it follows that every element in $H(B)$ is contained in $lB - B$ for some $l \geq 1$. Therefore,

$$H(B) = \bigcup_{l=1}^{\infty} (lB - B) \subseteq H(W) + B - B,$$

and so

$$|H(B)| \leq |H(W)||B - B| \leq r^{c^4} c^2 n.$$

This completes the proof.

7.7 Application: Essential components

For any set A of integers, let $A(m, n)$ denote the number of elements $a \in A$ such that $m < a \leq n$, and let $A(n) = A(0, n)$. The function $A(n)$ counts the number of positive elements of A not exceeding n. The *Shnirel'man density* of the set A is defined by

$$\sigma(A) = \inf \left\{ \frac{A(n)}{n} : n = 1, 2, 3, \ldots \right\}.$$

Then $0 \leq \sigma(A) \leq 1$ for every set A, and $1 \in A$ if $\sigma(A) > 0$.

The set B is called an *essential component* if $\sigma(A + B) > \sigma(A)$ for every set A with $0 < \sigma(A) < 1$. Let A and B be sets of integers with $1 \in A$ and $0 \in B$. Shnirel'man [118] proved the fundamental inequality

$$\sigma(A + B) \geq \sigma(A) + (1 - \sigma(A))\sigma(B).$$

If $0 < \sigma(A) < 1$ and $\sigma(B) > 0$, then $\sigma(A + B) > \sigma(A)$. Thus, any set B with $0 \in B$ and positive Shnirel'man density is an essential component.

There exist sets of density zero that are also essential components. Khinchin [75] proved that the set of squares forms an essential component. By Lagrange's theorem, the squares are a basis of order 4. Erdős [34] generalized Khinchin's result by proving that if B is any basis of order h and if A is any set of integers such that $0 < \sigma(A) < 1$, then

$$\sigma(A+B) \geq \sigma(A) + \frac{\sigma(A)(1-\sigma(A))}{2h} > \sigma(A).$$

Plünnecke applied his graph-theoretic method to obtain a considerable improvement of Erdős's theorem. Plünnecke's proof relies heavily on properties of the *impact function*, or *Wirkungsfunktion*, defined for $0 \leq \xi \leq 1$ and for any set B of nonnegative integers by

$$\phi(\xi, B) = \inf\{\sigma(A+B) : A \subseteq \mathbf{N}_0, \sigma(A) \geq \xi\}.$$

Lemma 7.5 *Let $0 \leq \xi \leq 1$, and let B be a set of nonnegative integers. Then*

$$\begin{aligned} \phi(\xi, B) &= \inf\left\{\frac{(A+B)(n)}{n} : A \subseteq \mathbf{N}_0, \sigma(A) \geq \xi, \right. \\ &\qquad \left. \text{and } \frac{A(n)}{n} \leq \frac{A(m)}{m} \quad \text{for } m = 1, \ldots, n\right\}. \end{aligned}$$

Proof. If $\sigma(A) \geq \xi$ and $n \geq 1$, then

$$\phi(\xi, B) \leq \sigma(A+B) \leq \frac{(A+B)(n)}{n}.$$

It suffices to show that for any $\varepsilon > 0$ there exist a set A and an integer n satisfying the conditions of the lemma such that

$$\phi(\xi, B) \leq \frac{(A+B)(n)}{n} < \phi(\xi, B) + \varepsilon.$$

It follows from the definition of the impact function that there exists a set A of nonnegative integers such that $\sigma(A) \geq \xi$ and

$$\phi(\xi, B) \leq \sigma(A+B) < \phi(\xi, B) + \frac{\varepsilon}{2}.$$

It follows from the definition of Shnirel'man density that there exists an integer $n \geq 1$ such that

$$\sigma(A+B) \leq \frac{(A+B)(n)}{n} < \sigma(A+B) + \frac{\varepsilon}{2},$$

and so

$$\phi(\xi, B) \leq \frac{(A+B)(n)}{n} < \phi(\xi, B) + \varepsilon. \tag{7.5}$$

Let n be the smallest positive integer such that inequality (7.5) is satisfied for some set A with $\sigma(A) \geq \xi$. For these choices of n and A, we have

$$\frac{(A+B)(n)}{n} < \phi(\xi, B) + \varepsilon \leq \frac{(A+B)(l)}{l}$$

for $l = 1, \ldots, n-1$, and so

$$\begin{aligned}(A+B)(l,n) &= (A+B)(n) - (A+B)(l) \\ &< (n-l)(\phi(\xi, B) + \varepsilon).\end{aligned}$$

We shall show that

$$\frac{A(n)}{n} \leq \frac{A(m)}{m}$$

for $m = 1, \ldots, n$. If not, there exists an integer l such that $1 \leq l \leq n-1$ and

$$\frac{A(l)}{l} = \min\left\{\frac{A(m)}{m} : m = 1, \ldots, n-1\right\} < \frac{A(n)}{n}.$$

Then for $i = 1, \ldots, n-l$ we have

$$\begin{aligned}A(l, l+i) &= A(l+i) - A(l) \\ &\geq \frac{(l+i)A(l)}{l} - A(l) \\ &= \frac{iA(l)}{l} \\ &\geq i\sigma(A) \\ &\geq i\xi.\end{aligned}$$

Let

$$A^* = \{a - l : a \in A, l < a \leq n\} \cup \{m \in \mathbf{N}_0 : m > n - l\}.$$

For $i = 1, \ldots, n-l$,

$$\begin{aligned}A^*(i) &= |\{a \in A : 0 < a - l \leq i\}| \\ &= |\{a \in A : l < a \leq l + i\}| \\ &= A(l, l+i) \\ &\geq i\xi.\end{aligned}$$

Since A^* contains all integers $m > n - l$, it follows for $i > n - l$ that

$$\begin{aligned}A^*(i) &= A^*(n-l) + (i - n + l) \\ &\geq (n-l)\xi + i - n + l \\ &= i\xi + (1-\xi)(i - n + l) \\ &\geq i\xi.\end{aligned}$$

It follows that

$$\sigma(A^*) \geq \xi.$$

Moreover,

$$\begin{aligned}
(A^* + B)(n-l) &= \left| \left\{ a^* + b : a^* \in A^*, b \in B, 0 < a^* + b \leq n - l \right\} \right| \\
&= \left| \left\{ a^* + b : a^* \in A^*, b \in B, l < a^* + l + b \leq n \right\} \right| \\
&\leq \left| \{ a + b : a \in A, b \in B, l < a + b \leq n \} \right| \\
&= (A + B)(l, n) \\
&< (n - l)(\phi(\xi, B) + \varepsilon).
\end{aligned}$$

Therefore,

$$\phi(\xi, B) \leq \frac{(A^* + B)(n-l)}{n-l} < \phi(\xi, B) + \varepsilon.$$

Comparing this with (7.5), we see that this contradicts the minimality of n (since $1 \leq n - l < n$), and so $A(n)/n \leq A(m)/m$ for all $m = 1, \ldots, n$. This completes the proof.

Lemma 7.6 *Let $0 < \xi \leq 1$, and let $h \geq 1$. If B is a set of nonnegative integers with $0 \in B$, then*

$$\phi(\xi, iB) \geq \xi^{1-i/h} \sigma(hB + 1)^{i/h}$$

for $i = 1, \ldots, h$.

Proof. Let A be a set of nonnegative integers such that

$$\sigma(A) \geq \xi > 0.$$

Then $1 \in A$. Let n be a positive integer such that

$$\frac{A(n)}{n} \leq \frac{A(m)}{m} \quad \text{for } m = 1, \ldots, n.$$

Let $\gamma = A(n)/n$. Then

$$\xi \leq \sigma(A) \leq \gamma.$$

Let G be the truncated addition graph constructed from A, B, and n as in Section 7.2. Its ith vertex set is

$$V_i = (A + iB) \cap [1, n].$$

Since $1 \in A$ and $0 \in B$, it follows that $1 \in V_i$ for $i = 0, 1, \ldots, h$. Let Z be a nonempty subset of $V_0 = A \cap [1, n]$. For $m = 0, 1, \ldots, n$, we have

$$\begin{aligned}
Z(m, n) &\leq A(m, n) \\
&= A(n) - A(m) \\
&= \gamma n - A(m) \\
&\leq \gamma(n - m),
\end{aligned}$$

and

$$|\mathrm{im}(Z, V_h)| = (Z + hB)(n)$$

in the graph G. Let z be the smallest positive element of Z. Then

$$\begin{aligned}
(Z + hB)(n) &\geq (\{z\} + hB)(n) \\
&= |\left\{b' \in hB : z \leq z + b' \leq n\right\}| \\
&= |\left\{b' \in hB : 1 \leq 1 + b' \leq n - z + 1\right\} \\
&= (hB + 1)(n - z + 1) \\
&\geq (n - z + 1)\sigma(hB + 1) \\
&\geq \gamma^{-1}Z(z - 1, n)\sigma(hB + 1) \\
&= \gamma^{-1}|Z|\sigma(hB + 1).
\end{aligned}$$

Thus,

$$\frac{|\mathrm{im}(Z, V_h)|}{|Z|} = \frac{(Z + hB)(n)}{|Z|} \geq \gamma^{-1}\sigma(hB + 1)$$

for all nonempty subsets Z of V_0, and so

$$D_h(G) \geq \gamma^{-1}\sigma(hB + 1).$$

Plünnecke's inequality implies that for $i = 1, \ldots, h$,

$$\begin{aligned}
\gamma^{-1}\sigma(hB + 1) &\leq D_h(G) \leq D_i(G)^{h/i} \\
&\leq \left(\frac{(A + iB)(n)}{A(n)}\right)^{h/i},
\end{aligned}$$

and so

$$\begin{aligned}
(A + iB)(n) &\geq A(n)(\gamma^{-1}\sigma(hB + 1))^{i/h} \\
&= n\gamma(\gamma^{-1}\sigma(hB + 1))^{i/h} \\
&= n\gamma^{1-i/h}\sigma(hB + 1)^{i/h} \\
&\geq n\xi^{1-i/h}\sigma(hB + 1)^{i/h}.
\end{aligned}$$

Equivalently,

$$\frac{(A + iB)(n)}{n} \geq \xi^{1-i/h}\sigma(hB + 1)^{i/h}.$$

It follows from Lemma 7.5 that

$$\phi(\xi, iB) \geq \xi^{1-i/h}\sigma(hB + 1)^{i/h}.$$

Theorem 7.10 (Plünnecke) *Let A be a set of nonnegative integers with $0 < \sigma(A) < 1$, and let B be a basis of order $h \geq 2$. Then*

$$\sigma(A + B) \geq \sigma(A)^{1-1/h}.$$

Proof. Since B is a basis of order $h \geq 2$, we must have $1 \in hB$, so $0 \in B \subseteq hB$. Therefore, hB is the set of all nonnegative integers and $hB + 1$ is the set of all positive integers. Thus, $\sigma(hB + 1) = 1$. Applying the previous Lemma with $i = 1$ gives

$$\phi(\xi, B) \geq \xi^{1-1/h}$$

for all $\xi \in (0, 1)$. Let $\xi = \sigma(A)$. Then

$$\sigma(A + B) \geq \phi(\xi, B) \geq \xi^{1-1/h} = \sigma(A)^{1-1/h}.$$

This completes the proof.

Corollary 7.2 (Erdős) *Let A be a set of nonnegative integers with $0 < \sigma(A) < 1$, and let B be a basis of order $h \geq 2$. Then*

$$\sigma(A + B) \geq \sigma(A) + \frac{\sigma(A)(1 - \sigma(A))}{h}.$$

(Erdős obtained a slightly weaker result in which the constant $1/h$ was replaced by $1/2h$.)

Proof. It suffices to prove that

$$x + \frac{x(1 - x)}{h} \leq x^{1-1/h}$$

for $0 \leq x \leq 1$, or, equivalently,

$$f(x) = x^{1/h}\left(1 + \frac{1 - x}{h}\right) \leq 1$$

for $0 \leq x \leq 1$. Since $f(0) = 0$ and $f(1) = 1$, it is enough to show that $f(x)$ is increasing on the unit interval, and this follows immediately by differentiation:

$$\begin{aligned} f'(x) &= \frac{x^{(1/h)-1}}{h}\left(1 + \frac{1 - x}{h}\right) - \frac{x^{1/h}}{h} \\ &= \frac{x^{(1/h)-1}}{h}\left(1 + \frac{1}{h}\right)(1 - x) \\ &> 0 \end{aligned}$$

for $0 < x < 1$. This completes the proof.

7.8 Notes

Plünnecke's work appeared in the monograph [104] and the papers [102, 103, 105]. These papers were ignored for many years, until Ruzsa rediscovered them and simplified Plünnecke's original proofs. In particular, the proof of Plünnecke's inequality (Theorem 7.4) comes from Ruzsa [112].

There are many proofs of Menger's theorem. The one is this chapter is due to Dirac [31]. McCuaig [86] has a very short proof.

The *Wirkungsfunktion* appears in Stöhr and Wirsing [122] and was studied by Plünnecke in great detail in [104]. The applications of Plünnecke's inequality in Section 7.6 are all due to Ruzsa [112, 114, 116]. The proof of Plünnecke's theorem on essential components (Theorem 7.10) is a simplified version, due to Malouf [83], of Plünnecke's original proof.

7.9 Exercises

1. Draw the addition graphs of level 3 determined by the following sets of integers:

   (a) $A = \{0\}$, $B = \{0, 1, 2\}$,

   (b) $A = \{0\}$, $B = \{0, 1, 3\}$,

   (c) $A = \{0, 1\}$, $B = \{0, 1, 4\}$.

2. Let G be the directed graph with vertex set

$$V(G) = \{a, b, x, y, z\}$$

   and edge set

$$E(G) = \{(a, x), (x, y), (x, z), (y, b), (z, b)\}.$$

   Show that $S_1 = \{x\}$ and $S_2 = \{y, z\}$ are minimal sets of vertices that separate a and b. Why does this not contradict Menger's theorem?

3. Let A and B be finite, nonempty subsets of an abelian group, and let G be the addition graph of level h determined by A and B. Prove that $d^+(v) = |B|$ for all $v \in V(G) \setminus V_h$.

4. Let G be an addition graph of level h. Prove directly that $D_i \geq 1$ for $i = 1, \ldots, h$.

5. Prove that the inverse of a Plünnecke graph is a Plünnecke graph.

6. Prove that the product of Plünnecke graphs is a Plünnecke graph.

7. Let B be a finite subset of an abelian group, and let $2 \leq i \leq h$. If $|B| = n$ and $|iB| \leq cn$, prove that

$$|hB| \leq c^{h/(i-1)}n.$$

8. Let B be a finite subset of an abelian group, and let $2 \leq i \leq h$. If $|B| = n$ and $|iB| \leq cn^{1+\delta}$, prove that

$$|hB| \leq c^{h/(i-1)}n^{1+\delta h/(i-1)}.$$

9. Let B be a finite subset of an abelian group. Let $1 \leq i \leq h$. Suppose that i divides h. Prove directly (without using Plünnecke's inequality) that $|hB| \leq |iB|^{h/i}$. What inequality can you obtain (without using Plünnecke's inequality) if i does not divide h?

10. In the following sequence of exercises, we construct a basis B of order h and a set A of positive Shnirel'man density that show that the exponent $1 - 1/h$ in Plünnecke's theorem on essential components (Theorem 7.10) is best possible.

    (a) Let $m \geq 2$. Prove that every nonnegative integer can be written uniquely in the form
    $$\sum_{i=0}^{\infty} u_i m^i,$$
    where $u_i \in \{0, 1, \ldots, m-1\}$ and $u_i \neq 0$ for only finitely many i.

    (b) Let $h \geq 2$. For $j = 0, 1, \ldots, h-1$, let B_j be the set of all nonnegative integers of the form
    $$\sum_{\substack{i=0 \\ i \equiv j \pmod{h}}}^{\infty} u_i m^i,$$
    where $u_i \in \{0, 1, \ldots, m-1\}$ and $u_i \neq 0$ for only finitely many i. Prove that
    $$\begin{aligned}&\{0, m^j, 2m^j, 3m^j, \ldots, (m-1)m^j\}\\ &\qquad \subseteq B_j \subseteq \{0, m^j, 2m^j, 3m^j, \ldots, (m-1)m^j\} + m^h * \mathbf{N}_0.\end{aligned}$$

    (c) Let
    $$B = B_0 \cup B_1 \cup \cdots \cup B_{h-1}.$$
    Prove that B is a basis of order h.

    (d) Let
    $$A = \{n \in \mathbf{N}_0 : n \equiv 1 \pmod{m^h}\}.$$
    Prove that $\sigma(A) = m^{-h}$.

    (e) Prove that
    $$A + B_j = \{1 + um^j : u = 0, 1, \ldots, m-1\} + m^h * \mathbf{N}_0$$
    and
    $$\begin{aligned} A + B &= \{1 + um^j : u = 0, 1, \ldots, m-1 \text{ and } j = 0, 1, \ldots, h-1\}\\ &\quad + m^h * \mathbf{N}_0.\end{aligned}$$

(f) Prove that

$$\sigma(A+B) \leq \frac{hm-h+1}{m^h} < hm^{1-h} = h\sigma(A)^{1-1/h}.$$

(g) Prove that the exponent $1-1/h$ in Theorem 7.10 cannot be replaced by $1-1/h-\varepsilon$ for any $\varepsilon > 0$.

8

Freiman's theorem

8.1 Multidimensional arithmetic progressions

A simple inverse theorem (Theorem 1.16) in additive number theory states that if A is a finite set of integers whose two-fold sumset is small in the sense that $|2A| \leq 3|A| - 4$, then A is a large subset of an ordinary arithmetic progression. Freiman discovered a deep generalization of this result. His theorem asserts that if A is a finite set of integers such that the sumset $2A$ is small, then A is a large subset of a multidimensional arithmetic progression. Freiman's inverse theorem can be stated as follows.

Theorem 8.1 (Freiman) *Let A be a finite set of integers such that $|2A| \leq c|A|$. There exist integers $a, q_1, \ldots, q_n, l_1, \ldots, l_n$ such that*

$$A \subseteq Q = \{a + x_1q_1 + \cdots + x_nq_n : 0 \leq x_i < l_i \text{ for } i = 1, \ldots, n\},$$

where $|Q| \leq c'|A|$ and n and c' depend only on c.

The object of this chapter is to present a beautiful proof due to I. Z. Ruzsa of a generalization of Freiman's theorem.

We begin with the following definition. Let $a, q_1, \ldots, q_n$ be elements of an abelian group G, and let $l_1, \ldots, l_n$ be positive integers. The set

$$\begin{aligned} Q &= Q(a; q_1, \ldots, q_n; l_1, \ldots, l_n) \\ &= \{a + x_1q_1 + \cdots + x_nq_n : 0 \leq x_i < l_i \text{ for } i = 1, \ldots, n\} \end{aligned}$$

is called an *n-dimensional arithmetic progression* in the group G. The *length* of Q is $l(Q) = l_1 \cdots l_n$. Clearly, $|Q| \leq l_1 \cdots l_n$, and Q is called *proper* if $|Q| = l(Q)$.

The representation of a set as a multidimensional arithmetic progression is not unique. A set can have more than one such representation, and these representations can have different dimensions and lengths.

Theorem 8.2 *Let G be an abelian group, and let Q and Q' be multidimensional arithmetic progressions in G of dimensions n and n' and lengths l and l', respectively. Then*

(i) $Q + Q'$ is a multidimensional arithmetic progression of dimension $n + n'$ and length ll'.

(ii) $Q - Q$ is an arithmetic progression of dimension n and length $l(Q - Q) < 2^n l$.

(iii) If Q is proper, then $l(hQ) < h^n |Q|$.

(iv) Every finite subset F of a group is a subset of an arithmetic progression of dimension $|F|$ and length $2^{|F|}$.

Proof. Let Q be the progression $Q(a; q_1, \ldots, q_n; l_1, \ldots, l_n)$, and let Q' be the progression $Q(a'; q'_1, \ldots, q'_{n'}; l'_1, \ldots, l'_{n'})$. Then

$$\begin{aligned} Q + Q' &= \{a + a' + x_1 q_1 + \cdots + x_n q_n + x'_1 q'_1 + \cdots + x'_{n'} q'_{n'} : \\ &\quad 0 \le x_i < l_i \text{ for } i = 1, \ldots, n \text{ and } 0 \le x'_j < l'_j \text{ for } j = 1, \ldots, n'\} \end{aligned}$$

is an arithmetic progression of dimension $n + n'$ and length

$$l(Q + Q') = l_1 \cdots l_n l'_1 \cdots l'_{n'} = l(Q) l(Q').$$

Similarly,

$$\begin{aligned} Q - Q &= \{y_1 q_1 + \cdots + y_n q_n : -l_i < y_i < l_i \text{ for } i = 1, \ldots, n\} \\ &= Q(b; q_1, \ldots, q_n; m_1, \ldots, m_n), \end{aligned}$$

where

$$b = -\sum_{i=1}^{n} (l_i - 1) q_i$$

and

$$m_i = 2l_i - 1,$$

and so $Q - Q$ is an arithmetic progression of dimension n and length

$$l(Q - Q) = m_1 \cdots m_n < 2^n l_1 \cdots l_n = 2^n l.$$

Since hQ can be represented in the form

$$hQ = Q(ha; q_1, \ldots, q_n; h(l_1 - 1) + 1, \ldots, h(l_n - 1) + 1).$$

it follows that if Q is proper, then

$$\begin{aligned} l(hQ) &\leq \prod_{i=1}^{n}(h(l_i - 1) + 1) \\ &< \prod_{i=1}^{n} hl_i \\ &= h^n |Q|. \end{aligned}$$

If F is a finite set of cardinality $|F| = n$, say, $F = \{f_1, \ldots, f_n\}$, then

$$\begin{aligned} F &\subseteq \{x_1 f_1 + \cdots + x_n f_n : 0 \leq x_i < 2 \text{ for } i = 1, \ldots, n\} \\ &= Q(0; f_1, \ldots, f_n; 2, \ldots, 2\} = Q, \end{aligned}$$

and Q is an n-dimensional arithmetic progression of length 2^n.

8.2 Freiman isomorphisms

Freiman introduced the important idea of a "local" isomorphism in additive number theory. Let G and H be abelian groups, and let $A \subseteq G$ and $B \subseteq H$. Let $h \geq 2$. The map $\phi : A \to B$ is called a *Freiman homomorphism of order* h if

$$\phi(a_1) + \cdots + \phi(a_h) = \phi(a_1') + \cdots + \phi(a_h')$$

for all $a_1, \ldots, a_h, a_1', \ldots, a_h' \in A$ such that

$$a_1 + \cdots + a_h = a_1' + \cdots + a_h'.$$

In this case, the induced map $\phi^{(h)} : hA \to hB$ defined by

$$\phi^{(h)}(a_1 + \cdots + a_h) = \phi(a_1) + \cdots + \phi(a_h)$$

is well defined.

If $\phi : A \to B$ is a one-to-one correspondence such that

$$a_1 + \cdots + a_h = a_1' + \cdots + a_h'$$

if and only if

$$\phi(a_1) + \cdots + \phi(a_h) = \phi(a_1') + \cdots + \phi(a_h'),$$

then ϕ is a *Freiman isomorphism of order* h, and the induced map $\phi^{(h)} : hA \to hB$ is also a one-to-one correspondence. A Freiman isomorphism of order 2 will be called simply a *Freiman isomorphism*.

Here are two examples. Let $A = [0, k-1] = \{0, 1, \ldots, k-1\}$, and let $B = \{a + xq_1 \mid 0 \leq x < k\}$. Then the map $\phi(x) = a + xq_1$ is a Freiman isomorphism of order h for all $h \geq 2$.

Let $A = \{(0,0), (1,0), (0,1)\} \subseteq \mathbf{Z}^2$, and let $B = \{0, 1, 3\} \subseteq \mathbf{Z}$. Define $\phi : A \to B$ by $\phi(0,0) = 0$, $\phi(1,0) = 1$, and $\phi(0,1) = 3$. Then ϕ is a Freiman isomorphism of order 2 but not of order 3.

If $\phi : A \to B$ is a Freiman homomorphism (resp. isomorphism) of order h and if $\psi : B \to C$ is a Freiman homomorphism (resp. isomorphism) of order h, then $\psi\phi : A \to C$ is a Freiman homomorphism (resp. isomorphism) of order h.

Let $\phi : A \to B$ be a Freiman isomorphism of order h. Then ϕ is also a Freiman isomorphism of order h' for every $h' \leq h$. If $A' \subseteq A$ and $B' = \phi(A')$, then the map $\phi : A' \to B'$ is also a Freiman isomorphism of order h.

If $f : G \to H$ is a group homomorphism, then f is a Freiman homomorphism of order h and $f^{(h)} = f$ for all $h \geq 2$. If f is a group isomorphism, then f is a Freiman isomorphism of order h for all $h \geq 2$.

If $\phi : G \to H$ is an affine map, that is, a map of the form $\phi(x) = a + f(x)$, where $a \in H$ and $f : G \to H$ is a group homomorphism (resp. isomorphism), then ϕ is a Freiman homomorphism (resp. isomorphism) of order h and $\phi^{(h)}(x) = ha(x)$ for all $h \geq 2$.

For example, let $q_1 \neq 0$, and let $Q = \{a + xq_1 \mid 0 \leq x < l\}$ be a 1-dimensional arithmetic progression (that is, an ordinary arithmetic progression) in the group G. Define $\phi : [0, l-1] \to Q$ by $\phi(x) = a + xq_1$. Then ϕ is the restriction of an affine map from $\mathbf{Z}$ into G and a Freiman isomorphism of order h for all $h \geq 2$.

Let $\mathbf{e}_1, \ldots, \mathbf{e}_n$ be the standard basis vectors in the Euclidean space $\mathbf{R}^n$, let $l_1, \ldots, l_n$ be positive integers, and let $\mathbf{P}$ be the fundamental parallelepiped of the lattice generated by the vectors $l_1\mathbf{e}_1, \ldots, l_n\mathbf{e}_n$. The *integer parallelepiped* $I(\mathbf{P})$ is defined by

$$I(\mathbf{P}) = \mathbf{P} \cap \mathbf{Z}^n = \{(x_1, \ldots, x_n) \in \mathbf{Z}^n : 0 \leq x_1 < l_i \text{ for } i = 1, \ldots, n\}.$$

Then $|I(\mathbf{P})| = l_1 \cdots l_n$. For $h \geq 2$ we have

$$hI(\mathbf{P}) = \{(x_1, \ldots, x_n) \in \mathbf{Z}^n : 0 \leq x_i < h(l_i - 1) + 1 \text{ for } i = 1, \ldots, n\}$$

and

$$|hI(\mathbf{P})| = \prod_{i=1}^{n} (h(l_i - 1) + 1) < h^n |I(\mathbf{P})|.$$

Let $a, q_1, \ldots, q_n \in \mathbf{Z}$, and let $Q = Q(a; q_1, \ldots, q_n; l_1, \ldots, l_n)$ be an n-dimensional integer arithmetic progression. Define the map $\phi : I(\mathbf{P}) \to Q$ by

$$\phi(x_1, \ldots, x_n) = a + x_1q_1 + \cdots + x_nq_n. \tag{8.1}$$

Since ϕ is the restriction of an affine map from $\mathbf{Z}^n$ into $\mathbf{Z}$, it follows that ϕ is a Freiman homomorphism of order h for all $h \geq 2$.

Theorem 8.3 *Let $h \geq 2$, and let $I(\mathbf{P})$ be the integer parallelepiped of dimension n determined by the integers $l_1, \ldots, l_n$. Then there exists an n-dimensional arithmetic progression Q such that $I(\mathbf{P})$ and Q are Freiman isomorphic of order h.*

Proof. Let a be any integer, and choose positive integers $q_1, \ldots, q_n$ so that

$$\sum_{j=1}^{k-1} h(l_j - 1)q_j < q_k \tag{8.2}$$

for $k = 2, \ldots, n$. Let

$$Q = Q(a; q_1, \ldots, q_n; l_1, \ldots, l_n).$$

Let $\phi : I(\mathbf{P}) \to Q$ be the affine map defined by (8.1). We shall prove that ϕ is a Freiman isomorphism of order h.

Let $\mathbf{x}_i = (x_{i1}, \ldots, x_{in}) \in I(\mathbf{P})$ and $\mathbf{y}_i = (y_{i1}, \ldots, y_{in}) \in I(\mathbf{P})$ for $i = i, \ldots, h$, and suppose that

$$\phi(\mathbf{x}_1) + \cdots + \phi(\mathbf{x}_h) = \phi(\mathbf{y}_1) + \cdots + \phi(\mathbf{y}_h).$$

This implies that

$$ha + \sum_{j=1}^{n} \sum_{i=1}^{h} x_{ij} q_j = ha + \sum_{j=1}^{n} \sum_{i=1}^{h} y_{ij} q_j.$$

Let

$$w_j = \sum_{i=1}^{h} x_{ij} - \sum_{i=1}^{h} y_{ij}.$$

Then

$$|w_j| \leq h(l_j - 1)$$

and

$$\sum_{j=1}^{n} w_j q_j = 0.$$

Suppose that $w_j \neq 0$ for some j, and let k be the greatest integer for which $w_k \neq 0$. Then

$$-w_k q_k = \sum_{j=1}^{k-1} w_j q_j$$

and so

$$q_k \leq |w_k q_k| = \left| \sum_{j=1}^{k-1} w_j q_j \right| \leq \sum_{j=1}^{k-1} h(l_j - 1) q_j < q_k,$$

which is absurd. Therefore, $w_j = 0$ for all j. It follows that

$$\mathbf{x}_1 + \cdots + \mathbf{x}_h = \mathbf{y}_1 + \cdots + \mathbf{y}_h.$$

Thus, every n-dimensional integer parallelepiped is Freiman isomorphic of order h to an n-dimensional arithmetic progression.

Corollary 8.1 *Let $h \geq 2$, and let A be a finite set of lattice points. Then A is Freiman isomorphic of order h to a set of integers.*

Proof. The set A is a subset of some n-dimensional integer parallelepiped $I(\mathbf{P})$, and $I(\mathbf{P})$ is Freiman isomorphic of order h to an n-dimensional arithmetic progression Q. Then A is Freiman isomorphic to its image under this isomorphism.

Corollary 8.2 *Let $h \geq 2$, and let A be a finite subset of a torsion-free abelian group. Then A is Freiman isomorphic of order h to a set of integers.*

Proof. Let G be the group generated by A. Since G is finitely generated, it follows from Theorem 6.10 that G is isomorphic to $\mathbf{Z}^n$ for some n, so there is a Freiman isomorphism of order h between A and some finite set of integer lattice points. By Corollary 8.1, this set is Freiman isomorphic of order h to a set of integers.

Theorem 8.4 *Let G and H be abelian groups, and let Q be an n-dimensional arithmetic progression contained in G. Let $h \geq 2$. If $\phi : Q \to H$ be a Freiman homomorphism of order $h \geq 2$, then $\phi(Q)$ is an n-dimensional arithmetic progression in H. If $\phi : Q \to \phi(Q)$ is a Freiman isomorphism, then Q is a proper n-dimensional arithmetic progression in G if and only if $\phi(Q)$ is also a proper n-dimensional arithmetic progression in H.*

Proof. Let $Q = \{a; q_1, \ldots, q_n; l_1, \ldots, l_n\}$. We define $a', q_1', \ldots, q_n' \in H$ by

$$\begin{aligned} a' &= \phi(a), \\ q_i' &= \phi(a + q_i) - \phi(a) \end{aligned}$$

for $i = 1, \ldots, n$. The set $Q' = \{a'; q_1', \ldots, q_n'; l_1, \ldots, l_n\}$ is an n-dimensional arithmetic progression in H. We shall show that $Q' = \phi(Q)$ and

$$\phi(a + x_1 q_1 + \cdots + x_n q_n) = a' + x_1 q_1' + \cdots + x_n q_n'$$

for all $a + x_1 q_1 + \cdots + x_n q_n \in Q$.

The proof is by induction on $m = \sum_{i=1}^n x_i$. It follows from the definition of $a', q_1', \ldots, q_n'$ that the statement is true for $m = 0$ and $m = 1$. Assume that the result holds for some $m \geq 1$. Let $r = a + x_1 q_1 + \cdots + x_n q_n \in Q$ with $\sum_{i=1}^n x_i = m + 1$. Choose j such that $x_j \geq 1$, and let $r' = r - q_j$. By the induction hypothesis for m, we have

$$\phi(r') = a' + x_1 q_1' + \cdots + x_{j-1} q_{j-1}' + (x_j - 1) q_j' + x_{j+1} q_{j+1}' + \cdots + x_n q_n'.$$

Since $r, a, r', a + q_j \in Q$, and

$$r + a = r' + (a + q_j),$$

and since a Freiman homomorphism of order h is also a Freiman homomorphism of order 2, it follows that

$$\phi(r) + \phi(a) = \phi(r') + \phi(a + q_j).$$

Therefore,

$$\begin{aligned}\phi(r) &= \phi(r') + \phi(a+q_j) - \phi(a) \\ &= \phi(r') + q'_j \\ &= a' + x_1 q'_1 + \cdots + x_n q'_n.\end{aligned}$$

Thus, the statement holds for all $m \geq 0$, and $\phi(Q) = Q'$. If ϕ is a Freiman isomorphism of order h, then $|Q| = |\phi(Q)|$, and so $\phi(Q)$ is proper if and only if Q is proper.

Theorem 8.5 *Let $h' = h(k+l)$, where h, k, and l are positive integers. Let G and H be abelian groups, and let $A \subseteq G$ and $B \subseteq H$ be nonempty, finite sets that are Freiman isomorphic of order h'. Then the difference sets $kA - lA$ and $kB - lB$ are Freiman isomorphic of order h.*

Proof. Let $\phi : A \rightarrow B$ be a Freiman isomorphism of order h', and let $\phi^{(k)} : kA \rightarrow kB$, $\phi^{(l)} : lA \rightarrow lB$, and $\phi^{(k+l)} : (k+l)A \rightarrow (k+l)B$ be the maps induced by ϕ. These maps are one-to-one correspondences, and

$$\begin{aligned}&\phi^{(k+l)}(a_1 + \cdots + a_k + a_{k+1} + \cdots + a_{k+l}) \\ &= \phi^{(k)}(a_1 + \cdots + a_k) + \phi^{(l)}(a_{k+1} + \cdots + a_{k+l})\end{aligned}$$

for all $a_1, \ldots, a_{k+l} \in A$. Let $d \in kA - lA$. If

$$d = u - v = u' - v',$$

where $u, u' \in kA$ and $v, v' \in lA$, then

$$u + v' = u' + v \in (k+l)A.$$

Since ϕ is a Freiman isomorphism of order $h' \geq k + l$, it follows that

$$\begin{aligned}\phi^{(k)}(u) + \phi^{(l)}(v') &= \phi^{(k+l)}(u + v') \\ &= \phi^{(k+l)}(u' + v) \\ &= \phi^{(k)}(u') + \phi^{(l)}(v),\end{aligned}$$

and so

$$\phi^{(k)}(u) - \phi^{(l)}(v) = \phi^{(k)}(u') - \phi^{(l)}(v').$$

This means that the map $\psi : kA - lA \rightarrow kB - lB$ defined by

$$\psi(d) = \psi(u - v) = \phi^{(k)}(u) - \phi^{(l)}(v)$$

is well defined. The map ψ is surjective since ϕ is surjective. Let $d = u - v \in kA - lA$ and $d' = u' - v' \in kA - lA$. If $\psi(d) = \psi(d')$, then

$$\phi^{(k)}(u) - \phi^{(l)}(v) = \phi^{(k)}(u') - \phi^{(l)}(v'),$$

and so

$$\phi^{(k+l)}(u+v') = \phi^{(k+l)}(u'+v).$$

This implies that $u+v' = u'+v$, and so $d = d'$. Thus, ψ is a one-to-one correspondence.

We shall prove that ψ is a Freiman isomorphism of order h. For $i = 1, \ldots, h$, let $d_i, d_i' \in kA - lA$ and let $d_i = u_i - v_i$ and $d_i' = u_i' - v_i'$, where $u_i, u_i' \in kA$ and $v_i, v_i' \in lA$. Then

$$d_1 + \cdots + d_h = d_1' + \cdots d_h'$$

if and only if

$$\begin{aligned} & u_1 + \cdots + u_h + v_1' + \cdots + v_h' \\ = \;& u_1' + \cdots + u_h' + v_1 + \cdots + v_h \in h(k+l)A \end{aligned}$$

if and only if

$$\begin{aligned} & \phi^{(h(k+l))}(u_1 + \cdots + u_h + v_1' + \cdots + v_h') \\ = \;& \phi^{(h(k+l))}(u_1' + \cdots + u_h' + v_1 + \cdots + v_h) \end{aligned}$$

if and only if

$$\begin{aligned} & \phi^{(k)}(u_1) + \cdots + \phi^{(k)}(u_h) + \phi^{(l)}(v_1') + \cdots + \phi^{(l)}(v_h') \\ & = \phi^{(k)}(u_1') + \cdots + \phi^{(k)}(u_h') + \phi^{(l)}(v_1) + \cdots + \phi^{(l)}(v_h) \end{aligned}$$

if and only if

$$\psi(d_1) + \cdots + \psi(d_h) = \psi(d_1') + \cdots + \psi(d_h').$$

This proves that ψ is a Freiman isomorphism of order h.

8.3 Bogolyubov's method

Let $m \geq 2$. If $x_i \equiv y_i \pmod{m}$ for $i = 1, \ldots, n$, then

$$(x_1, \ldots, x_n, m) = (y_1, \ldots, y_n, m),$$

and so the "greatest common divisor" of congruence classes modulo m is well defined. For $x \in \mathbf{R}$, let $\|x\|$ denote the distance from x to the nearest integer. Then $\|x\| \leq 1/4$ if and only if $\cos 2\pi x \geq 0$ if and only if $\Re(e^{2\pi i x}) \geq 0$. If $x, y \in \mathbf{Z}$ and $x \equiv y \pmod{m}$, then $\|x/m\| = \|y/m\|$. It follows that this "distance to the nearest integer" function is well defined on congruence classes modulo m. Similarly, the exponential function $e^{2\pi i x/m}$ is well defined on congruence classes modulo m. If $g \in \mathbf{Z}/m\mathbf{Z}$ and x is an integer in the congruence class g, we define $\|g/m\| = \|x/m\|$ and $e^{2\pi i g/m} = e^{2\pi i x/m}$. For $r_1, \ldots, r_k \in \mathbf{Z}/m\mathbf{Z}$ and $\varepsilon > 0$, we define the *Bohr neighborhood*

$$B(r_1, \ldots, r_n; \varepsilon) = \left\{ g \in \mathbf{Z}/m\mathbf{Z} : \left\| \frac{g r_i}{m} \right\| \leq \varepsilon \text{ for } i = 1, \ldots, n \right\}.$$

Note that $B(0; \varepsilon) = \mathbf{Z}/m\mathbf{Z}$ for every $\varepsilon > 0$.

Theorem 8.6 (Bogolyubov) *Let $m \geq 2$, and let A be a nonempty subset of $\mathbf{Z}/m\mathbf{Z}$. Define $\lambda \in (0, 1]$ by $|A| = \lambda m$. For some positive integer $n \leq \lambda^{-2}$, there exist pairwise distinct congruence classes $r_1, r_2, \ldots, r_n \in \mathbf{Z}/m\mathbf{Z}$ such that $r_1 = 0$ and*

$$B(r_1, \ldots, r_n; 1/4) \subseteq 2A - 2A.$$

Proof. Let $G = \mathbf{Z}/m\mathbf{Z}$. For $r \in G$, we consider the additive character $\chi_r : G \to \mathbf{C}$ defined by

$$\chi_r(g) = e^{2\pi i r g/m}.$$

This function is well defined on congruence classes r and g modulo m, and $\chi_0(g) = 1$ for all $g \in G$. Let

$$S_A(r) = \sum_{a \in A} \chi_r(a) = \sum_{a \in A} e^{2\pi i r a/m}.$$

Then

$$|S_A(r)| \leq S_A(0) = |A|$$

for all $r \in G$, and

$$\sum_{r \in G} |S_A(r)|^2 = \sum_{r \in G} \sum_{a, a' \in A} e^{2\pi i r(a-a')/m} = |G||A| = \lambda^{-1}|A|^2.$$

Let $g \in G$. Then

$$\sum_{r \in G} |S_A(r)|^4 \chi_r(g) = \sum_{r \in G} \sum_{a_1, a_2, a_3, a_4 \in A} e^{2\pi i r(g - a_1 - a_2 + a_3 + a_4)/m},$$

and this sum is nonzero if and only if g has at least one representation in the form $g = a_1 + a_2 - a_3 - a_4$, that is, if and only if g is in the difference set $2A - 2A$. Let

$$R_1 = \{r \in G : |S_A(r)| \geq \sqrt{\lambda}\, |A|\}$$

and

$$R_2 = \{r \in G : |S_A(r)| < \sqrt{\lambda}\, |A|\}.$$

Since $S_0 = |A| \geq \sqrt{\lambda}\, |A|$, it follows that $0 \in R_1$ and $R_2 \neq G$. Therefore,

$$\begin{aligned}
\left| \sum_{r \in R_2} |S_A(r)|^4 \chi_r(g) \right| &\leq \sum_{r \in R_2} |S_A(r)|^4 \\
&\leq \lambda |A|^2 \sum_{r \in R_2} |S_A(r)|^2 \\
&< \lambda |A|^2 \sum_{r \in G} |S_A(r)|^2 \\
&= \lambda |A|^2 \lambda^{-1} |A|^2 \\
&= |A|^4.
\end{aligned}$$

Let $R_1 = \{r_1, r_2, \ldots, r_n\}$, where $r_1 = 0$, and let $g \in B(r_1, \ldots, r_n; 1/4)$. Then $\|r_i g/m\| \leq 1/4$ for $i = 1, \ldots, n$, and so

$$\Re(\chi_{r_i}(g)) = \Re(e^{2\pi i r_i g/m}) = \cos(2\pi r_i g/m) \geq 0.$$

It follows that

$$\begin{aligned}
&\Re\left(\sum_{r\in G} |S_A(r)|^4 \chi_r(g)\right) \\
&= \Re\left(\sum_{r\in R_1} |S_A(r)|^4 \chi_r(g)\right) + \Re\left(\sum_{r\in R_2} |S_A(r)|^4 \chi_r(g)\right) \\
&= |A|^4 + \sum_{r\in R_1\setminus\{0\}} |S_A(r)|^4 \Re(\chi_r(g)) + \Re\left(\sum_{r\in R_2} |S_A(r)|^4 \chi_r(g)\right) \\
&\geq |A|^4 + \Re\left(\sum_{r\in R_2} |S_A(r)|^4 \chi_r(g)\right) \\
&\geq |A|^4 - \left|\sum_{r\in R_2} |S_A(r)|^4 \chi_r(g)\right| \\
&> 0.
\end{aligned}$$

Therefore,

$$\sum_{r\in G} |S_A(r)|^4 \chi_r(g) \neq 0$$

for all $g \in B(r_1, \ldots, r_n; 1/4)$, and so

$$B(r_1, \ldots, r_n; 1/4) \subseteq 2A - 2A.$$

Finally, we must estimate $n = |R_1|$. Since $|S_A(r)| \geq \sqrt{\lambda}\, |A|$ for all $r \in R_1$, it follows that

$$n\lambda |A|^2 \leq \sum_{r\in R_1} |S_A(r)|^2 \leq \sum_{r\in G} |S_A(r)|^2 = \lambda^{-1} |A|^2,$$

and so $n \leq \lambda^{-2}$. This completes the proof.

Theorem 8.7 *Let $m \geq 2$, and let $R = \{r_1, \ldots, r_n\}$ be a set of congruence classes modulo m. If $(r_1, \ldots, r_n, m) = 1$, then there exists a proper n-dimensional arithmetic progression Q in $\mathbf{Z}/m\mathbf{Z}$ such that Q is contained in the Bohr neighborhood $B(r_1, \ldots, r_n; 1/4)$ and*

$$|Q| > \frac{m}{(4n)^n}.$$

Proof. We shall apply results from the geometry of numbers that were obtained in Theorem 6.12. Let $\mathbf{u} = (u_1, \ldots, u_n)$ and $\mathbf{v} = (v_1, \ldots, v_n)$ be vectors in the lattice $\mathbf{Z}^n$. We write $\mathbf{u} \equiv \mathbf{v} \pmod{m}$ if $u_i \equiv v_i \pmod{m}$ for $i = 1, \ldots, n$. Let M be the lattice of all vectors $\mathbf{v} = (v_1, \ldots, v_n) \in \mathbf{Z}^n$ such that $v_i \equiv 0 \pmod{m}$ for $i = 1, \ldots, n$, that is, M consists of all vectors $\mathbf{v} \in \mathbf{Z}^n$ such that $\mathbf{v} \equiv \mathbf{0} \pmod{m}$. Then $M = (m\mathbf{Z})^n$ and $\det(M) = m^n$.

For $i = 1, \ldots, n$, let r_i also denote a fixed integer in the congruence class $r_i \in \mathbf{Z}/m\mathbf{Z}$. Let $\mathbf{r} = (r_1, \ldots, r_n) \in \mathbf{Z}^n$ for $i = 1, \ldots, n$. Then

$$(r_1, \ldots, r_n, m) = 1. \tag{8.3}$$

Let Λ be the set of all vectors $\mathbf{u} \in \mathbf{Z}^n$ such that $\mathbf{u} \equiv q\mathbf{r} \pmod{m}$ for some $q = 0, 1, \ldots, m-1$. Then

$$\Lambda = \bigcup_{q=0}^{m-1} (q\mathbf{r} + M).$$

The set Λ is a lattice, and M is a sublattice of Λ. Condition (8.3) implies that the m vectors $\mathbf{0}, \mathbf{r}, 2\mathbf{r}, \ldots, (m-1)\mathbf{r}$ are pairwise incongruent modulo m, and so the cosets $(q\mathbf{r} + M)$ are pairwise disjoint. It follows that the index of M in Λ is $[\Lambda : M] = m$ and, by Theorem 6.9,

$$\det(\Lambda) = \frac{\det(M)}{[\Lambda : M]} = m^{n-1}.$$

Let $K \subseteq \mathbf{R}^n$ be the cube consisting of all vectors $(x_1, \ldots, x_n)$ such that $|x_i| < 1/4$ for $i = 1, \ldots, n$. Then K is a convex body symmetric with respect to the origin, and $\mathrm{vol}(K) = 1/2^n$. Let $\lambda_1, \ldots, \lambda_n$ be the successive minima of K with respect to the lattice Λ, and let $\mathbf{b}_1, \ldots, \mathbf{b}_n$ be a corresponding set of linearly independent vectors in Λ. Then

$$\mathbf{b}_i = (b_{i1}, \ldots, b_{in}) \in \overline{\lambda_i K} \cap \Lambda$$

for $i = 1, \ldots, n$. It follows from Minkowski's second theorem (Theorem 6.6) that

$$\lambda_1 \cdots \lambda_n \leq \frac{2^n \det(\Lambda)}{\mathrm{vol}(K)} = 4^n m^{n-1}.$$

Since

$$\mathbf{b}_i \in \overline{\lambda_i K} = \left\{ (x_1, \ldots, x_n) \in \mathbf{R}^n : |x_i| \leq \frac{\lambda_i}{4} \text{ for } i = 1, \ldots, n \right\},$$

it follows that

$$|b_{ij}| \leq \frac{\lambda_i}{4}$$

for all $i, j = 1, \ldots, n$. Since

$$\mathbf{b}_i \in \Lambda,$$

it follows that

$$\mathbf{b}_i \equiv q_i \mathbf{r} \pmod{m}$$

for some integer $q_i \in [0, m-1]$. Then

$$b_{ij} \equiv q_i r_j \pmod{m}$$

for $i, j = 1, \ldots, n$. Let

$$l'_i = \left[\frac{m}{n\lambda_i}\right]$$

and

$$Q = \{x_1 q_1 + \cdots + x_n q_n : -l'_i \leq x_i \leq l'_i \text{ for } i = 1, \ldots, n\} \subseteq \mathbf{Z}/m\mathbf{Z}.$$

We shall show that $Q \subseteq B = B(r_1, \ldots, r_n; 1/4)$. Let

$$x = q_1 x_1 + \cdots + q_n x_n \in Q.$$

Then

$$x r_j = \sum_{i=1}^{n} x_i q_i r_j \equiv \sum_{i=1}^{n} x_i b_{ij} \pmod{m},$$

and so

$$\begin{aligned}
\left\| \frac{x r_j}{m} \right\| &= \left\| \sum_{i=1}^{n} \frac{x_i b_{ij}}{m} \right\| \\
&\leq \left| \sum_{i=1}^{n} \frac{x_i b_{ij}}{m} \right| \\
&\leq \sum_{i=1}^{n} \left| \frac{x_i b_{ij}}{m} \right| \\
&\leq \sum_{i=1}^{n} \frac{l'_i |b_{ij}|}{m} \\
&\leq \sum_{i=1}^{n} \frac{l'_i \lambda_i}{4m} \\
&\leq \sum_{i=1}^{n} \frac{1}{4n} \\
&= \frac{1}{4}.
\end{aligned}$$

This means that $x \in B$, and so $Q \subseteq B$.

Next we show that Q is a proper n-dimensional arithmetic progression. Suppose that

$$q_1 x_1 + \cdots + q_n x_n \equiv q_1 y_1 + \cdots + q_n y_n \pmod{m},$$

where $-l'_i \leq x_i, y_i \leq l'_i$ for $i = 1, \ldots, n$. Let $z_i = x_i - y_i$. Then $|z_i| \leq 2l'_i$ and

$$\sum_{i=1}^{n} q_i z_i \equiv 0 \pmod{m}.$$

It follows that

$$\sum_{i=1}^{n} q_i r_j z_i = \sum_{i=1}^{n} b_{ij} z_i \equiv 0 \pmod{m}$$

for $j = 1, \ldots, n$. Since

$$\begin{aligned}
\left|\sum_{i=1}^{n} b_{ij} z_i\right| &\leq \sum_{i=1}^{n} |b_{ij}||z_i| \\
&\leq \sum_{i=1}^{n} \left(\frac{\lambda_i}{4}\right)(2l_i') \\
&\leq \sum_{i=1}^{n} \frac{m}{2n} \\
&< m,
\end{aligned}$$

it follows that $\sum_{i=1}^{n} b_{ij} z_i = 0$ for $j = 1, \ldots, n$, and so

$$\sum_{i=1}^{n} z_i \mathbf{b}_i = \mathbf{0}.$$

Since the vectors $\mathbf{b_i}$ are linearly independent, we conclude that $z_i = 0$ for all i, and so

$$|Q| = (2l_1' + 1) \cdots (2l_n' + 1).$$

Let $l_i = 2l_i' + 1$ and $a = -\sum_{i=1}^{n} l_i' q_i$. Then Q is the proper n-dimensional arithmetic progression

$$Q(a; q_1, \ldots, q_n; l_1, \ldots, l_n).$$

Moreover,

$$\begin{aligned}
|Q| &= l_1 \cdots l_n \\
&\geq \prod_{i=1}^{n} (l_i' + 1) \\
&> \prod_{i=1}^{n} \frac{m}{n\lambda_i} \\
&= \left(\frac{m}{n}\right)^n \left(\prod_{i=1}^{n} \lambda_i\right)^{-1} \\
&\geq \left(\frac{m}{n}\right)^n (4^n m^{n-1})^{-1} \\
&= \frac{m}{(4n)^n}.
\end{aligned}$$

This completes the proof.

Theorem 8.8 *Let p be a prime, and let R be a nonempty set of congruence classes modulo p with $|R| = \lambda p$. For some positive integer $n \leq \lambda^{-2}$ there exists a proper n-dimensional arithmetic progression Q such that*

$$Q \subseteq 2R - 2R$$

and

$$l(Q) = |Q| > \delta p$$

where

$$\delta = \frac{1}{(4n)^n} \geq \left(\frac{\lambda^2}{4}\right)^{1/\lambda^2}.$$

Proof. If $\mathbf{Z}/p\mathbf{Z} = 2R - 2R$, let $n = 1$ and Q be the 1-dimensional arithmetic progression $Q = Q(0; 1; p) = \mathbf{Z}/p\mathbf{Z}$. Then $Q \subseteq 2R - 2R$ and $|Q| = p > \delta p$, where

$$\delta = \frac{1}{4} \geq \left(\frac{\lambda^2}{4}\right)^{1/\lambda^2}$$

for every $\lambda \in (0, 1]$.

Suppose that $2R - 2R \neq \mathbf{Z}/p\mathbf{Z}$. By Theorem 8.6, for some positive integer $n \leq \lambda^{-2}$ there exist pairwise distinct congruence classes $r_1, r_2, \ldots, r_n$ modulo p such that $r_1 = 0$ and

$$B = B(r_1, \ldots, r_n; 1/4) \subseteq 2R - 2R.$$

Since $B(0; 1/4) = \mathbf{Z}/p\mathbf{Z}$, we must have $n \geq 2$, and so

$$(r_1, \ldots, r_n, p) = 1.$$

By Theorem 8.7, there exists a proper n-dimensional arithmetic progression Q such that $Q \subseteq B$ and $|Q| > \delta p$, where

$$\delta = \frac{1}{(4n)^n} \geq \left(\frac{\lambda^2}{4}\right)^{1/\lambda^2}.$$

This completes the proof.

8.4 Ruzsa's proof, concluded

Theorem 8.9 (Ruzsa) *Let W be a finite nonempty set of integers. Let $h \geq 2$ and*

$$D = D_{h,h}(W) = hW - hW.$$

For every

$$m \geq 4h|D_{h,h}(W)| = 4h|D|,$$

there exists a set $W' \subseteq W$ such that

$$|W'| \geq \frac{|W|}{h},$$

and W' is Freiman isomorphic of order h to a set of congruence classes modulo m.

Proof. Let $m \geq 4h|D|$, and let p be a prime number such that

$$p > \max\{m, 2h \max_{w \in W} |w|\}.$$

Let $1 \leq q \leq p-1$. We shall construct a map $\phi_q : \mathbf{Z} \to \mathbf{Z}/m\mathbf{Z}$, and then prove that for some q there is a subset W' of W such that $|W'| \geq |W|/h$ and ϕ_q restricted to W' is a Freiman isomorphism of order h. The map ϕ_q will be the composition of four maps:

$$\mathbf{Z} \xrightarrow{\alpha} \mathbf{Z}/p\mathbf{Z} \xrightarrow{\beta_q} \mathbf{Z}/p\mathbf{Z} \xrightarrow{\gamma} \mathbf{Z} \xrightarrow{\delta} \mathbf{Z}/m\mathbf{Z},$$

where α, β_q, γ, and δ are defined below.

Let $\alpha : \mathbf{Z} \to \mathbf{Z}/p\mathbf{Z}$ be the natural map that sends w to $w + p\mathbf{Z}$. Since α is a group homomorphism, it is also a Freiman homomorphism of order h. Although α is not a group isomorphism, we can show that α restricted to W is a Freiman isomorphism of order h. Let $w_1, \ldots, w_h, w_1', \ldots, w_h' \in W$, and suppose that

$$\alpha(w_1) + \cdots + \alpha(w_h) = \alpha(w_1') + \cdots + \alpha(w_h').$$

Then

$$\alpha(w_1 + \cdots + w_h) = \alpha(w_1' + \cdots + w_h'),$$

and so

$$(w_1 + \cdots + w_h) - (w_1' + \cdots + w_h') \equiv 0 \pmod{p}.$$

Since

$$|(w_1 + \cdots + w_h) - (w_1' + \cdots + w_h')| \leq 2h \max_{w \in W} |w| < p,$$

it follows that

$$(w_1 + \cdots + w_h) - (w_1' + \cdots + w_h') = 0.$$

Therefore, $\alpha : W \to \alpha(W) \subseteq \mathbf{Z}/p\mathbf{Z}$ is a Freiman isomorphism of order h.

For $1 \leq q \leq p-1$, let $\beta_q : \mathbf{Z}/p\mathbf{Z} \to \mathbf{Z}/p\mathbf{Z}$ be the map that sends $w + p\mathbf{Z}$ to $wq + p\mathbf{Z}$. Since the map β_q is a group isomorphism, it is also a Freiman isomorphism of order h on every subset of $\mathbf{Z}/p\mathbf{Z}$.

Let $\gamma : \mathbf{Z}/p\mathbf{Z} \to \mathbf{Z}$ be the map that sends the congruence class $w + p\mathbf{Z}$ to its least nonnegative representative. The image of γ is the interval of integers $[0, p-1]$. The map γ is neither a group homomorphism nor a Freiman homomorphism of order h (see Exercise 9). We can, however, write $\mathbf{Z}/p\mathbf{Z}$ as the union of h subsets such that γ restricted to each of these subsets is a Freiman isomorphism of order h. For $i = 1, \ldots, h$, let

$$U_i = \gamma^{-1}\left[\frac{(i-1)(p-1)}{h}, \frac{i(p-1)}{h}\right] \subseteq \mathbf{Z}/p\mathbf{Z}.$$

Since

$$[0, p-1] = \bigcup_{i=1}^{h} \left[\frac{(i-1)(p-1)}{h}, \frac{i(p-1)}{h} \right],$$

it follows that

$$\mathbf{Z}/p\mathbf{Z} = \bigcup_{i=1}^{h} U_i.$$

Fix the set U_i, and let $u_j + p\mathbf{Z} \in U_i$ and $u'_j + p\mathbf{Z} \in U_i$ for $j = 1, \ldots, h$. If

$$u_1 + \cdots + u_h + p\mathbf{Z} = u'_1 + \cdots + u'_h + p\mathbf{Z}$$

in $\mathbf{Z}/p\mathbf{Z}$, then

$$\gamma(u_1 + p\mathbf{Z}) + \cdots + \gamma(u_h + p\mathbf{Z}) \equiv \gamma(u'_1 + p\mathbf{Z}) + \cdots + \gamma(u'_h + p\mathbf{Z}) \pmod{p}$$

in $\mathbf{Z}$. Since

$$\gamma(u_1 + p\mathbf{Z}) + \cdots + \gamma(u_h + p\mathbf{Z}) \in [(i-1)(p-1), i(p-1)]$$

and

$$\gamma(u'_1 + p\mathbf{Z}) + \cdots + \gamma(u'_h + p\mathbf{Z}) \in [(i-1)(p-1), i(p-1)],$$

it follows that

$$|(\gamma(u_1 + p\mathbf{Z}) + \cdots + \gamma(u_h + p\mathbf{Z})) - \left(\gamma(u'_1 + p\mathbf{Z}) + \cdots + \gamma(u'_h + p\mathbf{Z})\right)| \leq p-1,$$

and so

$$\gamma(u_1 + p\mathbf{Z}) + \cdots + \gamma(u_h + p\mathbf{Z}) = \gamma(u'_1 + p\mathbf{Z}) + \cdots + \gamma(u'_h + p\mathbf{Z}).$$

Thus, γ is a Freiman homomorphism of order h. Conversely, if

$$\gamma(u_1 + p\mathbf{Z}) + \cdots + \gamma(u_h + p\mathbf{Z}) = \gamma(u'_1 + p\mathbf{Z}) + \cdots + \gamma(u'_h + p\mathbf{Z})$$

in $\mathbf{Z}$, then

$$u_1 + \cdots + u_h + p\mathbf{Z} = u'_1 + \cdots + u'_h + p\mathbf{Z}$$

in $\mathbf{Z}/p\mathbf{Z}$, and so γ restricted to each set U_i is a Freiman isomorphism of order h.

Let

$$W_{i,q} = W \cap \alpha^{-1}(\beta_q^{-1}(U_i))$$

for $i = 1, \ldots, h$. Then

$$W = \cup_{i=1}^{h} W_{i,q},$$

and so

$$|W_{j,q}| \geq \frac{|W|}{h}$$

for some j. Let $W'_q = W_{j,q}$. Define $\theta_q : W \to \mathbf{Z}$ by $\theta_q = \gamma\beta_q\alpha$. Then

$$\theta_q(w) = wq - \left[\frac{wq}{p}\right] p \in [0, p-1]$$

for all $a \in W$. Let

$$V_q = \theta_q(W)$$

and

$$V_q' = \theta_q(W_q').$$

Then $\theta_q : W_q' \to V_q'$ is a Freiman isomorphism of order h.

Let $\delta : \mathbf{Z} \to \mathbf{Z}/m\mathbf{Z}$ be the natural map that sends w to $w + m\mathbf{Z}$. Then δ is a Freiman homomorphism of order h. We shall prove that there exists at least one $q \in [1, p-1]$ such that δ restricted to V_q is a Freiman isomorphism of order h.

Let $q \in [1, p-1]$, and suppose that $\delta : V_q \to \mathbf{Z}/m\mathbf{Z}$ is not a Freiman isomorphism of order h. Then there exist integers $v_1, \ldots, v_h, v_1', \ldots, v_h' \in V_q \subseteq [0, p-1]$ such that

$$v_1 + \cdots + v_h \neq v_1' + \cdots + v_h'$$

but

$$\begin{aligned} \delta(v_1 + \cdots + v_h) &= \delta(v_1) + \cdots + \delta(v_h) \\ &= \delta(v_1') + \cdots + \delta(v_h') \\ &= \delta(v_1' + \cdots + v_h'). \end{aligned}$$

Define

$$v^* = (v_1 + \cdots + v_h) - (v_1' + \cdots + v_h').$$

Since $m > 4h|D|$, we have

$$|v^*| \leq h(p-1) < hp < mp \tag{8.4}$$

and

$$v^* \equiv 0 \pmod{m}, \tag{8.5}$$

but

$$v^* \neq 0. \tag{8.6}$$

Choose $w_i, w_i' \in W$ such that $\theta_q(w_i) = v_i$ and $\theta_q(w_i') = v_i'$ for $i = 1, \ldots, h$. Define

$$w^* = (w_1 + \cdots + w_h) - (w_1' + \cdots + w_h').$$

Then $w^* \in D = hW - hW$. For $i = 1, \ldots, h$,

$$v_i \equiv w_i q \pmod{p}$$

and

$$v_i' \equiv w_i' q \pmod{p};$$

hence

$$v^* \equiv w^* q \pmod{p},$$

and so

$$v^* = \gamma(w^* q + p\mathbf{Z}) + xp \tag{8.7}$$

for some integer x. If $w^* \equiv 0 \pmod{p}$, then

$$v^* \equiv 0 \pmod{p}. \tag{8.8}$$

Since p is prime and $1 < m < p$, we have $(m, p) = 1$. Then congruences (8.5) and (8.8) imply that

$$v^* \equiv 0 \pmod{mp}.$$

Since $|v^*| < mp$ by (8.4), it follows that $v^* = 0$, which contradicts (8.6). Therefore,

$$w^* \not\equiv 0 \pmod{p}.$$

Recall that $\gamma(w^*q + p\mathbf{Z})$ is the least nonnegative integer in the congruence class $w^*q + p\mathbf{Z}$. By (8.5) and (8.7),

$$v^* = \gamma(w^*q + p\mathbf{Z}) + xp \equiv 0 \pmod{m},$$

and inequality (8.4) implies that

$$-h \leq x \leq h - 1.$$

Thus, if $q \in [1, p-1]$ and $\delta : V_q \to \mathbf{Z}/m\mathbf{Z}$ is not a Freiman isomorphism of order h, then there exist integers $w^* \in D$ and $x \in [-h, h-1]$ such that $w^* \not\equiv 0 \pmod{p}$ and

$$\gamma(w^*q + p\mathbf{Z}) + xp \equiv 0 \pmod{m}. \tag{8.9}$$

Let us count the number of triples

$$(q, w^*, x)$$

that satisfy these conditions. Choose an integer $x \in [-h, h-1]$. There are $2h$ such choices. Since $p > m$, the congruence

$$y + xp \equiv 0 \pmod{m}$$

has at most

$$\left(\frac{p-1}{m}\right) + 1 \leq \frac{2(p-1)}{m}$$

solutions $y \in [1, p-1]$. Choose an integer $w^* \in D$ such that $w^* \not\equiv 0 \pmod{p}$. Since $0 \in D$, there are at most $|D| - 1$ such choices. Since $w^* \not\equiv 0 \pmod{p}$, for each integer $y \in [1, p-1]$ there is a unique integer $q \in [1, p-1]$ such that $y = \gamma(w^*q + p\mathbf{Z})$. Thus, for each of the permissible choices of x and w^* there are at most $2(p-1)/m$ choices of $q \in [1, p-1]$ such that the triple (q, w^*, x) yields a solution of the congruence (8.9). Since $m \geq 4h|D|$, the number of triples is at most

$$2h\left(\frac{2(p-1)}{m}\right)(|D| - 1) < \frac{4h|D|(p-1)}{m} \leq p - 1.$$

Therefore, at least one integer $q \in [1, p-1]$ occurs in none of the triples, and for this q the map

$$\delta : V_q = \theta_q(W) \to \mathbf{Z}/m\mathbf{Z}$$

is a Freiman isomorphism of order h. Let $W' = W'_q$. Since

$$\theta_q : W' \to V'_q \subseteq V_q \subseteq \mathbf{Z}$$

is also a Freiman isomorphism of order h, it follows that there is a Freiman isomorphism of order h from W' into $\mathbf{Z}/m\mathbf{Z}$. Moreover, $|W'| \geq |W|/h$. This completes the proof.

Theorem 8.10 *Let c, c_1, and c_2 be positive real numbers. Let $k \geq 1$, and let A and B be finite subsets of a torsion-free abelian group such that*

$$c_1 k \leq |A|, |B| \leq c_2 k$$

and

$$|A + B| \leq ck.$$

Then A is a subset of an n-dimensional arithmetic progression of length at most lk, where n and l depend only on c, c_1, and c_2.

Proof. Let G be the group generated by A. Since G is a finitely generated torsion-free abelian group, it follows from Corollary 8.2 with $h = 32$ that there is a Freiman isomorphism of order 32 between A and some set W of integers. Since $32 = 2(8+8)$, Theorem 8.5 implies that the difference sets $D_{8,8}(A) = 8A - 8A$ and $D_{8,8}(W) = 8W - 8W$ are Freiman isomorphic of order 2, and so $|D_{8,8}(W)| = |D_{8,8}(A)|$. Let $c_3 = c/c_1$. Since

$$|A + B| \leq ck \leq (c/c_1)|A| = c_3|A|,$$

it follows from Theorem 7.8 that

$$|D_{8,8}(W)| = |D_{8,8}(A)| \leq c_3^{16}|A| = c_3^{16}|W|.$$

Bertrand's postulate in elementary number theory states that for every positive integer n there exists a prime number between n and $2n$. Let $n = 32|D_{8,8}(W)|$. Then there is a prime number p such that

$$|W| < 32|W| \leq 32|D_{8,8}(W)| < p < 64|D_{8,8}(W)| \leq 64c_3^{16}|W|.$$

By Theorem 8.9 with $h = 8$, there exists a set $W' \subseteq W$ such that $|W'| \geq |W|/8$ and W' is Freiman isomorphic of order 8 to a set R of congruence classes modulo p. Define $\lambda \in (0, 1]$ by $\lambda p = |R|$. Then

$$\lambda p = |R| = |W'| \geq \frac{|W|}{8} > \frac{p}{8 \cdot 64c_3^{16}},$$

and so

$$\lambda > 2^{-9}c_3^{-16}.$$

By Theorem 8.8, the difference set $2R - 2R$ contains a proper n_1-dimensional arithmetic progression Q' of length

$$l(Q') = |Q'| > \delta p > \delta|W| = \delta|A|,$$

where

$$n_1 \leq \lambda^{-2} < 2^{18}c_3^{32}$$

and

$$\delta = (4n_1)^{-n_1} > (2^{20}c_3^{32})^{-2^{18}c_3^{32}}.$$

Since $8 = 2(2+2)$, it follows from Theorem 8.5 that the difference sets $2R - 2R$ and $2W' - 2W'$ are Freiman isomorphic of order 2.

The sets W and A are Freiman isomorphic of order 32 and thus also of order 8. Let A' be the image of W' under this isomorphism. By Theorem 8.5, the difference sets $2W' - 2W'$ and $2A' - 2A'$ are Freiman isomorphic of order 2, and so $2R - 2R$ and $2A' - 2A'$ are Freiman isomorphic. Let Q_1 be the image of Q' under this isomorphism. By Theorem 8.4, the set Q_1 is a proper n_1-dimensional arithmetic progression such that

$$Q_1 \subseteq 2A' - 2A' \subseteq 2A - 2A \subseteq G$$

and

$$\delta|A| < |Q'| = |Q_1| = l(Q_1) \leq |2A - 2A| \leq c_3^4|A|.$$

Let $A^* = \{a_1, \ldots, a_{n_2}\}$ be a maximal set of elements in A such that the sets $a_i + Q_1$ are pairwise disjoint. Since

$$\bigcup_{i=1}^{n_2}(a_i + Q_1) = A^* + Q_1 \subseteq A + Q_1 \subseteq 3A - 2A,$$

it follows from Theorem 7.8 that

$$\begin{aligned} n_2|Q_1| &= \sum_{i=1}^{n_2}|a_i + Q_1| \\ &= \left|\bigcup_{i=1}^{n_2}(a_i + Q_1)\right| \\ &= |A^* + Q_1| \\ &\leq |3A - 2A| \\ &\leq c_3^5|A|. \end{aligned}$$

Then

$$n_2 \leq \frac{c_3^5|A|}{|Q_1|}$$

$$
\begin{aligned}
&< \frac{c_3^5|A|}{\delta|A|} \\
&= c_3^5(4n_1)^{n_1}.
\end{aligned}
$$

The set A^* is a subset of the n_2-dimensional arithmetic progression

$$Q_2 = \{x_1a_1 + \cdots + x_{n_2}a_{n_2} : 0 \leq x_i < 2 \text{ for } i = 1, \ldots, n_2\}$$

of length $l(Q_2) = 2^{n_2}$. Since the set A^* is maximal, for every $a \in A$ there is an $a_i \in A^*$ such that

$$(a + Q_1) \cap (a_i + Q_1) \neq \emptyset,$$

and so there exist integers $q, q' \in Q_1$ such that $a + q = a_i + q'$. Then

$$a = a_i + q' - q \in A^* + Q_1 - Q_1 \subseteq Q_2 + Q_1 - Q_1.$$

Let $Q = Q_2 + Q_1 - Q_1$. Then

$$A \subseteq Q.$$

By Theorem 8.2, $Q_1 - Q_1$ is an n_1-dimensional arithmetic progression of length

$$l(Q_1 - Q_1) < 2^{n_1} l(Q_1) \leq 2^{n_1} c_3^4 |A| \leq 2^{n_1} \left(\frac{c}{c_1}\right)^4 c_2 k,$$

and so $Q = Q_2 + (Q_1 - Q_1)$ is an arithmetic progression of dimension

$$n = n_1 + n_2$$

whose length satisfies

$$l = l(Q) \leq l(Q_2) l(Q_1 - Q_1) < 2^{n_2} 2^{n_1} \left(\frac{c}{c_1}\right)^4 c_2 k = 2^n \left(\frac{c}{c_1}\right)^4 c_2 k,$$

where n and l depend only on c, c_1, and c_2. This completes the proof.

Freiman's Theorem 8.1 is the special case of the preceding result, when A is a finite set of integers and $B = A$.

8.5 Notes

There are important unsolved problems related to Freiman's theorem. Let A be a finite set of integers. If $|2A| \leq c|A|$, then Theorem 8.1 states that A is a subset of an n-dimensional arithmetic progression Q such that $|Q| \leq c'|A|$, where the numbers n and c' depend only on c. It is not known how small these constants can be made simultaneously. Another important question, the "Proper Conjecture," will be discussed in Section 9.6.

If the cardinality of the sumset $2A$ is somewhat greater than $c|A|$, is there a structure theorem? For example, what do we know about the set A if $|2A| \leq c|A|(\log|A|)^{\delta}$ for some $\delta > 0$?

Let $h \geq 3$. It is not known how to extend Freiman's theorem to the h-fold sum hA. We would like to find a condition of the form $|hA| \leq c|A|^u$ that implies something about the structure of A. If $u = 1$, then $|hA| \leq c|A|$ and $|2A| \leq |hA|$ imply that $|2A| \leq c|A|$, so Freiman's theorem applies directly. On the other hand, the condition $|hA| \leq c|A|^{h-1}$ is not sufficient to restrict the structure of the set A (see Exercise 12). We can formulate many open problems of this kind. For example, does there exist $\delta > 1$ such that if $|3A| \leq c|A|^{1+\delta}$, then the structure of A is in some sense determined?

Freiman's theorem first appeared in a short paper [53] in 1964 and in a 1966 monograph [54]. Freiman published a revised proof in the proceedings of the New York Number Theory Seminar [56]. Bilu [9] has a different version of Freiman's original proof. Ruzsa's proof appears in [113] and [115]. Bogolyubov's theorem is part of his "arithmetic" proof [11] that every almost periodic function can be uniformly approximated by trigonometric polynomials. Expositions of Bogolyubov's method appear in Jessen [72] and Maak [82].

8.6 Exercises

1. Let $\phi : A \to B$ be a *Freiman isomorphism of order* h, and let $A' \subseteq A$ and $B' = \phi(A')$. Prove that $\phi : A' \to B'$ is also a Freiman isomorphism of order h.

2. Let $h' \leq h$, and let $\phi : A \to B$ be a Freiman isomorphism of order h. Prove that ϕ is also a Freiman isomorphism of order h'.

3. Prove that the composition of Freiman isomorphisms of order h is a Freiman isomorphism of order h.

4. (Freiman [54]) Let $A = \{0, 1, h+1\}$ and $B = \{0, 1, h+2\}$. Prove that A and B are Freiman isomorphic of order h but not of order $h+1$.

5. Let $P = [0, 35]$. Represent P as an n-dimensional arithmetic progression for $n = 1, 2, 3, 4, 6, 9$.

6. Let P and P' be multidimensional arithmetic progressions of dimensions n and n', respectively. Show that $P - P'$ is an arithmetic progression of dimension $n + n'$.

7. Let Ω be some condition such that if the finite subset A of an abelian group satisfies Ω, then $A \subseteq Q$, where Q is an n-dimensional arithmetic progression such that $|Q| \leq c'|A|$, where n and c' depend only on Ω. Prove that $|2A| \leq 2^n c'|A|$.

8. Let A be a finite subset of an abelian group, and let $k = |A|$. Suppose that

$$c_1 k \log k < |2A| < c_2 k \log k.$$

Show that A is not a large subset of a multidimensional arithmetic progression, that is, show that there do not exist numbers $n = n(c_1, c_2)$ and $l = l(c_1, c_2)$ such that $A \subseteq Q$, where Q is an n-dimensional arithmetic progression and $|Q| \leq l|A|$.

9. Let $\phi : \mathbf{Z}/m\mathbf{Z} \rightarrow \mathbf{Z}$ be the map that sends each congruence class to its least nonnegative residue. Show that ϕ is not a Freiman homomorphism of order 2.

10. Let A be a finite set of lattice points in the plane whose elements do not all lie on a line. Prove that there does not exist a set B of integers such that A and B are Freiman isomorphic of order h for all $h \geq 2$.

11. Let A be a set of integers. Let $h \geq 2$, and let $D = hA - hA$. Prove that there exists a set $A'' \subseteq A$ such that $|A''| \geq |A|/h^2$ and A'' is Freiman isomorphic of order h to a subset of $[1, 2|D|]$.
(Hint: Use Theorem 8.9.)

12. Let the set B of nonnegative integers be a basis of order 2, and let $A = B \cap [0, n]$. Prove that
$$|3A| \leq c_1|A|^2,$$
where the constant c_1 depends on B but not on n. Suppose that B is a thin basis, that is, $|B \cap [0, n]| \leq cn^{1/2}$ for all $n \geq 1$. Prove that
$$|3A| \geq c_2|A|^2,$$
where the constants c_1, c_2 depend only on B.

9

Applications of Freiman's theorem

9.1 Combinatorial number theory

Many theorems and conjectures in combinatorial number theory are about sets of integers that contain finite arithmetic progressions. In this chapter we use Freiman's theorem to solve two such problems. The first is to prove that if A is a sufficiently large set of integers such that $|2A| \leq c|A|$, then A contains a long arithmetic progression. This is done in the next section. The second problem is to prove that a set of integers that contains many three–term arithmetic progressions must contain a long arithmetic progression. The proof uses a beautiful theorem of Balog and Szemerédi that is a kind of "density" version of Freiman's theorem.

9.2 Small sumsets and long progressions

The most famous result about arithmetic progressions is the following theorem of Szemerédi. Let $\delta > 0$ and $t \geq 3$. There exists an integer $l_0(\delta, t)$ such that if $l \geq l_0(\delta, t)$ and A is a subset of $[0, l-1]$ with $|A| \geq \delta l$, then A contains an arithmetic progression of length t. The following lemma is a simple consequence of this result.

Lemma 9.1 *Let $\delta > 0$ and $t \geq 3$. There exists an integer $l_0(\delta, t)$ such that if Q is an arithmetic progression of length l in a torsion-free abelian group and B is a subset of Q with $|B| \geq \delta l$ and $l \geq l_0(\delta, t)$, then B contains an arithmetic progression of length t.*

Proof. Let $l_0(\delta, t)$ be the integer determined by Szemerédi's theorem. Since Q is a 1-dimensional arithmetic progression, there exist group elements a and $q \neq 0$ such that

$$Q = \{a + xq : 0 \leq x < l\}.$$

Let

$$A = \{x \in [0, l-1] : a + xq \in B\}.$$

Then A is a set of integers, and $|A| = |B| \geq \delta l$. By Szemerédi's theorem, A contains an arithmetic progression of length t, and so there exist integers a' and $q' \neq 0$ such that $a' + yq' \in A$ for $0 \leq y < t$. Let $a'' = a + a'q$ and $q'' = q'q$. Then $q'' \neq 0$ since the group is torsion-free, and

$$a + (a' + yq')q = a'' + yq'' \in B$$

for $0 \leq y < t$. Thus, B contains an arithmetic progression of length t.

Theorem 9.1 *Let $c \geq 2$ and $t \geq 3$. There exists an integer $k_0(c, t)$ such that if A is a subset of a torsion-free abelian group, $|A| \geq k_0(c, t)$, and $|2A| \leq c|A|$, then A contains an arithmetic progression of length at least t.*

Proof. Let $|A| = k$. By Freiman's theorem (Theorem 8.10), there exist integers $n = n(c)$ and $l = l(c)$ such that A is a subset of an n-dimensional arithmetic progression Q of length

$$l(Q) \leq lk.$$

Let

$$Q = \{a + x_1q_1 + \cdots + x_nq_n : 0 \leq x_i < l_i \text{ for } i = 1, \ldots, n\}.$$

Then

$$l(Q) = l_1l_2 \cdots l_n.$$

We can assume without loss of generality that

$$l_1 \leq l_2 \leq \cdots \leq l_n.$$

It follows that

$$k = |A| \leq |Q| \leq l(Q) = l_1l_2 \cdots l_n \leq l_n^n,$$

and so

$$l_n \geq k^{1/n}.$$

Let Y be the set of all lattice points

$$\mathbf{y} = (y_1, \ldots, y_{n-1}) \in \mathbf{Z}^{n-1}$$

such that

$$0 \leq y_i < l_i$$

for $i = 1, \ldots, n-1$. Then

$$|Y| = l_1l_2 \cdots l_{n-1}.$$

For each $\mathbf{y} \in Y$, the set

$$L(\mathbf{y}) = \{a + y_1 q_1 + \cdots + y_{n-1} q_{n-1} + x_n q_n : 0 \leq x_n < l_n\}$$

is an arithmetic progression of length l_n in the group. Since

$$A \subseteq Q = \bigcup_{\mathbf{y} \in Y} L(\mathbf{y}),$$

it follows that

$$A = \bigcup_{\mathbf{y} \in Y} (L(\mathbf{y}) \cap A)$$

and so

$$k = |A| \leq \sum_{\mathbf{y} \in Y} |L(\mathbf{y}) \cap A|.$$

We compute a lower bound for the average cardinality of the intersections of the arithmetic progressions $L(\mathbf{y})$ with the set A as follows:

$$\frac{\sum_{\mathbf{y} \in Y} |L(\mathbf{y}) \cap A|}{|Y|} \geq \frac{k}{l_1 \cdots l_{n-1}} \geq \frac{l_n}{l} \geq \frac{k^{1/n}}{l}.$$

It follows that there exists some $\mathbf{y} \in Y$ such that

$$|L(\mathbf{y}) \cap A| \geq \frac{k^{1/n}}{l}.$$

Let

$$k_0(c, t) = l_0(1/l, t)^n,$$

where $l_0(1/l, t)$ is the integer constructed in Lemma 9.1. If $|A| = k \geq k_0(c, t)$, then $L(\mathbf{y})$ is an arithmetic progression of length

$$l_n \geq k^{1/n} \geq k_0(c, t)^{1/n} = l_0(1/l, t),$$

and so $L(\mathbf{y}) \cap A$ contains an arithmetic progression of length at least t. This completes the proof.

9.3 The regularity lemma

Let A and B be nonempty subsets of a torsion-free abelian group, and let $|A| = |B| = k$. For $W \subseteq A \times B$, let

$$S(W) = \{a + b : (a, b) \in W\}.$$

In particular,

$$S(A \times B) = A + B.$$

Freiman's theorem states that if $|S(A \times B)| \leq ck$, then A is contained in an n-dimensional arithmetic progression Q whose length $l(Q)$ satisfies $l(Q) \leq lk$, where n and l are parameters that depend only on c.

Balog and Szemerédi [6] proved a Freiman–type result for large subsets of $A \times B$. They showed that if $W \subseteq A \times B$, $|W| \geq c_1 k^2$, and $|S(W)| \leq c_2 k$, then there exists a set $A' \subseteq A$ such that $|A'| \geq c_1' k$ and $|2A'| \leq c_2' k$, where c_1' and c_2' are positive constants that depend only on c_1 and c_2. It follows that

$$|2A'| \leq c_2'|k| = \left(\frac{c_2'}{c_1'}\right) c_1'|k| \leq c|A'|,$$

where $c = c_2'/c_1'$. Applying Freiman's theorem, we conclude that A' is a "large" subset of a multidimensional arithmetic progression. The proof of the Balog–Szemerédi theorem uses an important result in graph theory, also discovered by Szemerédi, known as the regularity lemma.

The regularity lemma asserts the existence of a remarkable class of partitions of the vertex set of a graph. Let $G = (V, E)$ be a *graph*, or, more precisely, an undirected graph with loops but with no multiple edges. Then V is a finite set of vertices and E is the set of edges, where each edge is a set $\{v, v'\}$ of not necessarily distinct elements of V. The vertices v, v' are called the *endpoints* of the edge $e = \{v, v'\}$, and an edge e is *adjacent* to a vertex v if $v \in e$. The *degree* of a vertex v is the number of edges adjacent to v. There is at most one edge between any two vertices of the graph.

Let A and B be subsets of V. We denote by $e(A, B)$ the number of edges with one endpoint in A and the other endpoint in B. If A and B are disjoint nonempty sets, we define the *density of edges* between A and B by

$$d(A, B) = \frac{e(A, B)}{|A||B|},$$

where $|A|$ denotes the cardinality of the set A. Since $0 \leq e(A, B) \leq |A||B|$, it follows that $0 \leq d(A, B) \leq 1$.

We require the following lemmas.

Lemma 9.2 *Let $G = (V, E)$ be a graph, and let A and B be disjoint nonempty subsets of V. If $A' \subseteq A$ and $B' \subseteq B$ satisfy*

$$\begin{aligned} |A'| &\geq (1 - \delta)|A|, \\ |B'| &\geq (1 - \delta)|B|, \end{aligned}$$

where $0 < \delta < 1$, then

$$\begin{aligned} |d(A, B) - d(A', B')| &\leq \frac{2\delta}{(1 - \delta)^2}, \\ |d(A, B)^2 - d(A', B')^2| &\leq \frac{4\delta}{(1 - \delta)^2}. \end{aligned}$$

In particular, if $\delta \leq 1/2$, *then*

$$|d(A, B) - d(A', B')| \leq 8\delta$$

and

$$|d(A, B)^2 - d(A', B')^2| \leq 16\delta.$$

Proof. Let $A'' = A \setminus A'$ and $B'' = B \setminus B'$. Then $|A''| = |A| - |A'| \leq \delta|A|$ and $|B''| \leq \delta|B|$, and

$$\begin{aligned} e(A, B) &= e(A', B') + e(A', B'') + e(A'', B') + e(A'', B'') \\ &= e(A', B') + e(A', B'') + e(A'', B) \\ &\leq e(A', B') + e(A, B'') + e(A'', B) \\ &\leq e(A', B') + |A||B''| + |A''||B| \\ &\leq e(A', B') + 2\delta|A||B|. \end{aligned}$$

It follows that

$$\begin{aligned} d(A, B) &= \frac{e(A, B)}{|A||B|} \\ &\leq \frac{e(A', B')}{|A||B|} + 2\delta \\ &\leq \frac{e(A', B')}{|A'||B'|} + 2\delta \\ &= d(A', B') + 2\delta \\ &\leq d(A', B') + \frac{2\delta}{(1-\delta)^2}, \end{aligned}$$

and so

$$d(A, B) - d(A', B') \leq \frac{2\delta}{(1-\delta)^2}.$$

Similarly,

$$\begin{aligned} d(A', B') &= \frac{e(A', B')}{|A'||B'|} \\ &\leq \frac{e(A, B)}{|A'||B'|} \\ &\leq \frac{e(A, B)}{(1-\delta)^2|A||B|} \\ &= \frac{d(A, B)}{(1-\delta)^2} \end{aligned}$$

and so

$$d(A', B') - d(A, B) \leq d(A, B)\left(\frac{1}{(1-\delta)^2} - 1\right)$$

$$\begin{aligned} &\le \frac{1}{(1-\delta)^2} - 1 \\ &\le \frac{2\delta}{(1-\delta)^2}. \end{aligned}$$

Therefore,

$$|d(A,B) - d(A',B')| \le \frac{2\delta}{(1-\delta)^2}$$

and

$$\begin{aligned} |d(A,B)^2 - d(A',B')^2| &= |d(A,B) + d(A',B')||d(A,B) - d(A',B')| \\ &\le 2|d(A,B) - d(A',B')| \\ &\le \frac{4\delta}{(1-\delta)^2}. \end{aligned}$$

If $0 < \delta \le 1/2$, then $1/(1-\delta)^2 \le 4$. This completes the proof.

Lemma 9.3 (Schwarz's inequality) *Let $x_1, \ldots, x_n$ be real numbers. Then*

$$\sum_{i=1}^{n} x_i^2 \ge \frac{1}{n}\left(\sum_{i=1}^{n} x_i\right)^2. \tag{9.1}$$

For $m = 1, \ldots, n-1$, let

$$\Delta = \frac{1}{m}\left(\sum_{i=1}^{m} x_i\right) - \frac{1}{n}\left(\sum_{i=1}^{n} x_i\right).$$

Then

$$\sum_{i=1}^{n} x_i^2 \ge \frac{1}{n}\left(\sum_{i=1}^{n} x_i\right)^2 + \frac{mn\Delta^2}{n-m}. \tag{9.2}$$

Proof. Let $S_1(n) = \sum_{i=1}^{n} x_i$ and $S_2(n) = \sum_{i=1}^{n} x_i^2$. Since

$$\begin{aligned} 0 &\le \sum_{i=1}^{n}\left(x_i - \frac{S_1(n)}{n}\right)^2 \\ &= \sum_{i=1}^{n}\left(x_i^2 - \frac{2x_iS_1(n)}{n} + \frac{S_1(n)^2}{n^2}\right) \\ &= S_2(n) - \frac{2S_1(n)^2}{n} + \frac{nS_1^2}{n^2} \\ &= S_2(n) - \frac{S_1(n)^2}{n}, \end{aligned}$$

it follows that

$$S_2(n) \ge \frac{S_1(n)^2}{n},$$

which is inequality (9.1). This implies that

$$\begin{aligned} S_2(n) - S_2(m) &= \sum_{i=m+1}^{n} x_i^2 \\ &\geq \frac{1}{n-m}\left(\sum_{i=m+1}^{n} x_i\right)^2 \\ &= \frac{(S_1(n) - S_1(m))^2}{n-m}, \end{aligned}$$

and so

$$\begin{aligned} S_2(n) &= S_2(m) + (S_2(n) - S_2(m)) \\ &\geq \frac{S_1(m)^2}{m} + \frac{(S_1(n) - S_1(m))^2}{n-m} \\ &= \frac{S_1(n)^2}{n-m} - \frac{2S_1(n)S_1(m)}{n-m} + \frac{nS_1(m)^2}{m(n-m)} \\ &= \frac{S_1(n)^2}{n} + \frac{mS_1(n)^2}{n(n-m)} - \frac{2S_1(n)S_1(m)}{n-m} \\ &\quad + \frac{nS_1(m)^2}{m(n-m)} \\ &= \frac{S_1(n)^2}{n} + \frac{mn}{n-m}\left(\frac{S_1(n)}{n} - \frac{S_1(m)}{m}\right)^2 \\ &= \frac{S_1(n)^2}{n} + \frac{mn\Delta^2}{n-m}. \end{aligned}$$

This proves inequality (9.2).

Lemma 9.4 *Let $G = (V, E)$ be a graph, and let*

$$A = \bigcup_{i=1}^{q} A_i \subseteq V$$

and

$$B = \bigcup_{j=1}^{r} B_j \subseteq V,$$

where

$$|A_i| = a \geq 1$$

for $i = 1, \ldots, q$,

$$|B_j| = b \geq 1$$

for $j = 1, \ldots, r$, and

$$A_i \cap B_j = \emptyset$$

for all i and j. Then

$$\frac{1}{qr}\sum_{i=1}^{q}\sum_{j=1}^{r} d(A_i, B_j)^2 \geq d(A, B)^2. \tag{9.3}$$

Let $0 < \theta < 1$, *and let* q' *and* r' *be integers such that* $0 < \theta q \leq q' < q$ *and* $0 < \theta r \leq r' < r$. *Let* $A' = \cup_{i=1}^{q'} A_i$ *and* $B' = \cup_{j=1}^{r'} B_j$. *Then*

$$\frac{1}{qr}\sum_{i=1}^{q}\sum_{j=1}^{r} d(A_i, B_j)^2 \geq d(A, B)^2 + \theta^2(d(A, B) - d(A', B'))^2. \tag{9.4}$$

Proof. Since $|A| = qa$ and $|B| = rb$, it follows that

$$\begin{aligned}
\frac{1}{qr}\sum_{i=1}^{q}\sum_{j=1}^{r} d(A_i, B_j) &= \frac{1}{qr}\sum_{i=1}^{q}\sum_{j=1}^{r} \frac{e(A_i, B_j)}{|A_i||B_j|} \\
&= \frac{1}{qarb}\left(\sum_{i=1}^{q}\sum_{j=1}^{r} e(A_i, B_j)\right) \\
&= \frac{e(A, B)}{|A||B|} \\
&= d(A, B).
\end{aligned}$$

Similarly,

$$\frac{1}{q'r'}\sum_{i=1}^{q'}\sum_{j=1}^{r'} d(A_i, B_j) = d(A', B').$$

Then

$$\begin{aligned}
\Delta &= \frac{1}{qr}\sum_{i=1}^{q}\sum_{j=1}^{r} d(A_i, B_j) - \frac{1}{q'r'}\sum_{i=1}^{q'}\sum_{j=1}^{r'} d(A_i, B_j) \\
&= d(A, B) - d(A', B').
\end{aligned}$$

It follows from (9.1) of Lemma 9.3 that

$$\begin{aligned}
\frac{1}{qr}\sum_{i=1}^{q}\sum_{j=1}^{r} d(A_i, B_j)^2 &\geq \left(\frac{1}{qr}\right)^2\left(\sum_{i=1}^{q}\sum_{j=1}^{r} d(A_i, B_j)\right)^2 \\
&= d(A, B)^2.
\end{aligned}$$

This proves the first inequality.

The conditions $q' \geq \theta q$ and $r' \geq \theta r$ imply that

$$\frac{q'r'}{qr - q'r'} \geq \frac{qr\theta^2}{qr - q'r'} \geq \theta^2.$$

To obtain the second inequality, we apply (9.2) of Lemma 9.3 with $n = qr$ and $m = q'r'$. This gives

$$\begin{aligned}\frac{1}{qr}\sum_{i=1}^{q}\sum_{j=1}^{r} d(A_i, B_j)^2 &\geq \left(\frac{1}{qr}\right)^2\left(\sum_{i=1}^{q}\sum_{j=1}^{r} d(A_i, B_j)\right)^2 \\ &\quad + \frac{q'r'\Delta^2}{qr - q'r'} \\ &\geq d(A, B)^2 + \theta^2\Delta^2 \\ &= d(A, B)^2 + \theta^2(d(A, B) - d(A', B'))^2.\end{aligned}$$

This completes the proof.

Let $G = (V, E)$ be a graph, and let $\mathcal{P}$ be a partition of the vertex set V into $m + 1$ sets $C_0, C_1, \ldots, C_m$. The partition $\mathcal{P}$ will be called *equitable* if $|C_s| = |C_t|$ for $1 \leq s < t \leq m$. The set C_0 is called the *exceptional set* of the partition. The *partition density* of the equitable partition $\mathcal{P}$ is defined by

$$d(\mathcal{P}) = \frac{1}{m^2}\sum_{1\leq s<t\leq m} d(C_s, C_t)^2.$$

Since there are $m(m-1)/2$ summands and each summand satisfies $0 \leq d(C_s, C_t) \leq 1$, it follows that

$$0 \leq d(\mathcal{P}) < \frac{1}{2}.$$

Let A and B be disjoint nonempty subsets of the vertex set V, and let $\varepsilon > 0$. The pair (A, B) will be called *ε-regular* if the conditions

$$X \subseteq A, |X| \geq \varepsilon|A|$$

and

$$Y \subseteq B, |Y| \geq \varepsilon|B|$$

imply that

$$|d(A, B) - d(X, Y)| < \varepsilon.$$

An equitable partition of V into $m + 1$ pairwise disjoint sets $C_0, C_1, \ldots, C_m$ will be called *ε-regular* if

$$|C_0| \leq \varepsilon|V|$$

and if the pair (C_s, C_t) is ε-regular for all but at most εm^2 pairs (C_s, C_t) with $1 \leq s < t \leq m$.

The following result is the heart of the proof of the regularity lemma.

Lemma 9.5 *Let $0 < \varepsilon < 1$, and let the integer m satisfy*

$$4^m \geq 2^{10}\varepsilon^{-5}.$$

Let $G = (V, E)$ be a graph with k vertices, and let $\mathcal{P}$ be an equitable partition of V into $m + 1$ classes $C_0, C_1, \ldots, C_m$ such that

$$|C_s| \geq 4^{2m}$$

for $s = 1, \ldots, m$. If the number of ε-irregular pairs (C_s, C_t) with $1 \leq s < t \leq m$ is greater than εm^2, then there is an equitable partition $\mathcal{P}'$ of V into $m4^m + 1$ classes such that the exceptional class has cardinality less than

$$|C_0| + \frac{k}{4^m},$$

and the partition density of $\mathcal{P}$ satisfies

$$d(\mathcal{P}') \geq d(\mathcal{P}) + \frac{\varepsilon^5}{32}.$$

Proof. Let $q = 4^m$, and let

$$e = \left[\frac{|C_s|}{q}\right]$$

for $s = 1, \ldots, m$. Then

$$q^2 = 4^{2m} \leq |C_s| = eq + r < (e+1)q,$$

where

$$0 \leq r < q \leq e.$$

It follows that

$$q = \left[\frac{|C_s|}{e}\right].$$

Let $1 \leq s < t \leq m$. If the pair (C_s, C_t) is ε-irregular, then we choose sets $X_s(t) \subseteq C_s$ and $X_t(s) \subseteq C_t$ such that

$$|X_s(t)| \geq \varepsilon |C_s|, \tag{9.5}$$

$$|X_t(s)| \geq \varepsilon |C_t|, \tag{9.6}$$

and

$$|d(C_s, C_t) - d(X_s(t), X_t(s))| \geq \varepsilon. \tag{9.7}$$

If the pair (C_s, C_t) is ε-regular, let $X_s(t) = X_t(s) = \emptyset$. Thus, for each $s = 1, \ldots, m$, we have constructed $m - 1$ sets

$$X_s(1), \ldots, X_s(s-1), X_s(s+1), \ldots, X_s(m)$$

contained in C_s. These sets determine a partition of C_s into 2^{m-1} pairwise disjoint sets (some of which may be empty) in the following way. Let $\Lambda = \{0, 1\}^{m-1}$ be the set of all $(m-1)$-tuples of 0s and 1s. Then $|\Lambda| = 2^{m-1}$. For $\lambda = (\lambda_1, \ldots, \lambda_{s-1}, \lambda_{s+1}, \ldots, \lambda_m) \in \Lambda$, let $Y_s(\lambda)$ consist of all $v \in C_s$ such that $v \in X_s(j)$ if $\lambda_j = 1$ and $v \notin X_s(j)$ if $\lambda_j = 0$.

We shall use the 2^{m-1} sets $Y_s(\lambda)$ to construct an equitable partition $\mathcal{P}'$ that will consist of $m4^m$ sets of cardinality e and an exceptional set of cardinality less than $|C_0| + k/4^m$.

In each set $Y_s(\lambda)$, we choose

$$q_\lambda = \left[\frac{|Y_s(\lambda)|}{e}\right]$$

pairwise disjoint sets, each of cardinality exactly e. Let $Y_s'(\lambda)$ be the union of these q_λ sets. Then

$$|Y_s(\lambda) \setminus Y_s'(\lambda)| < e.$$

Since the set

$$C_s \setminus \cup_{\lambda\in\Lambda} Y_s'(\lambda)$$

has cardinality exactly

$$|C_s| - \sum_{\lambda\in\Lambda} |Y_s'(\lambda)| = (eq + r) - e\sum_{\lambda\in\Lambda} q_\lambda = e\left(q - \sum_{\lambda\in\Lambda} q_\lambda\right) + r,$$

we can choose an additional

$$q - \sum_{\lambda\in\Lambda} q_\lambda$$

pairwise disjoint sets of cardinality e in C_s. This construction produces $q = 4^m$ pairwise disjoint sets, which we shall denote by $C_s'(i)$ for $i = 1, \ldots, q$. Let

$$C_s' = \cup_{i=1}^{q} C_s'(i) \subseteq C_s.$$

Then

$$C_0' = V \setminus \cup_{s=1}^{m} C_s' \supseteq V \setminus \cup_{s=1}^{m} C_s = C_0$$

and

$$|C_s \setminus C_s'| = r < e \leq \frac{|C_s|}{q};$$

hence

$$\begin{aligned} |C_0'| &= |C_0| + \sum_{s=1}^{m} |C_s \setminus C_s'| \\ &< |C_0| + \frac{1}{q}\sum_{s=1}^{m} |C_s| \\ &\leq |C_0| + \frac{|V \setminus C_0|}{q} \\ &\leq |C_0| + \frac{k}{4^m}. \end{aligned}$$

Let $\mathcal{P}'$ be the partition of V that consists of the $mq + 1 = m4^m + 1$ sets $C'_s(i)$ for $1 \leq s \leq m$ and $1 \leq i \leq q$, together with the exceptional set C'_0. It remains to prove the partition density inequality

$$d(\mathcal{P}') \geq d(\mathcal{P}) + \frac{\varepsilon^5}{32}.$$

Since

$$|C_s \setminus C'_s| = r < e \leq \frac{|C_s|}{q}$$

for $s = 1, \ldots, m$, it follows that

$$|C'_s| > \left(1 - \frac{1}{q}\right)|C_s|,$$

where $0 < 1/q < 1/2$. By Lemma 9.2,

$$|d(C_s, C_t) - d(C'_s, C'_t)| \leq \frac{8}{q} \leq \frac{\varepsilon^5}{2^7} < \frac{\varepsilon}{4}$$

and

$$|d(C_s, C_t)^2 - d(C'_s, C'_t)^2| \leq \frac{16}{q} \leq \frac{\varepsilon^5}{64}$$

for $1 \leq s < t \leq m$. Applying inequality (9.3) of Lemma 9.4 with $q = r$, $A_i = C'_s(i)$, and $B_j = C'_t(j)$, we obtain

$$\begin{aligned} \frac{1}{q^2}\sum_{i=1}^{q}\sum_{j=1}^{q} d(C'_s(i), C'_t(j))^2 &\geq d(C'_s, C'_t)^2 \\ &> d(C_s, C_t)^2 - \frac{\varepsilon^5}{64} \end{aligned}$$

for all pairs (C_s, C_t) with $1 \leq s < t \leq m$.

Now let (C_s, C_t) be an ε-irregular pair, and let $X_s = X_s(t) \subseteq C_s$ and $X_t = X_t(s) \subseteq C_t$ be the sets that were chosen to satisfy conditions (9.5)–(9.7). In particular, $|X_s| \geq \varepsilon|C_s|$. Recall that the set C_s was partitioned into 2^{m-1} sets $Y_s(\lambda)$, where $\lambda = (\lambda_1, \ldots, \lambda_{s-1}, \lambda_{s+1}, \ldots, \lambda_m) \in \Lambda$. The set $X_s = X_s(t)$ is the union of the 2^{m-2} sets $Y_s(\lambda)$ with $\lambda_t = 1$. Each of these sets $Y_s(\lambda)$ contains $[|Y_s(\lambda)|/e]$ pairwise disjoint sets belonging to the partition $\mathcal{P}'$. Let $C'_s(1), \ldots, C'_s(q')$ denote the sets in the partition $\mathcal{P}'$ that are contained in $Y_s(\lambda)$ for some $Y_s(\lambda) \subseteq X_s$, and let

$$X'_s = \bigcup_{i=1}^{q'} C'_s(i).$$

Let $\sum$ denote the sum over all $\lambda \in \Lambda$ with $\lambda_t = 1$. Then

$$|X'_s| = q'e = \sum \left[\frac{|Y_s(\lambda)|}{e}\right] e.$$

Since

$$\begin{aligned}
|X_s \setminus X'_s| &= \sum |Y_s(\lambda)| - |X'_s| \\
&= \sum |Y_s(\lambda)| - \sum \left[\frac{|Y_s(\lambda)|}{e}\right] e \\
&= \sum \left(|Y_s(\lambda)| - \left[\frac{|Y_s(\lambda)|}{e}\right] e \right) \\
&< \sum e \\
&= 2^{m-2} e \\
&\le \frac{2^{m-2}|C_s|}{q} \\
&= 2^{-m-2}|C_s| \\
&\le 2^{-m-2}\varepsilon^{-1}|X_s| \\
&= q^{-1/2}2^{-2}\varepsilon^{-1}|X_s| \\
&\le (\varepsilon^5 2^{-10})^{1/2} 2^{-2}\varepsilon^{-1}|X_s| \\
&< 2^{-7}\varepsilon|X_s|,
\end{aligned}$$

it follows that

$$|X'_s| > (1 - 2^{-7}\varepsilon)|X_s|$$

and

$$\begin{aligned}
q' &= \frac{|X'_s|}{e} \\
&> (1 - 2^{-7}\varepsilon)\frac{|X_s|}{e} \\
&\ge (1 - 2^{-7}\varepsilon)\varepsilon\frac{|C_s|}{e} \\
&> (1 - 2^{-7})\varepsilon q \\
&> \frac{\varepsilon q}{2}.
\end{aligned}$$

Similarly, let X'_t be the union of the r' sets $C'_t(1), \ldots, C'_t(r')$ of the partition $\mathcal{P}'$ that are contained in X_t. Then

$$|X'_t| > (1 - 2^{-7}\varepsilon)|X_t|$$

and

$$r' \ge \frac{\varepsilon q}{2}.$$

Therefore,

$$|d(X_s, X_t) - d(X'_s, X'_t)| < 8\varepsilon 2^{-7} = \frac{\varepsilon}{16} < \frac{\varepsilon}{4}.$$

Since the pair (C_s, C_t) is not ε-regular, we obtain

$$\begin{aligned}
\varepsilon &\leq |d(C_s, C_t) - d(X_s, X_t)| \\
&\leq |d(C_s, C_t) - d(C_s', C_t')| + |d(C_s', C_t') - d(X_s', X_t')| \\
&\quad +|d(X_s', X_t') - d(X_s, X_t)| \\
&< \frac{\varepsilon}{4} + |d(C_s', C_t') - d(X_s', X_t')| + \frac{\varepsilon}{4},
\end{aligned}$$

and so

$$|d(C_s', C_t') - d(X_s', X_t')| > \frac{\varepsilon}{2}.$$

Applying inequality (9.4) of Lemma 9.4 with $r = q$ and $\theta = \varepsilon/2$, we see that if (C_s, C_t) is an ε-irregular pair, then

$$\begin{aligned}
q^{-2}\sum_{i,j=1}^{q} d(C_s'(i), C_t'(j))^2 &\geq d(C_s', C_t')^2 + (\varepsilon/2)^2(d(C_s', C_t') - d(X_s', X_t'))^2 \\
&> d(C_s', C_t')^2 + \varepsilon^4/16 \\
&> d(C_s, C_t)^2 - \varepsilon^5/64 + \varepsilon^4/16.
\end{aligned}$$

For $1 \leq s < t \leq m$, let $\chi(s,t) = 1$ if the pair (C_s, C_t) is ε-irregular, and let $\chi(s,t) = 0$ if the pair (C_s, C_t) is ε-regular. Then

$$\sum_{1\leq s<t\leq m} \chi(s,t) \geq \varepsilon m^2$$

and

$$q^{-2}\sum_{i,j=1}^{q} d(C_s'(i), d(C_t'(j))^2 > d(C_s, C_t)^2 - \varepsilon^5/64 + \chi(s,t)\varepsilon^4/16.$$

We can now estimate the partition density $d(\mathcal{P}')$:

$$\begin{aligned}
d(\mathcal{P}') &= (mq)^{-2}\left(\sum_{1\leq s<t\leq m}\sum_{i,j=1}^{q} d(C_s'(i), C_t'(j))^2 + \sum_{s=1}^{m}\sum_{1\leq i<j\leq q} d(C_s'(i), C_s(j))^2\right) \\
&\geq m^{-2}\sum_{1\leq s<t\leq m} q^{-2}\sum_{i,j=1}^{q} d(C_s'(i), C_t'(j))^2 \\
&\geq m^{-2}\sum_{1\leq s<t\leq m}\left(d(C_s, C_t)^2 - \varepsilon^5/64 + \chi(s,t)\varepsilon^4/16\right) \\
&\geq m^{-2}\left(\sum_{1\leq s<t\leq m} d(C_s, C_t)^2 - \binom{m}{2}\varepsilon^5/64 + (\varepsilon m^2)(\varepsilon^4/16)\right) \\
&\geq m^{-2}\sum_{1\leq s<t\leq m}(d(C_s, C_t)^2 - m^{-2}\binom{m}{2}\varepsilon^5/64) + m^{-2}m^2\varepsilon^5/16 \\
&> d(\mathcal{P}) + \varepsilon^5/32.
\end{aligned}$$

This completes the proof of the Lemma.

Theorem 9.2 (the regularity lemma) *Let $0 < \varepsilon < 1$ and $m' \geq 1$. There exist numbers $K = K(\varepsilon, m')$ and $M = M(\varepsilon, m')$ such that, if $G = (V, E)$ is a graph with $|V| \geq K$ vertices, then there exists an ε-regular partition of V into $m+1$ sets, where $m' \leq m \leq M$.*

Proof. Let $t' = [16\varepsilon^{-5}]$. We construct a sequence $m_0, m_1, m_2, \ldots$ of integers as follows. Let m_0 be an integer such that

$$m_0 \geq m'$$

and

$$4^{m_0} \geq \max\{2^{10}\varepsilon^{-5}, 2^{t'+2}\varepsilon^{-1}\},$$

and let

$$m_{t+1} = m_t 4^{m_t}$$

for $t = 0, 1, 2, 3, \ldots$. We define numbers M and K by

$$M = M(\varepsilon, m') = m_{t'}$$

and

$$K = K(\varepsilon, m') = \max\{2m_0/\varepsilon, m_{t'}16^{m_{t'}}/(1-\varepsilon)\}.$$

Let $G = (V, E)$ be a graph with $|V| = k \geq K$. Let T be the set of nonnegative integers t with the property that there exists an equitable partition $\mathcal{P}$ of V into $m_t + 1$ sets such that

$$d(\mathcal{P}) \geq t\varepsilon^5/32 \tag{9.8}$$

and the exceptional set C_0 of the partition has cardinality

$$|C_0| < \varepsilon k(1 - 2^{-t-1}). \tag{9.9}$$

Consider any partition $\mathcal{P}_0$ of V that consists of m_0 pairwise disjoint sets of size $[k/m_0]$, together with an exceptional set C_0 of cardinality less than m_0. Then

$$|C_0| < m_0 = \varepsilon K/2 \leq \varepsilon k/2 = \varepsilon k(1 - 1/2).$$

Since $d(\mathcal{P}_0) \geq 0$, it follows that $t = 0$ satisfies conditions (9.8) and (9.9). Thus, $0 \in T$.

Since $d(\mathcal{P}) < 1/2$, if t satisfies condition (9.8), then

$$t \leq t' = [16\varepsilon^{-5}].$$

It follows that condition (9.8) is satisfied for only finitely many positive integers. Thus, the set T is finite and there exists a greatest integer $t \leq t'$ such that conditions (9.8) and (9.9) are satisfied for some partition $\mathcal{P}$ of V into $m_t + 1$ sets. Let $\mathcal{P} = \{C_0, C_1, \ldots, C_{m_t}\}$. Then

$$|C_s| = \frac{|V| - |C_0|}{m_t}$$

$$\begin{aligned} &> \frac{k(1-\varepsilon)}{m_t} \\ &\geq \frac{K(1-\varepsilon)}{m_{t'}} \\ &\geq 16^{m_{t'}} \\ &\geq 16^{m_t} \end{aligned}$$

for $s = 1, \ldots, m_t$ and

$$4^{m_t} \geq 4^{m_0} > 2^{10}\varepsilon^{-5}.$$

Since the exceptional set of the partition $\mathcal{P}$ satisfies $|C_0| < \varepsilon k$, it follows that, if the partition $\mathcal{P}$ is not ε-regular, then the number of ε-irregular pairs (C_s, C_t) is greater that εm^2. Lemma 9.5 implies that there exists an equitable partition $\mathcal{P}'$ of V into $m_t 4^{m_t} + 1 = m_{t+1} + 1$ sets such that

$$d(\mathcal{P}') \geq d(\mathcal{P}) + \varepsilon^5/32 \geq (t+1)\varepsilon^5/32$$

and the exceptional set C_0' of $\mathcal{P}'$ satisfies

$$\begin{aligned} |C_0'| &< |C_0| + k/4^{m_t} \\ &< \varepsilon k(1 - 2^{-t-1}) + k/4^{m_t} \\ &\leq \varepsilon k(1 - 2^{-t-1} + \varepsilon^{-1}4^{-m_t}) \\ &\leq \varepsilon k(1 - 2^{-t-1} + \varepsilon^{-1}4^{-m_0}) \\ &\leq \varepsilon k(1 - 2^{-t-1} + 2^{-t'-2}) \\ &\leq \varepsilon k(1 - 2^{-t-1} + 2^{-t-2}) \\ &= \varepsilon k(1 - 2^{-t-2}). \end{aligned}$$

This implies that $t+1$ satisfies conditions (9.8) and (9.9), which contradicts the maximality of t. Therefore, $\mathcal{P}$ is ε-regular. This completes the proof.

9.4 The Balog–Szemerédi theorem

Theorem 9.3 *Let δ, σ, λ, and μ be positive real numbers. There exist positive numbers c_1', c_2' and K that depend only on δ, σ, λ, and μ with the following property. Let $k \geq K$, and let A and B be finite subsets of an abelian group such that*

$$\lambda k \leq |A| \leq \mu k$$

and

$$\lambda k \leq |B| \leq \mu k.$$

Let W be a subset of $A \times B$ such that

$$|W| \geq \delta k^2$$

and

$$S = S(W) = \{a + b : (a, b) \in W\}$$

satisfies

$$|S| \leq \sigma k. \tag{9.10}$$

Then there exists a set $A' \subseteq A$ such that

$$|A'| \geq c_1' k$$

and

$$|2A'| \leq c_2' k.$$

In particular,

$$|2A'| \leq c|A'|,$$

where $c = c_2'/c_1'$.

Proof. Let

$$0 < \varepsilon < \min\left\{1, \frac{\delta^2}{16\mu\sigma(12\mu^2 + 2\mu\sigma)}, \frac{\delta^2}{64\mu^3\sigma}\right\} \tag{9.11}$$

and

$$m' = \left[\frac{1}{\varepsilon}\right] + 1 > \frac{1}{\varepsilon}.$$

Let $M(\varepsilon, m')$ and $K(\varepsilon, m')$ be the numbers constructed in the regularity lemma. Let $M = M(\varepsilon, m')$ and $K = K(\varepsilon, m')/\lambda$. Then M and K depend only on δ, σ, λ, and μ.

For $s \in S$, let $r(s)$ denote the number of representations of s in the form $a + b$, where $(a, b) \in W$, that is,

$$r(s) = |\{(a, b) \in W : a + b = s\}|.$$

For each $a \in A$ there is at most one $b \in B$ such that $a + b = s$, and so

$$1 \leq r(s) \leq |A| \leq \mu k$$

for all $s \in S$. Let

$$c_3 = \frac{\delta}{2\sigma} > 0$$

and

$$S' = \{s \in S : r(s) > c_3 k\}.$$

Then

$$\begin{aligned}
\delta k^2 &\leq |W| \\
&= \sum_{s \in S} r(s) \\
&= \sum_{s \in S \setminus S'} r(s) + \sum_{s \in S'} r(s) \\
&\leq c_3 k |S \setminus S'| + \mu k |S'| \\
&\leq c_3 k |S| + \mu k |S'| \\
&\leq \sigma c_3 k^2 + \mu k |S'|,
\end{aligned}$$

and so

$$|S'| \geq \frac{(\delta - \sigma c_3)k}{\mu} = \frac{\delta k}{2\mu}.$$

Let

$$W' = \{(a, b) \in W : a + b \in S'\}.$$

Then

$$|W'| = \sum_{s \in S'} r(s) > c_3 k |S'| \geq \frac{\delta c_3 k^2}{2\mu} = c_4 k^2,$$

where

$$c_4 = \frac{\delta c_3}{2\mu} = \frac{\delta^2}{4\mu\sigma}.$$

Let $G = (V, E)$ be the graph with vertex set

$$V = A \cup B$$

and edge set

$$E = \{\{a, b\} : (a, b) \in W'\}.$$

The sets A and B are not necessarily disjoint, and

$$\lambda k \leq |A| \leq |V| \leq |A| + |B| \leq 2\mu k.$$

If $e = \{a, b\} \in E$, then the set W' contains at least one and possibly both of the ordered pairs (a, b) and (b, a), and so

$$\frac{c_4 k^2}{2} \leq \frac{|W'|}{2} \leq |E| \leq |W'|.$$

Since $|V| \geq \lambda k \geq \lambda K = K(\varepsilon, m')$, the regularity lemma implies that there exists a partition of V into $m + 1$ pairwise disjoint sets $C_0, C_1, \ldots, C_m$, where

$$m' \leq m \leq M,$$

$$|C_0| \leq \varepsilon |V| \leq \varepsilon 2\mu k,$$

and

$$\begin{aligned} \frac{(1-\varepsilon)\lambda k}{m} &\leq \frac{(1-\varepsilon)|V|}{m} \\ &\leq |C_i| \\ &= \frac{|V| - |C_0|}{m} \\ &\leq \frac{|V|}{m} \\ &\leq \frac{2\mu k}{m} \end{aligned}$$

for $i = 1, \ldots, m$. Moreover, there are at most εm^2 pairs (C_i, C_j) with $1 \leq i < j \leq m$ that fail to be ε-regular.

We shall construct a subset E' of the edge set E by deleting the following four classes of edges from E:

1. Delete all edges with at least one endpoint in C_0. Since the degree of a vertex is at most $|V|$, it follows that the number of edges removed from E is at most
$$|C_0||V| \leq \varepsilon|V|^2.$$

2. For $i = 1, \ldots, m$, delete all edges with both endpoints in the same set C_i. The number of edges removed from E is at most
$$\sum_{i=1}^{m} |C_i|^2 \leq m\left(\frac{|V|}{m}\right)^2 = \frac{|V|^2}{m} \leq \frac{|V|^2}{m'} < \varepsilon|V|^2.$$

3. Let $1 \leq i < j \leq m$. If (C_i, C_j) is not an ε-regular pair, delete all edges between C_i and C_j. Since $e(C_1, C_j) \leq |C_i||C_j| \leq |V|^2/m^2$ and there are at most εm^2 irregular pairs, the number of edges removed from E in this way does not exceed
$$\frac{\varepsilon m^2|V|^2}{m^2} = \varepsilon|V|^2.$$

4. We use the set S' to "color" the edges of E by assigning to the edge $\{a, b\} \in E$ the color $s = a + b \in S'$. For each $i = 1, \ldots, m$ and for each color $s \in S'$, we consider the number of edges of color s with at least one endpoint in the set C_i. If this number is less than $\varepsilon|C_i|$, then we delete all these edges. The total number of edges of all colors deleted from the sets $C_1, \ldots, C_m$ is at most
$$|S'|\sum_{i=1}^{m} \varepsilon|C_i| \leq \frac{m|S|\varepsilon|V|}{m} = \varepsilon|S||V|.$$

Let E' be the set of edges remaining after the four deletions just described. Then
$$|E \setminus E'| < 3\varepsilon|V|^2 + \varepsilon|S||V| \leq (12\mu^2 + 2\mu\sigma)\varepsilon k^2,$$
and so
$$\begin{aligned}
|E'| &= |E| - |E \setminus E'| \\
&> \frac{c_4k^2}{2} - (12\mu^2 + 2\mu\sigma)\varepsilon k^2 \\
&= \left(\frac{c_4}{2} - (12\mu^2 + 2\mu\sigma)\varepsilon\right)k^2 \\
&= \left(\frac{\delta^2}{8\mu\sigma} - (12\mu^2 + 2\mu\sigma)\varepsilon\right)k^2 \\
&> \frac{\delta^2k^2}{16\mu\sigma} \\
&= c_5k^2,
\end{aligned}$$
since
$$\varepsilon < \frac{\delta^2}{16\mu\sigma(12\mu^2 + 2\mu\sigma)}$$

by (9.11), and

$$c_5 = \frac{\delta^2}{16\mu\sigma} > 0.$$

All of the edges in E' are edges between ε-regular pairs (C_s, C_t), where $1 \leq s < t \leq m$. Let $e'(C_s, C_t)$ (resp. $e(C_s, C_t)$) denote the number of edges in E' (resp. in E) with one endpoint in C_s and the other endpoint in C_t. Since there are at most $\binom{m}{2}$ ε-regular pairs, there must exist some ε-regular pair, say, (C_1, C_2), such that the number of edges in E' between C_1 and C_2 is

$$e'(C_1, C_2) \geq \frac{|E'|}{\binom{m}{2}} \geq \frac{c_5 k^2}{\binom{m}{2}} > \frac{c_5 k^2}{m^2}.$$

Therefore, by (9.11),

$$\begin{aligned} d(C_1, C_2) &= \frac{e(C_1, C_2)}{|C_1||C_2|} \\ &\geq \frac{e'(C_1, C_2)}{|C_1||C_2|} \\ &\geq \left(\frac{c_5 k^2}{m^2}\right)\left(\frac{2\mu k}{m}\right)^{-2} \\ &= \frac{c_5}{4\mu^2} \\ &= \frac{\delta^2}{64\mu^3\sigma} \\ &> \varepsilon. \end{aligned}$$

Let S'' be the set of colors of edges in E' between C_1 and C_2:

$$S'' = \{a + b : \{a, b\} \in e'(C_1, C_2)\}.$$

Since $r(s) \leq \mu k$ for every color $s \in S$, it follows that

$$|S''| \geq \frac{e'(C_1, C_2)}{\mu k} \geq \frac{c_5 k}{\mu m^2} \geq \frac{c_5 k}{\mu M^2} = c_6 k.$$

We shall prove that the sumset $2S'' = S'' + S''$ is small. Let $s^* \in 2S''$. We fix some representation $s^* = s_1 + s_2 \in 2S''$, where $s_1 \in S''$ and $s_2 \in S''$. Since s_1 is the color of an edge that has "survived" step 4 of the deletion process just described, there are at least $\varepsilon|C_1|$ edges in E' of color s_1 with exactly one endpoint in the set C_1. Let X_1 be the set of these endpoints. Then

$$X_1 \subseteq C_1$$

and

$$|X_1| \geq \varepsilon|C_1|.$$

Similarly, there are at least $\varepsilon|C_2|$ edges in E' of color s_2 with exactly one endpoint in the set C_2. Let X_2 be the set of these endpoints. Then

$$X_2 \subseteq C_2$$

and

$$|X_2| \geq \varepsilon|C_2|.$$

Since the pair (C_1, C_2) is ε-regular, it follows that

$$d(X_1, X_2) \geq d(C_1, C_2) - \varepsilon \geq \left(\frac{\delta^2}{64\mu^3\sigma}\right) - \varepsilon = c_7 > 0.$$

Therefore,

$$\begin{aligned} e(X_1, X_2) &= d(X_1, X_2)|X_1||X_2| \\ &\geq c_7\varepsilon^2|C_1||C_2| \\ &\geq \frac{c_7\varepsilon^2(1-\varepsilon)^2\lambda^2k^2}{m^2} \\ &\geq \frac{c_7\varepsilon^2(1-\varepsilon)^2\lambda^2k^2}{M^2} \\ &= c_8k^2. \end{aligned}$$

Let Ω denote the set of all ordered triples of the form

$$(s, v_1, v_2),$$

where $s \in S$ and $v_1, v_2 \in V$. Then

$$|\Omega| = |S||V|^2 \leq 4\sigma\mu^2k^3.$$

Let $\{v_1, v_2\} \in E$ be an edge between the sets X_1 and X_2, where $v_1 \in X_1$ and $v_2 \in X_2$. Then there exist vertices $v_1' \in V$ and $v_2' \in V$ such that $\{v_1, v_1'\} \in E'$ with $v_1 + v_1' = s_1 \in S'$ and $\{v_2, v_2'\} \in E'$ with $v_2 + v_2' = s_2 \in S'$. Let $s = v_1 + v_2 \in S$. To the edge $\{v_1, v_2\}$ we associate the ordered triple

$$(s, v_1', v_2') \in \Omega.$$

Observe that

$$s + v_1' + v_2' = (v_1 + v_1') + (v_2 + v_2') = s_1 + s_2 = s^* \in 2S''.$$

Conversely, if $(s, v_1', v_2') \in \Omega$ is a triple constructed in this way from an edge $\{v_1, v_2\}$ between X_1 and X_2, then $v_1 = s_1 - v_1'$ and $v_2 = s_2 - v_2'$. Therefore, the number of distinct triples $(s, v_1', v_2') \in \Omega$ such that $s + v_1' + v_2' = s_1 + s_2 = s^*$ is at least $e(X_1, X_2) \geq c_8k^2$. Applying this construction to each element $s^* \in 2S''$, we obtain $c_8k^2|2S''|$ distinct triples in Ω. Since

$$c_8k^2|2S''| \leq |\Omega| \leq 4\sigma\mu^2k^3,$$

it follows that

$$|2S''| \leq \frac{4\sigma\mu^2 k}{c_8} = c_2' k.$$

For $b \in B$, let $R(b)$ denote the number of $s \in S''$ such that $s = a + b$ for some $a \in A$. Choose $b' \in B$ such that $R(b') = \max\{R(b) : b \in B\}$. Since $s \in S'' \subseteq S'$, we have $r(s) > c_3 k$ and

$$\begin{aligned}
c_3 c_6 k^2 &\leq c_3 k |S''| \\
&< \sum_{s \in S''} r(s) \\
&= |\{(a, b) \in W : a + b \in S''\}| \\
&\leq |\{(a, b) \in A \times B : a + b \in S''\}| \\
&= \sum_{b \in B} R(b) \\
&\leq |B| R(b') \\
&\leq \mu k R(b').
\end{aligned}$$

Therefore,

$$R(b') \geq \frac{c_3 c_6 k}{\mu} = c_1' k.$$

Let

$$A' = \{a \in A : a + b' \in S''\}.$$

Then

$$A' + \{b'\} \subseteq S'',$$

$$|A'| = R(b') \geq c_1' k,$$

and

$$|2A'| = |2(A' + \{b\})| \leq |2S''| \leq c_2' k.$$

This completes the proof of the theorem.

Theorem 9.4 *Let δ, σ, λ, and μ be positive real numbers. There exist positive numbers c_1'', c_2'', n, and K that depend only on δ, σ, λ, and μ with the following property. Let $k \geq K$, and let A and B be finite subsets of a torsion-free abelian group such that*

$$\lambda k \leq |A| \leq \mu k$$

and

$$\lambda k \leq |B| \leq \mu k.$$

Let W be a subset of $A \times B$ such that

$$|W| \geq \delta k^2$$

and

$$S(W) = \{a + b : (a, b) \in W\}$$

satisfies

$$|S(W)| \leq \sigma k.$$

Then there exists an n-dimensional arithmetic progression Q such that

$$|A \cap Q| \geq c_1''|A|$$

and

$$|A \cap Q| \geq c_2''|Q|.$$

Proof. This follows immediately from the preceding result and Freiman's theorem.

9.5 A conjecture of Erdős

Erdős conjectured that a set of integers that contains "many" three-term arithmetic progressions must contain a "long" arithmetic progression. In this section, we shall apply Theorems 9.1 and 9.3 to prove a quantitative version of this statement.

An arithmetic progression of length three is a set $\{a, b, c\}$ such that $b - a = c - b \neq 0$. Two such arithmetic progressions $\{a, b, c\}$ and $\{a', b', c'\}$ are *distinct* if $\{a, b, c\} \neq \{a', b', c'\}$. A set of k integers contains at most k^2 pairwise distinct arithmetic progressions of length three.

Theorem 9.5 *Let $\delta > 0$ and $t \geq 3$. There exists an integer $k_1(\delta, t)$ such that if A is a set of $k \geq k_1(\delta, t)$ integers that contains at least δk^2 distinct arithmetic progressions of length three, then A contains an arithmetic progression of length t.*

Proof. Let A be a set of k integers, and let

$$\{\{a_i, b_i, c_i\} : i \in I\}$$

be a family of

$$|I| \geq \delta k^2$$

three term arithmetic progressions in A such that

$$b_i - a_i = c_i - b_i > 0.$$

Then

$$a_i + c_i = 2b_i$$

for all $i \in I$. Let

$$W = \{(a_i, c_i) : i \in I\} \subseteq A \times A.$$

Then

$$|W| = I \geq \delta k^2$$

and

$$S(W) = \{2b_i : i \in I\} \subseteq \{2b : b \in A\} = 2 * A.$$

It follows that

$$|S(W)| \leq |2 * A| = k.$$

We apply Theorem 9.3 with $A = B$ and $\lambda = \mu = \sigma = 1$. If $k \geq K = K(\delta)$, then there exists a set $A' \subseteq A$ such that $|A'| \geq c_1' k$ and $|2A'| \leq c|A'|$, where c depends only on δ. By Theorem 9.1, if $c_1' k \geq k_0(c, t)$, then A' contains an arithmetic progression of length t. This completes the proof.

9.6 The proper conjecture

The n-dimensional arithmetic progression

$$\begin{aligned} Q &= Q(a; q_1, \ldots, q_n; l_1, \ldots, l_n) \\ &= \{a + x_1 q_1 + \cdots + x_n q_n \mid 0 \leq x_i < l_i \text{ for } i = 1, \ldots, n\} \end{aligned}$$

is called *proper* if $|Q| = l_1 \cdots l_n = l(Q)$. This means that every element of Q has a *unique* representation. If Q is proper, then

$$hQ = \{a + x_1 q_1 + \cdots + x_n q_n \mid 0 \leq x_i \leq h(l_i - 1) \text{ for } i = 1, \ldots, n\}$$

and so $|hQ| < h^n l_1 \cdots l_n = h^n |Q|$. Let $\{\mathbf{e}_1, \ldots, \mathbf{e}_n\}$ be the standard basis for $\mathbf{R}^n$, let $l_1, \ldots, l_n$ be positive integers, and consider the parallelepiped

$$P = \{x_1 \mathbf{e}_1 + \cdots + x_n \mathbf{e}_n : 0 \leq x_i < l_i \text{ for } i = 1, \ldots, n\}.$$

For every $h \geq 2$ there is a Freiman isomorphism of order h between the set of lattice points $\mathbf{Z}^n \cap P$ and a proper n-dimensional arithmetic progression.

Every 1-dimensional arithmetic progression is proper, but it is easy, for every $n \geq 2$, to construct examples of n-dimensional arithmetic progressions that fail to be proper.

By Freiman's theorem, every finite set A of integers with a small sumset $2A$ is a large subset of a multidimensional arithmetic progression, but it is not known if A must be a large subset of a *proper* multidimensional arithmetic progression. This can be called the "proper conjecture."

Conjecture 9.1 *Let c, c_1, and c_2 be positive real numbers. Let $k \geq 1$, and let A and B be finite subsets of a torsion-free abelian group such that*

$$c_1 k \leq |A|, |B| \leq c_2 k$$

and

$$|A + B| \leq ck.$$

Then A is a subset of a proper n-dimensional arithmetic progression of length at most lk, where n and l depend only on c, c_1, and c_2, and n and l are both "small."

9.7 Notes

Let $c \geq 2$ and $\varepsilon > 0$. Is it true that if A is a finite set of integers with $|A| = k$ sufficiently large and $|2A| \geq ck$, then A contains an arithmetic progression of length k^{ε}? This would significantly strengthen Theorem 9.1.

Freiman [54, 55] and Ruzsa [113] studied sets with small sumsets that contain three-term arithmetic progressions.

It is not known whether Theorem 9.3 remains true if inequality (9.10) is replaced with the weaker condition

$$|S| \leq \sigma k^{2-\delta}$$

for some $\delta > 0$. Nor is it known whether the theorem can be generalized to the sum of $h \geq 3$ finite subsets of the integers or of $h \geq 3$ finite subsets of an arbitrary abelian group.

The regularity lemma for ε-regular partitions of the vertices of a graph is due to Szemerédi [124]. He used a variant of this result in his proof [123] that every infinite set of integers of positive upper density contains arbitrarily long finite arithmetic progressions. Chung [20] and Frankl and Rödl [48] have generalized the regularity lemma to hypergraphs. The version of Freiman's theorem for large subsets of $A \times B$ was proved by Balog and Szemerédi [6] in the case that $A = B$ is a finite set of integers. The somewhat more general result given in this chapter requires only minor modifications of their proof.

Laczkovich and Ruzsa [77] have applied Freiman's theorem to combinatorial geometry. Let a, b, c, and d be four distinct points in the plane. Their cross ratio is

$$(a, b, c, d) = \frac{\frac{c-a}{c-b}}{\frac{d-a}{d-b}}.$$

Let P be a finite set of complex numbers. Let A be a set of n complex numbers, and let $S(P, A)$ denote the number of subsets $X \subseteq A$ such that $|X| = |P|$ and X is homothetic, or affinely equivalent, to P in the sense that

$$X = aP + B$$

for some $a, b \in \mathbf{C}$. Let

$$S(P, n) = \max\{S(P, A) : |A| = n\}.$$

Laczkovich and Ruzsa proved that

$$S(P, n) \geq cn^2$$

for some constant $c > 0$ and all $n \geq |P|$ if and only if the cross ratio of every four–element subset of P is algebraic. This generalizes a result of Elekes and Erdős [33].

9.8 Exercises

1. Let $\delta > 0$ and let n and t be integers such that $n \geq 1$ and $t \geq 3$. Let Q be an n-dimensional arithmetic progression in an abelian group. Prove that there exists an integer $m_1(\delta, n, t)$ such that if the length of Q satisfies $l(Q) \geq m_1(\delta, n, t)$ and if B is a subset of Q such that $|B| \geq \delta l(Q)$, then B contains an arithmetic progression of length t.

2. Let A and B be finite subsets of an abelian group such that $|A| = |B| = k$. Let $S = A + B$. Let $r(s)$ denote the number of ordered pairs $(a, b) \in A \times B$ such that $a + b = s$. Suppose that
$$\sum_{s \in S} r(s)^2 \geq \delta k^3.$$
   (a) Prove that there exist constants $d_1 > 0$ and $d_2 > 0$ such that, if $S' = \{s \in S : r(s) \geq d_1 k^{1/2}\}$, then $|S'| \geq d_2 k$.
   (b) Suppose that $|S| \leq \sigma k$. Prove that there exist constants $d_3 > 0$ and $d_4 > 0$ such that, if $S' = \{s \in S : r(s) \geq d_3 k\}$, then $|S'| \geq d_4 k$.

3. Let $h \geq 2$, and let $A_1, \ldots, A_h$ be finite subsets of an abelian group such that $|A_i| = k$ for $i = 1, 2, \ldots, h$. Let $S = A_1 + \cdots + A_h$. Let $r(s)$ denote the number of ordered pairs $(a_1, \ldots, a_h) \in A_1 \times \cdots \times A_h$ such that $a_1 + \cdots + a_h = s$. Prove the following:
   (a) $k \leq |S| \leq k^h$.
   (b) $r(s) \leq k^{h-1}$ for all $s \in S$.
   (c) $\sum_{s \in S} r(s) = k^h$.

4. Let $h \geq 2$, and let $A_1, \ldots, A_h$ be finite subsets of an abelian group such that $|A_i| = k$ for $i = 1, 2, \ldots, h$. Let $S = A_1 + \cdots + A_h$. Let $r(s)$ denote the number of ordered pairs $(a_1, \ldots, a_h) \in A_1 \times \cdots \times A_h$ such that $a_1 + \cdots + a_h = s$. Suppose that
$$\sum_{s \in S} r(s)^m \geq \delta k^{h+\alpha}$$
and
$$|S| \leq \sigma k^{h-\delta}.$$
Let
$$S' = \left\{ s \in S : r(s) \geq \left(\frac{\delta}{2\sigma} \right) k^{(\alpha+\delta)/m} \right\}.$$
Prove that $|S'| \geq (\delta/2) k^{\beta}$, where $\beta = h + \alpha - (h-1)m$.

5. Let $x_1, \ldots, x_n$ be real numbers, and let $p_1, \ldots, p_n$ be positive real numbers such that $p_1 + \cdots + p_n = 1$. Prove that
$$\sum_{i=1}^{n} p_i x_i^2 \geq \left(\sum_{i=1}^{n} p_i x_i \right)^2.$$

6. Let $x_1, \ldots, x_n$ be real numbers, and let $p_1, \ldots, p_n$ be positive real numbers such that $p_1 + \cdots + p_n = 1$. For $m = 1, \ldots, n-1$, let $P(m) = \sum_{i=1}^{m} p_i$ and

$$\Delta = \frac{1}{P(m)} \left(\sum_{i=1}^{m} p_i x_i \right) - \sum_{i=1}^{n} p_i x_i .$$

Prove that

$$\sum_{i=1}^{n} p_i x_i^2 \geq \left(\sum_{i=1}^{n} p_i x_i \right)^2 + \frac{P(m)\Delta^2}{1 - P(m)} .$$

7. Let $\delta > 0$. Let A be a set of k integers that contains at least δk^2 arithmetic progressions of length 3. Prove that A contains a subset A' such that $|A'| \geq c_1' k$ and $|2A| \leq c_2' k$, where c_1' and c_2' depend only on δ.

References

[1] N. Alon and M. Dubiner. Zero-sum sets of prescribed size. In *Combinatorics: Paul Erdős is Eighty*, Colloq. Math. Soc. János Bolyai, pages 33–50. North-Holland Publishing Co., Amsterdam, 1993.

[2] N. Alon, M. B. Nathanson, and I. Z. Ruzsa. Adding distinct congruence classes modulo a prime. *Am. Math. Monthly*, 102:250–255, 1995.

[3] N. Alon, M. B. Nathanson, and I. Z. Ruzsa. The polynomial method and restricted sums of congruence classes. *J. Number Theory*, 56:404–417, 1996.

[4] G. E. Andrews. *The Theory of Partitions*. Addison–Wesley, Reading, Mass., 1976.

[5] C. Bailey and R. B. Richter. Sum zero, mod n, size n subsets of integers. *Am. Math. Monthly*, 96:240–242, 1989.

[6] A. Balog and E. Szemerédi. A statistical theorem of set addition. *Combinatorica*, 14:263–268, 1994.

[7] R. P. Bambah, A. Woods, and H. Zassenhaus. Three proofs of Minkowski's second inequality in the geometry of numbers. *J. Austral. Math. Soc.*, 5:453–462, 1965.

[8] A. Bialostocki and M. Lotspeich. Some developments of the Erdős-Ginzburg-Ziv theorem. In G. Halász, L. Lovász, D. Miklós, and T. Szőnyi, editors, *Sets, Graphs, and Numbers*, volume 60 of *Colloq. Math. Soc. János Bolyai*, pages 97–117. North-Holland Publishing Co., Amsterdam, 1992.

[9] Y. Bilu. Structure of sets with small sumsets. Mathématiques Stochastiques, Univ. Bordeaux 2, Preprint 94–10, 1994.

[10] Y. Bilu. Addition of sets of integers of positive density. Preprint, 1996.

[11] N. N. Bogolyubov. Sur quelques propriétés arithmétiques des presque-périodes. *Ann. Chaire Math. Phys. Kiev*, 4:185–194, 1939.

[12] L. V. Brailovsky and G. A. Freiman. On a product of finite subsets in a torsion-free group. *J. Algebra*, 130:462–476, 1990.

[13] W. Brakemeier. Eine Anzahlformel von Zahlen modulo n. *Monat. Math.*, 85:277–282, 1978.

[14] J. W. S. Cassels. *An Introduction to the Geometry of Numbers*. Springer–Verlag, Berlin, 1959.

[15] J. W. S. Cassels. On the representations of integers as the sums of distinct summands taken from a fixed set. *Acta Sci. Math. (Szeged)*, 21:111–124, 1960.

[16] A. L. Cauchy. Recherches sur les nombres. *J. École polytech.*, 9:99–116, 1813.

[17] J. Cherly and J.-M. Deshouillers. Un théoréme d'addition dans $F_q[x]$. *J. Number Theory*, 34:128–131, 1990.

[18] I. Chowla. A theorem on the addition of residue classes: Application to the number $\Gamma(k)$ in Waring's problem. *Proc. Indian Acad. Sci., Section A*, 1:242–243, 1935.

[19] S. Chowla, H. B. Mann, and E. G. Straus. Some applications of the Cauchy-Davenport theorem. *Det Kongelige Norske Videnskabers Selskabs*, 32(13):74–80, 1959.

[20] F. R. K. Chung. Regularity lemmas for hypergraphs and quasi-randomness. *Random Structures and Algorithms*, 2:241–252, 1991.

[21] I. Danicic. An elementary proof of Minkowski's second inequality. *J. Austral. Math. Soc.*, 10:177–181, 1969.

[22] H. Davenport. On the addition of residue classes. *J. London Math. Soc.*, 10:30–32, 1935.

[23] H. Davenport. Minkowski's inequality for the minima associated with a convex body. *Q. J. Math.*, 10:119–121, 1939.

[24] H. Davenport. The geometry of numbers. *Math. Gazette*, 31:206–210, 1947.

[25] H. Davenport. A historical note. *J. London Math. Soc.*, 22:100–101, 1947.

[26] J.-M. Deshouillers, G. A. Freiman, V. Sós, and M. Temkin. On the structure of sum-free sets, 2. In *Conference on the Structure Theory of Set Addition*, pages 79–97. CIRM, Marseille, 1993.

[27] J.-M. Deshouillers and E. Wirsing. untitled. Preprint, 1977.

[28] J. A. Dias da Silva and Y. O. Hamidoune. A note on the minimal polynomial of the Kronecker sum of two linear operators. *Linear Algebra and its Applications*, 141:283–287, 1990.

[29] J. A. Dias da Silva and Y. O. Hamidoune. Cyclic spaces for Grassmann derivatives and additive theory. *Bull. London Math. Soc.*, 26:140–146, 1994.

[30] G. T. Diderich. On Kneser's addition theorem in groups. *Proc. Am. Math. Soc.*, 38:443–451, 1973.

[31] G. A. Dirac. Short proof of Menger's graph theorem. *Mathematika*, 13:42–44, 1966.

[32] H. G. Eggleston. *Convexity*. Cambridge University Press, Cambridge, 1958.

[33] G. Elekes and P. Erdős. Similar configurations and pseudogrids. Preprint, 1993.

[34] P. Erdős. On the arithmetical density of the sum of two sequences, one of which forms a basis for the integers. *Acta Arith.*, 1:197–200, 1936.

[35] P. Erdős. On an elementary proof of some asymptotic formulas in the theory of partitions. *Annals Math.*, 43:437–450, 1942.

[36] P. Erdős. On the addition of residue classes (mod p). In *Proceedings of the 1963 Number Theory Conference at the University of Colorado*, pages 16–17, Boulder, 1963. University of Colorado.

[37] P. Erdős. Some problems in number theory. In A. O. L. Atkin and B. J. Birch, editors, *Computers in Number Theory*, pages 405–414. Academic Press, New York, 1971.

[38] P. Erdős. Problems and results on combinatorial number theory III. In M. B. Nathanson, editor, *Number Theory Day, New York 1976*, volume 626 of *Lecture Notes in Mathematics*, pages 43–72, Berlin, 1977. Springer–Verlag.

[39] P. Erdős and G. Freiman. On two additive problems. *J. Number Theory*, 34:1–12, 1990.

[40] P. Erdős, A. Ginzburg, and A. Ziv. Theorem in the additive number theory. *Bull. Research Council Israel*, 10F:41–43, 1961.

[41] P. Erdős, B. Gordon, L. A. Rubel, and E. G. Straus. Tauberian theorems for sum sets. *Acta Arith.*, 9:177–189, 1964.

[42] P. Erdős and R. L. Graham. *Old and New Problems and Results in Combinatorial Number Theory*. L'Enseignement Mathématique, Geneva, 1980.

[43] P. Erdős and H. Heilbronn. On the addition of residue classes mod p. *Acta Arith.*, 9:149–159, 1964.

[44] P. Erdős and E. Szemerédi. On sums and products of integers. In P. Erdős, L. Alpár, G. Halász, and A. Sárközy, editors, *Studies in Pure Mathematics, To the Memory of Paul Turán*, pages 213–218. Birkhäuser Verlag, Basel, 1983.

[45] P. C. Fishburn. On a contribution of Freiman to additive number theory. *J. Number Theory*, 35:325–334, 1990.

[46] J. Folkman. On the representation of integers as sums of distinct terms from a fixed sequence. *Canadian J. Math.*, 18:643–655, 1966.

[47] K. B. Ford. Sums and products from a finite set. Preprint, 1996.

[48] P. Frankl and V. Rödl. The uniformity lemmas for hypergraphs. *Graphs and Combinatorics*, 8:309–312, 1992.

[49] G. A. Freiman. On the addition of finite sets. I. *Izv. Vysh. Zaved. Matematika*, 13(6):202–213, 1959.

[50] G. A. Freiman. Inverse problems of additive number theory. On the addition of sets of residues with respect to a prime modulus. *Doklady Akad. Nauk SSSR*, 141:571–573, 1961.

[51] G. A. Freiman. Inverse problems of additive number theory. On the addition of sets of residues with respect to a prime modulus. *Soviet Math.–Doklady*, 2:1520–1522, 1961.

[52] G. A. Freiman. Inverse problems of additive number theory. VI. on the addition of finite sets. III. *Izv. Vysh. Ucheb. Zaved. Matematika*, 28(3):151–157, 1962.

[53] G. A. Freiman. Addition of finite sets. *Doklady Akad. Nauk SSSR*, 158:1038–1041, 1964.

[54] G. A. Freiman. *Foundations of a Structural Theory of Set Addition*. Kazan Gos. Ped. Inst., Kazan, 1966.

[55] G. A. Freiman. *Foundations of a Structural Theory of Set Addition*, volume 37 of *Translations of Mathematical Monographs*. American Mathematical Society, Providence, R.I., 1973.

[56] G. A. Freiman. What is the structure of K if $K + K$ is small? In D. V. Chudnovsky, G. V. Chudnovsky, H. Cohn, and M. B. Nathanson, editors, *Number Theory, New York 1984–85*, volume 1240 of *Lecture Notes in Mathematics*, pages 109–134, New York, 1987. Springer–Verlag.

[57] G. A. Freiman. On the structure and number of sum-free sets. In *Proceedings of the Journées Arithmétiques de Genève*, volume 209, pages 195–201. Astérique, 1992.

[58] G. A. Freiman. Sumsets and powers of 2. In G. Halász, L. Lovász, D. Miklós, and T. Szőnyi, editors, *Sets, Graphs, and Numbers*, volume 60 of *Colloq. Math. Soc. János Bolyai*, pages 279–286. North-Holland Publishing Co., Amsterdam, 1992.

[59] G. A. Freiman, L. Low, and J. Pitman. The proof of Paul Erdős' conjecture of the addition of different residue classes modulo prime number. In *Structure Theory of Set Addition, 7-11 June 1993*, pages 99–108, Marseille, 1993. CIRM.

[60] P. M. Gruber and C. G. Lekkerkerker. *Geometry of Numbers*. North-Holland, Amsterdam, 1987.

[61] Y. O. Hamidoune. A generalization of an addition theorem of Shatrowsky. *Europ. J. Combin.*, 13:249–255, 1992.

[62] Y. O. Hamidoune. On a subgroup contained in words with a bounded length. *Discrete Math.*, 103:171–176, 1992.

[63] Y. O. Hamidoune. An isoperimetric method in additive theory. Preprint, 1994.

[64] Y. O. Hamidoune. On inverse additive problems. Preprint, 1994.

[65] Y. O. Hamidoune. Subsets with small sums in abelian groups I. The Vosper's property. Preprint, 1994.

[66] Y. O. Hamidoune. Subsets with small sums in abelian groups II. The critical pair problem. Preprint, 1994.

[67] Y. O. Hamidoune, O. Ordaz, and A. Ortunio. On a combinatorial theorem of Erdős, Ginzburg and Ziv. Preprint, 1995.

[68] Y. O. Hamidoune and Ö. J. Rödseth. On bases for σ-finite groups. *Math. Scand.*, 1996. To appear.

[69] B. Hornfeck. Ein Satz über die Primzahlmenge. *Mat. Annalen*, 60:271–273, 1954.

[70] B. Hornfeck. Berichtigung zur Arbeit: Ein Satz über Primzahlmenge. *Mat. Annalen*, 62:502, 1955.

[71] C. G. J. Jacobi. *Gesammelte Werke*. Chelsea Publishing Company, New York, 1969.

[72] B. Jessen. Review of paper by N. N. Bogolyubov. *Math. Reviews*, 8:512, 1947.

[73] X.-D. Jia and M. B. Nathanson. Addition theorems for σ-finite groups. In *Proceedings of the Hans Rademacher Centenary Conference*, volume 166 of *Contemporary Mathematics*, pages 275–284, Providence, 1994. American Mathematical Society.

[74] J. H. B. Kemperman. On small subsets of an abelian group. *Acta Math.*, 103:63–88, 1960.

[75] A. Khinchin. Über ein metrisches Problem der additiven Zahlentheorie. *Mat. Sbornik N.S.*, 10:180–189, 1933.

[76] M. Kneser. Abschätzungen der asymptotischen Dichte von Summenmengen. *Math. Z.*, 58:459–484, 1953.

[77] M. Laczkovich and I. Z. Ruzsa. The number of homothetic sets. Preprint, 1993.

[78] S. Lang. *Algebra.* Addison–Wesley, Reading, Mass., 3rd edition, 1993.

[79] V. F. Lev. Representing powers of 2 by a sum of four integers. *Combinatorica*, 16:1–4, 1996.

[80] V. F. Lev. Structure theorem for multiple addition and the Frobenius problem. *J. Number Theory*, 58:79–88, 1996.

[81] V. F. Lev and P. Y. Smeliansky. On addition of two different sets of integers. *Acta Arith.*, 70:85–91, 1995.

[82] W. Maak. *Fastperiodische Funktionen.* Springer–Verlag, Heidelberg, 2nd edition, 1967.

[83] J. L. Malouf. The density of the sum of a basis and a set of positive density; a bound of Helmut Plünnecke. *J. Number Theory*, 1994. To appear.

[84] H. B. Mann. *Addition Theorems.* Wiley–Interscience, New York, 1965.

[85] R. Mansfield. How many slopes in a polygon? *Israel J. Math.*, 39:265–272, 1981.

[86] W. McCuaig. A simple proof of Menger's theorem. *Journal of Graph Theory*, 8:427–429, 1984.

[87] G. Mohanty. *Lattice Path Counting and Applications.* Academic Press, New York, 1979.

[88] W. Moran and A.D. Pollington. On a result of Freiman. Preprint, 1992.

[89] T. V. Narayana. *Lattice Path Combinatorics, with Statistical Applications.* University of Toronto Press, Toronto, 1979.

[90] M. B. Nathanson. Additive number theory: Extremal problems and the combinatorics of sumsets. In preparation.

[91] M. B. Nathanson. Sums of finite sets of integers. *Am. Math. Monthly*, 79:1010–1012, 1972.

[92] M. B. Nathanson. The simplest inverse problems in additive number theory. In A. Pollington and W. Moran, editors, *Number Theory with an Emphasis on the Markoff Spectrum*, pages 191–206, New York, 1993. Marcel Dekker.

[93] M. B. Nathanson. Ballot numbers, alternating products, and the Erdős-Heilbronn conjecture. In R. L. Graham and J. Nesetril, editors, *The Mathematics of Paul Erdős*. Springer–Verlag, Heidelberg, 1994.

[94] M. B. Nathanson. An inverse theorem for sums of sets of lattice points. *J. Number Theory*, 46:29–59, 1994.

[95] M. B. Nathanson. Inverse theorems for subset sums. *Trans. Am. Math. Soc.*, 347:1409–1418, 1995.

[96] M. B. Nathanson. *Additive Number Theory: The Classical Bases*, volume 164 of *Graduate Texts in Mathematics*. Springer–Verlag, New York, 1996.

[97] M. B. Nathanson. On sums and products of integers. *Proc. Am. Math. Soc.*, 1996. To appear.

[98] M. B. Nathanson and A. Sárközy. Sumsets containing long arithmetic progressions and powers of 2. *Acta Arith.*, 54:147–154, 1989.

[99] M. B. Nathanson and G. Tenenbaum. Inverse theorems and the number of sums and products. To appear, 1996.

[100] H.-H. Ostmann. Untersuchen über den Summenbegriff in der additiven Zahlentheorie. *Mat. Annalen*, 120:165–196, 1948.

[101] S. S. Pillai. Generalization of a theorem of Davenport on the addition of residue classes. *Proc. Indian Acad. Sci. Ser. A*, 6:179–180, 1938.

[102] H. Plünnecke. Über ein metrisches Problem der additiven Zahlentheorie. *J. reine angew. Math.*, 197:97–103, 1957.

[103] H. Plünnecke. Über die Dichte der Summe zweier Mengen, deren eine die dichte null hat. *J. reine angew. Math.*, 205:1–20, 1960.

[104] H. Plünnecke. *Eigenschaften und Abschätzungen von Wirkungsfunktionen*, volume 22. Berichte der Gesellschaft für Mathematik und Datenverarbeitung, Bonn, 1969.

[105] H. Plünnecke. Eine zahlentheoretische Anwendung der Graphtheorie. *J. reine angew. Math.*, 243:171–183, 1970.

[106] J. M. Pollard. A generalization of a theorem of Cauchy and Davenport. *J. London Math. Soc.*, 8:460–462, 1974.

[107] J. M. Pollard. Additive properties of residue classes. *J. London Math. Soc.*, 11:147–152, 1975.

[108] L.P. Postnikova. Fluctuations in the distribution of fractional parts. *Doklady Akad. Nauk SSSR*, 161:1282–1284, 1965.

[109] L. Pyber. On the Erdős-Heilbronn conjecture. Personal Communication.

[110] U.-W. Rickert. *Über eine Vermutung in der additiven Zahlentheorie*. PhD thesis, Technical University of Braunschweig, 1976.

[111] Ö. J. Rödseth. Sums of distinct residues mod p. *Acta Arith.*, 65:181–184, 1993.

[112] I. Z. Ruzsa. An application of graph theory to additive number theory. *Scientia, Ser. A*, 3:97–109, 1989.

[113] I. Z. Ruzsa. Arithmetic progressions and the number of sums. *Periodica Math. Hungar.*, 25:105–111, 1992.

[114] I. Z. Ruzsa. An analog of Freiman's theorem in groups. Preprint, 1993.

[115] I. Z. Ruzsa. Generalized arithmetic progressions and sumsets. *Acta Math. Hungar.*, 65(4):379–388, 1994.

[116] I. Z. Ruzsa. Sums of finite sets. In D. V. Chudnovsky, G. V. Chudnovsky, and M. B. Nathanson, editors, *Number Theory: New York Seminar*. Springer–Verlag, New York, 1996.

[117] L. Shatrovskii. A new generalization of Davenport's-Pillai's theorem on the addition of residue classes. *Doklady Akad. Nauk CCCR*, 45:315–317, 1944.

[118] L. G. Shnirel'man. Über additive Eigenschaften von Zahlen. *Mat. Annalen*, 107:649–690, 1933.

[119] C. L. Siegel. *Lectures on the Geometry of Numbers*. Springer–Verlag, Berlin, 1989.

[120] R. Spigler. An application of group theory to matrices and to ordinary differential equations. *Linear Algebra and its Applications*, 44:143–151, 1982.

[121] J. Steinig. On G. A. Freiman's theorems concerning the sum of two finite sets of integers. In *Conference on the Structure Theory of Set Addition*, pages 173–186. CIRM, Marseille, 1993.

[122] A. Stöhr and E. Wirsing. Beispiele von wesentlichen Komponenten, die keine Basen sind. *J. reine angew. Math.*, 196:96–98, 1956.

[123] E. Szemerédi. On sets of integers containing no k elements in arithmetic progression. *Acta Arith.*, 27:199–245, 1975.

[124] E. Szemerédi. Regular partitions of graphs. In J.-C. Bermond et al., editors, *Problèmes Combinatoires et Théorie des Graphes*, number 260 in Colloques internationaux C.N.R.S., pages 399–401, Paris, 1978. C.N.R.S.

[125] A. I. Vinogradov and Yu. V. Linnik. Estimate of the sum of the number of divisors in a short segment of an arithmetic progression. *Uspekhi Mat. Nauk (N.S.)*, 12:277–280, 1957.

[126] A. G. Vosper. The critical pairs of subsets of a group of prime order. *J. London Math. Soc.*, 31:200–205, Addendum 280–282, 1956.

[127] H. Weyl. On geometry of numbers. *Proc. London Math. Soc.*, 47:268–289, 1942. Reprinted in *Gesammelte Abhandlungen*, volume IV, pages 75–96, Springer–Verlag, Berlin, 1968.

[128] E. Wirsing. Ein metrischer Satz über Mengen ganzer Zahlen. *Archiv Math.*, 4:392–398, 1953.

[129] D. Zeilberger. André's reflection proof generalized to the many-candidate ballot problem. *Discrete Math.*, 44:325–326, 1983.

Index

Graduate Texts in Mathematics

continued from page ii

119 ROTMAN. An Introduction to Algebraic Topology.
120 ZIEMER. Weakly Differentiable Functions: Sobolev Spaces and Functions of Bounded Variation.
121 LANG. Cyclotomic Fields I and II. Combined 2nd ed.
122 REMMERT. Theory of Complex Functions. *Readings in Mathematics*
123 EBBINGHAUS/HERMES et al. Numbers. *Readings in Mathematics*
124 DUBROVIN/FOMENKO/NOVIKOV. Modern Geometry—Methods and Applications. Part III.
125 BERENSTEIN/GAY. Complex Variables: An Introduction.
126 BOREL. Linear Algebraic Groups.
127 MASSEY. A Basic Course in Algebraic Topology.
128 RAUCH. Partial Differential Equations.
129 FULTON/HARRIS. Representation Theory: A First Course. *Readings in Mathematics*
130 DODSON/POSTON. Tensor Geometry.
131 LAM. A First Course in Noncommutative Rings.
132 BEARDON. Iteration of Rational Functions.
133 HARRIS. Algebraic Geometry: A First Course.
134 ROMAN. Coding and Information Theory.
135 ROMAN. Advanced Linear Algebra.
136 ADKINS/WEINTRAUB. Algebra: An Approach via Module Theory.
137 AXLER/BOURDON/RAMEY. Harmonic Function Theory.
138 COHEN. A Course in Computational Algebraic Number Theory.
139 BREDON. Topology and Geometry.
140 AUBIN. Optima and Equilibria. An Introduction to Nonlinear Analysis.
141 BECKER/WEISPFENNING/KREDEL. Gröbner Bases. A Computational Approach to Commutative Algebra.
142 LANG. Real and Functional Analysis. 3rd ed.
143 DOOB. Measure Theory.
144 DENNIS/FARB. Noncommutative Algebra.
145 VICK. Homology Theory. An Introduction to Algebraic Topology. 2nd ed.
146 BRIDGES. Computability: A Mathematical Sketchbook.
147 ROSENBERG. Algebraic K-Theory and Its Applications.
148 ROTMAN. An Introduction to the Theory of Groups. 4th ed.
149 RATCLIFFE. Foundations of Hyperbolic Manifolds.
150 EISENBUD. Commutative Algebra with a View Toward Algebraic Geometry.
151 SILVERMAN. Advanced Topics in the Arithmetic of Elliptic Curves.
152 ZIEGLER. Lectures on Polytopes.
153 FULTON. Algebraic Topology: A First Course.
154 BROWN/PEARCY. An Introduction to Analysis.
155 KASSEL. Quantum Groups.
156 KECHRIS. Classical Descriptive Set Theory.
157 MALLIAVIN. Integration and Probability.
158 ROMAN. Field Theory.
159 CONWAY. Functions of One Complex Variable II.
160 LANG. Differential and Riemannian Manifolds.
161 BORWEIN/ERDÉLYI. Polynomials and Polynomial Inequalities.
162 ALPERIN/BELL. Groups and Representations.
163 DIXON/MORTIMER. Permutation Groups.
164 NATHANSON. Additive Number Theory: The Classical Bases.
165 NATHANSON. Additive Number Theory: Inverse Problems and the Geometry of Sumsets.
166 SHARPE. Differential Geometry: Cartan's Generalization of Klein's Erlangen Programme.
167 MORANDI. Field and Galois Theory.